CHEMISTRY
foundations and applications

CHEMISTRY

foundations and applications

J. J. Lagowski, Editor in Chief

MACMILLAN
REFERENCE
USA™

New York • Detroit • San Diego • San Francisco • Cleveland • New Haven, Conn. • Waterville, Maine • London • Munich

Chemistry: Foundations and Applications
J. J. Lagowski, Editor in Chief

For more information, contact
Macmillan Reference USA
27500 Drake Rd.
Farmington Hills, MI 48331-3535
Or you can visit our Internet site at
http://www.gale.com

Library of Congress Cataloging-in-Publication Data

Chemistry : foundations and applications / edited by J.J. Lagowski.
p. cm.
Includes index.
ISBN 0-02-865721-7 (set hardcover : alk. paper) — ISBN 0-02-865722-5 (v. 1 : alk. paper) — ISBN 0-02-865723-3 (v. 2 : alk. paper) — ISBN 0-02-865724-1 (v. 3 : alk. paper) — ISBN 0-02-865725-X (v. 4 : alk. paper)
1. Chemistry—Encyclopedias. I. Lagowski, J. J.
QD4.C48 2004
540'.3—dc22
2003021038

This title is also available as an e-book.
ISBN 0-02-865913-9
Contact your Gale sales representative for ordering information.

Printed in the United States of America
10 9 8 7 6 5 4 3 2

Table of Contents

Volume 1

Volume 2

Volume 3

Volume 4

Preface

Chemistry. The word conjures up mystery—perhaps magic—smoke, fireworks, explosions, unpleasant odors. But it could evoke "smokeless burning," which would be invisible, fluorescent lights, "neon" signs, the quiet crumbling of rocks under the pressure of freezing water, the slow and quiet formation of caves in limestone, and the delightful scents of perfumes or fruit aromas. There is no magic, only knowledge and understanding. We offer this *Encyclopedia* as a contribution to help readers gain knowledge and understanding of chemistry.

Chemistry was manifested as an art at the beginnings of civilization. The early decorative chemical arts included the preparation of pigments such as the Egyptian blue applied to King Tutankhamen's golden death mask; the various bronze alloys that were used to make vases in the ancient world of the Middle East as well as in China; and the glass objects that have been found in Mesopotamia (now known as Iraq). Those chemical arts became a science in the eighteenth century when Antoine Laurent Lavoisier (1743–1794) led what has been called "the chemical revolution." Using accurate measurements of primarily mass, early chemists began to make order out of the myriad of substances that are found in the natural world. This order was eventually expressed in a number of chemical concepts that include the laws of chemical composition (constant composition, mass conservation, multiple proportions), periodicity, the nature of atoms, chemical bonding, and a variety of concepts involving chemical structures. The early symbiosis of chemistry with civilization remains. Chemistry is still a useful science in the advancement of civilization. Chemists have developed and refined the core concepts of chemistry to the point where they have become powerful tools to assist humankind in the acquisition of materials of practical use to extend and preserve civilization. Humans now have available a broader array of substances with a remarkable spectrum of properties than was available before chemistry became a science. Light emitting diodes (LEDs) produce more light than the individual torches, candles, and oil lamps of the distant past—indeed, than the incandescent light bulbs of the immediate past—more efficiently and with less pollution. Polymeric materials or composites can be produced with virtually any property desired—from stretching clingy Saran Wrap to Kevlar used in bullet proof vests; from nonstick Teflon to optical fibers; from rubber objects that are impervious to oil and gasoline to tires that can be used for 100,000 miles before needing replacement; from fibers that compete with (in some cases surpass) natural materials to plastics that have more desirable optical properties than glass. In a word, chemistry is *everywhere.*

There is no magic, only knowledge and understanding.

These volumes are a contribution to assist readers in their understanding of chemistry and chemical ideas and concepts. The 509 articles have been carefully chosen to provide basic information on a broad range of topics. For those readers who desire to expand their knowledge of a topic, we have included bibliographic references to readily accessible sources.

The continual evolution of the discipline of chemistry is reflected in our treatment of the elements. The careful reader will note that we have included articles for the first 104 elements; the remainder of the elements are "recently" discovered or exist only as short-lived species and, accordingly, are not readily available for the usual chemical studies that reveal, for example, their bulk properties or reactivity and much of the "standard chemistry" that is of interest. Much of what little we know about the elements beyond 104 permits us to place these elements in their appropriate places in the periodic table, which nevertheless still turns out to be quite insightful from a chemical point of view.

Entries in the *Encyclopedia* are in alphabetic sequence. Cross-references appear in small capitals at the ends of the articles to help readers locate related discussions. Articles range from brief, but concise, definitions to extensive overviews that treat key concepts in larger contexts. A list of common abbreviations and symbols, and a list of the known elements, as well as a modern version of the periodic table are included in the For Your Reference section at the beginning of each volume. A glossary and a comprehensive index appear at the end of each volume. Contributors are listed alphabetically, together with their academic and professional affiliations, at the beginning of each volume.

Following this preface, we offer a topical arrangement of the articles in the *Encyclopedia.* This outline provides a general overview of the principal parts of the subject of chemistry and is arranged in alphabetical order.

Many individuals have contributed greatly and in many ways to this *Encyclopedia.* The associate editors—Alton J. Banks, Thomas Holme, Doris Kolb, and Herbert Silber—carried the major responsibility in shaping the intellectual content of the *Encyclopedia.* The authors of the articles executed that plan admirably and we thank them for that effort.

The staff at Macmillan Reference USA—Marie-Claire Antoine, Hélène Potter, Ray Abruzzi, Gloria Lam, and Christine Slovey—have been outstanding in their dedication and contributions to bring this *Encyclopedia* from its initial concept to the current reality. Without their considerable input, insightful guidance, and effort this *Encyclopedia* would never have seen the light of day. I take this opportunity to thank them personally and publicly. I am particularly grateful to Rita Wilkinson, my administrative assistant for her persistent and careful attention to details that kept the editorial office and my office connected for the smooth transmission of numerous critical details. I am especially grateful to Christine Slovey who, through her determined efforts and dedication, made a potentially difficult and tedious task far less onerous and, indeed, enjoyable.

J. J. Lagowski

Topical Outline

Analytical Chemistry Applications

Adhesives
Agricultural Chemistry
Analytical Chemistry
Bleaches
Ceramics
Chemical Engineering
Chemical Informatics
Coal
Cosmetics
Cryogenics
Detergents
Disposable Diapers
Dyes
Explosions
Fertilizer
Fibers
Food Preservatives
Forensic Chemistry
Formulation Chemistry
Freons
Gardening
Gasoline
Gemstones
Genetic Engineering
Glass
Hair Dyes and Hair Treatments
Herbicides
Industrial Chemistry, Inorganic
Industrial Chemistry, Organic
Insecticides
Irradiated Foods
Materials Science
Nanochemistry
Nylon
Pesticides
Pigments
Polymers, Synthetic
Recycling
Rocketry
Superconductors
Zeolites

Aqueous Chemistry

Acid-Base Chemistry
Bases
Bleaches
Chemical Reactions
Colloids
Corrosion
Equilibrium
Solution Chemistry
Water

Astrochemistry

Astrochemistry

Biochemistry

Acetylcholine
Active Site
Allosteric Enzymes
Amino Acid
Antibiotics
Artificial Sweeteners
Base Pairing
Bioluminescence
Caffeine
Carbohydrates
Cellulose
Chemiluminescence
Cholecalciferol
Cholesterol
Chromosome
Clones
Codon
Coenzyme
Cofactor
Collagen
Cortisone
Denaturation
Deoxyribonucleic Acid
Disaccharides
DNA Replication
Dopamine
Double Helix
Endorphins
Enzymes
Epinephrine
Estrogen
Fats and Fatty Acids
Fibrous Protein
Genes
Genetic Engineering
Genome
Globular Protein
Glycolysis
Glycoprotein
Hemoglobin
Hydrolase
Hydrolysis
Ion Channels
Kinase
Krebs Cycle
Lipid Bilayers
Lipids
Low Density Lipoprotein (LDL)
Membrane
Methylphenidate
Mutagen
Mutation
Neurochemistry
Neurotoxins
Neurotransmitters
Nicotinamide
Nicotinamide Adenine Dinucleotide (NAD)
Nicotine
Norepinephrine
Nucleic Acids
Nucleotide
Peptide Bond
Phospholipids

Photosynthesis
Polymerase Chain Reaction (PCR)
Polymers, Natural
Polysaccharides
Proteins
Primary Structure
Protein Solubility
Protein Synthesis
Protein Translation
Quaternary Structure
Residue
Restriction Enzymes
Retinol
Rhodium
Ribonucleic Acid
RNA Synthesis
Secondary Structure
Starch
Steroids
Stimulants
Storage Protein
Substrate
Taste Receptors
Teratogen
Tertiary Structure
Testosterone
Thiamin
Toxicity
Transmembrane Protein
Transport Protein
Triglycerides
Venom
Zwitterion

Biographies

Al Razi, Abu Bakr Muhammed ibn Zakariya
Anfinsen, Christian
Arrhenius, Svante
Avery, Oswald
Avogadro, Amedeo
Baekeland, Leo
Balmer, Johann Jakob
Bardeen, John
Becquerel, Antoine-Henri
Berg, Paul
Berthollet, Claude-Louis
Berzelius, Jöns Jakob
Black, Joseph
Bohr, Niels
Boltzmann, Ludwig
Boyle, Robert
Bragg, William Henry
Bragg, William Lawrence
Brønsted, Johannes Nicolaus
Bunsen, Robert
Caldwell, Mary
Calvin, Melvin
Cannizzaro, Stanislao
Carnot, Sadi
Carothers, Wallace
Carver, George Washington
Cavendish, Henry
Chadwick, James
Chappelle, Emmett
Chardonnet, Hilaire
Charles, Jacques
Chevreul, Michel
Cleve, Per Theodor
Cori, Carl and Gerty
Cottrell, Frederick
Coulomb, Charles
Curie, Marie Sklodowska
Dalton, John
Davy, Humphry
de Broglie, Louis
Dewar, James
du Pont, Éleuthère Irénée
Ehrlich, Paul
Einstein, Albert
Elion, Gertrude Belle
Fahrenheit, Gabriel
Faraday, Michael
Fermi, Enrico
Fischer, Emil Hermann
Fleming, Alexander
Franklin, Rosalind
Gadolin, Johan
Gay-Lussac, Joseph-Louis
Gibbs, Josiah Willard
Goodyear, Charles
Haber, Fritz
Hall, Charles
Heisenberg, Werner
Hess, Germain Henri
Heyrovsky, Jaroslav
Hodgkin, Dorothy
Julian, Percy
Kekulé, Friedrich August
Kelsey, Frances Kathleen Oldham
Khorana, Har Gobind
Krebs, Hans Adolf
Lavoisier, Antoine
Lawrence, Ernest
Le Bel, Joseph-Achille
Leblanc, Nicolas
LeChâtelier, Henri
Leclanché, Georges
Leloir, Luis
Lewis, Gilbert N.
Liebig, Justus von
Lister, Joseph
Lonsdale, Kathleen
Lucretius
Marsden, Ernest
Martin, Archer John Porter
Maxwell, James Clerk
Meitner, Lise
Mendeleev, Dimitri
Menten, Maud
Meyer, Lothar
Millikan, Robert
Morgan, Agnes Fay
Moseley, Henry
Nernst, Walther Hermann
Newton, Isaac
Nobel, Alfred Bernhard
Northrop, John
Oppenheimer, Robert
Ostwald, Friedrich Wilhelm
Paracelsus
Pasteur, Louis
Pauli, Wolfgang
Pauling, Linus
Perkin, William Henry
Planck, Max
Priestley, Joseph
Raman, Chandrasekhara
Ramsay, William
Remsen, Ira
Robinson, Robert
Röntgen, Wilhelm
Rutherford, Ernest
Rydberg, Johannes
Sanger, Frederick
Scheele, Carl
Schrödinger, Erwin
Seaborg, Glenn Theodore
Seibert, Florence
Soddy, Frederick
Stanley, Wendell
Staudinger, Hermann
Strutt, John (Lord Rayleigh)
Sumner, James
Svedberg, Theodor
Synge, Richard Laurence Millington
Szent-Györgyi, Albert
Thomson, Joseph John
Todd, Alexander
Travers, Morris
Urey, Harold
van der Waals, Johannes
Van Helmont, Johann Baptista
Van't Hoff, Jacobus Hendricus
Volta, Alessandro
Waksman, Selman
Watson, James Dewey
Weizmann, Chaim

Chemical Substances

Computing

Elements

Energy

Environmental Chemistry

For Your Reference

TABLE 1. SELECTED METRIC CONVERSIONS

WHEN YOU KNOW	MULTIPLY BY	TO FIND
Temperature		
Celsius (˚C)	1.8 (˚C) +32	Fahrenheit (˚F)
Celsius (˚C)	˚C +273.15	Kelvin (K)
degree change (Celsius)	1.8	degree change (Fahrenheit)
Fahrenheit (˚F)	[(˚F) −32] / 1.8	Celsius (˚C)
Fahrenheit (˚F)	[(˚F −32) / 1.8] +273.15	Kelvin (K)
Kelvin (K)	K −273.15	Celsius (˚C)
Kelvin (K)	1.8(K −273.15) +32	Fahrenheit (˚F)

WHEN YOU KNOW	MULTIPLY BY	TO FIND
Distance/Length		
centimeters	0.3937	inches
kilometers	0.6214	miles
meters	3.281	feet
meters	39.37	inches
meters	0.0006214	miles
microns	0.000001	meters
millimeters	0.03937	inches

WHEN YOU KNOW	MULTIPLY BY	TO FIND
Capacity/Volume		
cubic kilometers	0.2399	cubic miles
cubic meters	35.31	cubic feet
cubic meters	1.308	cubic yards
cubic meters	8.107×10^{-4}	acre-feet
liters	0.2642	gallons
liters	33.81	fluid ounces

WHEN YOU KNOW	MULTIPLY BY	TO FIND
Area		
hectares (10,000 square meters)	2.471	acres
hectares (10,000 square meters)	107,600	square feet
square meters	10.76	square feet
square kilometers	247.1	acres
square kilometers	0.3861	square miles

WHEN YOU KNOW	MULTIPLY BY	TO FIND
Weight/Mass		
kilograms	2.205	pounds
metric tons	2205	pounds
micrograms (μg)	10^{-6}	grams
milligrams (mg)	10^{-3}	grams
nanograms (ng)	10^{-9}	grams

TABLE 2. ALPHABETIC TABLE OF THE ELEMENTS

Symbol	Element	Atomic Number	Atomic Mass*	Symbol	Element	Atomic Number	Atomic Mass*
Ac	Actinium	89	(227)	Mt	Meitnerium	109	(266)
Al	Aluminum	13	26.982	Md	Mendelevium	101	(258)
Am	Americium	95	(243)	Hg	Mercury	80	200.59
Sb	Antimony	51	121.75	Mo	Molybdenum	42	95.94
Ar	Argon	18	39.948	Nd	Neodymium	60	144.24
As	Arsenic	33	74.922	Ne	Neon	10	20.180
At	Astatine	85	(210)	Np	Neptunium	93	237.048
Ba	Barium	56	137.33	Ni	Nickel	28	58.69
Bk	Berkelium	97	(247)	Nb	Niobium	41	92.908
Be	Beryllium	4	9.012	N	Nitrogen	7	14.007
Bi	Bismuth	83	208.980	No	Nobelium	102	(259)
Bh	Bohrium	107	(262)	Os	Osmium	76	190.2
B	Boron	5	10.811	O	Oxygen	8	15.999
Br	Bromine	35	79.904	Pd	Palladium	46	106.42
Cd	Cadmium	48	112.411	P	Phosphorus	15	30.974
Ca	Calcium	20	40.08	Pt	Platinum	78	195.08
Cf	Californium	98	(251)	Pu	Plutonium	94	(244)
C	Carbon	6	12.011	Po	Polonium	84	(209)
Ce	Cerium	58	140.115	K	Potassium	19	39.1
Cs	Cesium	55	132.90	Pr	Praseodymium	59	140.908
Cl	Chlorine	17	35.453	Pm	Promethium	61	(145)
Cr	Chromium	24	51.996	Pa	Protactinium	91	231.036
Co	Cobalt	27	58.933	Ra	Radium	88	226.025
Cu	Copper	29	63.546	Rn	Radon	86	(222)
Cm	Curium	96	(247)	Re	Rhenium	75	186.207
Ds	Darmastadtium	110	(269)	Rh	Rhodium	45	102.906
Db	Dubnium	105	(262)	Rb	Rubidium	37	85.47
Dy	Dysprosium	66	162.50	Ru	Ruthenium	44	101.07
Es	Einsteinium	99	(252)	Rf	Rutherfordium	104	(261)
Er	Erbium	68	167.26	Sm	Samarium	62	150.36
Eu	Europium	63	151.965	Sc	Scandium	21	44.966
Fm	Fermium	100	(257)	Sg	Seaborgium	106	(263)
F	Fluorine	9	18.998	Se	Selenium	34	78.96
Fr	Francium	87	(223)	Si	Silicon	14	28.086
Gd	Gadolinium	64	157.25	Ag	Silver	47	107.868
Ga	Gallium	31	69.723	Na	Sodium	11	22.990
Ge	Germanium	32	72.61	Sr	Strontium	38	87.62
Au	Gold	79	196.967	S	Sulfur	16	32.066
Hf	Hafnium	72	178.49	Ta	Tantalum	73	180.948
Hs	Hassium	108	(265)	Tc	Technetium	43	(98)
He	Helium	2	4.003	Te	Tellurium	52	127.60
Ho	Holmium	67	164.93	Tb	Terbium	65	158.925
H	Hydrogen	1	1.008	Tl	Thallium	81	204.383
In	Indium	49	114.82	Th	Thorium	90	232.038
I	Iodine	53	126.905	Tm	Thulium	69	168.934
Ir	Iridium	77	192.22	Sn	Tin	50	118.71
Fe	Iron	26	55.847	Ti	Titanium	22	47.88
Kr	Krypton	36	83.80	W	Tungsten	74	183.85
La	Lanthanum	57	138.906	U	Uranium	92	238.029
Lr	Lawrencium	103	(260)	V	Vanadium	23	50.942
Pb	Lead	82	207.2	Xe	Xenon	54	131.29
Li	Lithium	3	6.941	Yb	Ytterbium	70	173.04
Lu	Lutetium	71	174.967	Y	Yttrium	39	88.906
Mg	Magnesium	12	24.305	Zn	Zinc	30	65.38
Mn	Manganese	25	54.938	Zr	Zirconium	40	91.224

*Atomic masses are based on the relative atomic mass of $^{12}C=12$. These values apply to the elements as they exist in materials of terrestrial origin and to certain artificial elements. Values in parenthesis are the mass number of the isotope of the longest half-life.

TABLE 3. COMMON ABBREVIATIONS, SYMBOLS, AND ACRONYMS

′	minute (of arc); single prime
″	second (of arc); double prime
+	plus
$^+$	positive charge
−	minus
$^-$	negative charge
±	plus-or-minus
∓	minus-or-plus
×	multiplied by
·	multiplied by
÷	divided by
=	equals
≠	not equal to
∽	about, approximately
≅	congruent to; approximately equal to
≈	approximately equal to
≡	identical to; equivalent to
<	less than
≤	less than or equal to
>	greater than
≥	greater than or equal to
%	percent
°	degree (temperature; angle of arc)
@	at
—	single bond
═	double bond
: :	double bond
≡	triple bond
: : :	triple bond
∞	infinity
∝	variation
∂	partial derivative or differential
α	proportional to, alpha
√	square root
Δ	delta; increment of a variable
ϵ_0	dielectric constant; permittivity
θ	plane angle
λ	wavelength
μ	magnetic moment; micro
μA	microampere
μC	microcoulomb
μF	microfarad
μg	microgram
μg/ml	microgram per milliliter
μK	microkelvin
μm	micrometer (also called micron)
μmol;	micromole
μs, μsec	microsecond
ν	frequency
v	velocity
π or pi	ratio of the circumference of a circle to its diameter; double as in double bond
σ	single as in single bond; Stefan-Boltzmann constant
Σ	summation
φ	null set
ψ	amplitude of a wave (as in *wave,* or *psi, function)*
→	reaction to right
←	reaction to left
↔	connecting resonance forms
⇆	equilibrium reaction beginning at right
⇄	equilibrium reaction beginning at left
⇌	reversible reaction beginning at left
⇋	reversible reaction beginning at right
↑	elimination
⇓	absorption
a	acceleration
A	area
a_o	Bohr Unit
AAS	atomic absorption spectroscopy
ABS	alkylbenzene sulfate
ACS	American Chemical Society
ADH	alcohol dehydrogenase
ADP	adenosine diphosphate
AEC	Atomic Energy Commission
AES	atomic emission spectroscopy
AFM	atomic force microscope; atomic force microscopy
AFS	atomic fluorescence spectroscopy
ALDH	aldehyde dehydronase
amp	ampere
AMS	accelerator mass spectrometry
AMU	atomic mass unit
atm.	standard atmosphere (unit of pressure)
ATP	adenosine triphosphate
β	beta
b.p.	boiling point
Btu	British thermal unit
c	centi-; speed of light
C	carbon; Celsius; centigrade; coulomb
C	heat capacity; electric capacitance

TABLE 3. COMMON ABBREVIATIONS, SYMBOLS, AND ACRONYMS [continued]

Abbreviation	Meaning
Ci	Curies
cm	centimeter
CT	computed tomography
d	d-orbital
D	dipole moment
DC	direct current
deg	degree
dr	diastereomer ratio
e	elementary charge
E	electric field strength; energy
E_a	activation energy
E_g	bandgap energy
EA	electron affinity
er	enantiomer ratio
eV	electron volts
f	f-orbital
F	Fahrenheit; Faraday's constant; fluorine
F	force
g	gram
g	g-orbital; gas
h	hour
h	Planck's constant
Hz	hertz
i	i-orbital
IUPAC	International Union of Pure and Applied Chemistry
J	joule
J	electric current density
k	k-orbital
K	degrees Kelvin; Kelvin; potassium
K_a	acidity constant for the dissociation of weak acid (the weaker the acid, the lower the K_a value
k_B	Boltzmann's constant
Kg	kilogram
kHz	kilohertz
kJ	kilojoule
kJ mol	kilojoule mole
km	kilometer
K_m	Michaelis constant
l	length; liquid
L	lambert; liter
L	length; Avogadro's constant
LD	lethal dose
L/mole	liters per mole
ln	natural logarithm
log	logarithm
m	meter; milli-; molal (concentration)
m	mass
M	molar (concentration)
m_e	electron mass
mA	milliamperes
mg	milligram
mg/L	milligrams per liter
MHz	megahertz
min	minute
ml	milliliter
MO	molecular orbital
p	p-orbital
ω	omega
ppb	parts per billion
ppm	parts per million
ppt	parts per trillion
psi	per square inch; English for ψ
q	quantity
REM	Roentgen Equivalent Man (radiation-dose unit of measure)
s	solid; s-orbital
S	entropy
sec	second; secant
SEM	scanning electron microscope
SI	Système Internationale (International System of Measurements)
SPM	scanning probe microscope
STM	scanning tunneling microscope
STP	standard temperature and pressue (°C, 1 atm)
Sv	sievert unit (1 Sv = 100 REM; used to measure radiation dose)
t	time
T	moment of force, thermodynamic temperature (in degrees Kelvin); torque
T_c	critical temperature
TEM	transmission electron microscope
u	unified atomic mass unit
U	electric potential
V	electric potential; vanadium; volume
V	volt
vap.	vaporization
VB	valence bond
vel.	velocity
VSEPR	valence shell electron pair repulsion
Z	atomic number

PERIODIC TABLE OF THE ELEMENTS*†

s

I	II
3 LITHIUM **Li** 6.941	4 BERYLLIUM **Be** 9.012
11 SODIUM **Na** 22.990	12 MAGNESIUM **Mg** 24.305
19 POTASSIUM **K** 39.1	20 CALCIUM **Ca** 40.08
37 RUBIDIUM **Rb** 85.47	38 STRONTIUM **Sr** 87.62
55 CESIUM **Cs** 132.90	56 BARIUM **Ba** 137.33
87 FRANCIUM **Fr** (223)	88 RADIUM **Ra** 226.025

1 HYDROGEN **H** 1.008

d

21 SCANDIUM **Sc** 44.966	22 TITANIUM **Ti** 47.88	23 VANADIUM **V** 50.942	24 CHROMIUM **Cr** 51.996	25 MANGANESE **Mn** 54.938	26 IRON **Fe** 55.847	27 COBALT **Co** 58.933	28 NICKEL **Ni** 58.69	29 COPPER **Cu** 63.546	30 ZINC **Zn** 65.38
39 YTTRIUM **Y** 88.906	40 ZIRCONIUM **Zr** 91.224	41 NIOBIUM **Nb** 92.908	42 MOLYBDENIUM **Mo** 95.94	43 TECHNETIUM **Tc** (98)	44 RUTHENIUM **Ru** 101.07	45 RHODIUM **Rh** 102.906	46 PALLADIUM **Pd** 106.42	47 SILVER **Ag** 107.868	48 CADMIUM **Cd** 112.411
71 LUTETIUM **Lu** 174.967	72 HAFNIUM **Hf** 178.49	73 TANTALUM **Ta** 180.948	74 TUNGSTEN **W** 183.85	75 RHENIUM **Re** 186.207	76 OSMIUM **Os** 190.2	77 IRIDIUM **Ir** 192.22	78 PLATINUM **Pt** 195.08	79 GOLD **Au** 196.967	80 MERCURY **Hg** 200.59
103 LAWRENCIUM **Lr** (260)	104 RUTHERFORDIUM **Rf** (261)	105 DUBNIUM **Db** (262)	106 SEABORGIUM **Sg** (263)	107 BOHRIUM **Bh** (262)	108 HASSIUM **Hs** (265)	109 MEITNERIUM **Mt** (266)	110 DARMASTADTIUM **Ds** (269)	111 UNUNUNIUM **Uuu** (272)	112 (?)

p

III	IV	V	VI	VII	VIII
					2 HELIUM **He** 4.003
5 BORON **B** 10.811	6 CARBOM **C** 12.011	7 NITROGEN **N** 14.007	8 OXYGEN **O** 15.999	9 FLUORINE **F** 18.998	10 NEON **Ne** 20.180
13 ALUMINUM **Al** 26.982	14 SILICON **Si** 28.086	15 PHOSPHORIUS **P** 30.974	16 SULFUR **S** 32.066	17 CHLORINE **Cl** 35.453	18 ARGON **Ar** 39.948
31 GALLIUM **Ga** 69.73	32 GERMANIUM **Ge** 72.61	33 ARSENIC **As** 74.922	34 SELENIUM **Se** 78.96	35 BROMINE **Br** 79.904	36 KRYPTON **Kr** 83.80
49 INDIUM **In** 114.82	50 TIN **Sn** 118.71	51 ANTIMONY **Sb** 121.75	52 TELLURIUM **Te** 127.60	53 IODINE **I** 126.905	54 XENON **Xe** 131.29
81 THALLIUM **Tl** 204.383	82 LEAD **Pb** 207.2	83 BISMUTH **Bi** 208.980	84 POLONIUM **Po** (209)	85 ASTATINE **At** (210)	86 RADON **Rn** (222)

f

57 LANTHANUM **La** 138.906	58 CERIUM **Ce** 140.15	59 PRAESEODYMIUM **Pr** 140.908	60 NEODYMIUM **Nd** 144.24	61 NEODYMIUM **Pm** (145)	62 SAMARIUM **Sm** 150.36	63 EUROPIUM **Eu** 151.965	64 GADOLINIUM **Gd** 157.25	65 TERBIUM **Tb** 158.925	66 DYSPROSIUM **Dy** 162.50	67 HOLMIUM **Ho** 164.93	68 ERBIUM **Er** 167.26	69 THULIUM **Tm** 168.934	70 YTTERBIUM **Yb** 173.04
89 ACTINIUM **Ac** 227.03	90 THORIUM **Th** 232.038	91 PROTACTINIUM **Pa** 231.036	92 URANIUM **U** 238.029	93 NEPTUNIUM **Np** 237.048	94 PLUTONIUM **Pu** (244)	95 AMERICIUM **Am** (243)	96 CURIUM **Cm** (247)	97 BERKELIUM **Bk** (247)	98 CALIFORNIUM **Cf** (251)	99 EINSTEINIUM **Es** (252)	100 FERMIUM **Fm** (257)	101 MENDELEVIUM **Md** (258)	102 NOBELIUM **No** (259)

*Each element in the table is listed with (from top to bottom) its atomic number, its name, its symbol, and its atomic mass. Atomic mass numbers in parentheses are the mass numbers of the longest-lived isotope. Other atomic mass numbers are the average mass number of the naturally occurring isotopes.

†The names and labels for elements beyond number 103 are controversial. IUPAC initially ruled in favor of Latin names based on atomic number, but in 1994 a set of specific names and symbols was suggested. After considerable debate, a revised final list of names for elements 104-109 was issued on August 30, 1997. Temporary names were also assigned for elements 110 and 111. Various groups have suggested alternatives names for some of these elements. Additional elements continue to be synthesized, though with increasing difficulty, with no definite upper atomic-number limit yet established.

Contributors

Catalina Achim
Carnegie Mellon University
Pittsburgh, Pennsylvania

David T. Allen
University of Texas at Austin
Austin, Texas

Carolyn J. Anderson
Washington University School of Medicine
St. Louis, Missouri

Ronald D. Archer
Professor Emeritus
University of Massachusetts, Amherst
Amherst, Massachusetts

D. Eric Aston
University of Idaho
Moscow, Idaho

Peter Atkins
University of Oxford
Oxford, United Kingdom

Cynthia Atterholt
Western Carolina University
Cullowhee, North Carolina

Michèle Auger
Laval University
Quebec, Quebec, Canada

Karl W. Böer
Professor Emeritus
University of Delaware
Newark, Delaware

Catherine H. Banks
Peace College
Raleigh, North Carolina

Joseph Bariyanga
University of Wisconsin
Milwaukee, Wisconsin

Nathan J. Barrows
University of Northern Colorado
Greeley, Colorado

David A. Bassett
Brockton, Massachusetts

James D. Batteas
CUNY - College of Staten Island
Staten Island, New York

Bruce Beaver
Duquesne University
Pittsburgh, Pennsylvania

C. Larry Bering
Clarion University
Clarion, Pennsylvania

Bill Bertoldi
Kingsford, Michigan

W. E. Billups
Rice University
Houston, Texas

Koen Binnemans
Catholic University of Leuven
Leuven, Belgium

James P. Birk
Arizona State University
Tempe, Arizona

Jeffrey R. Bloomquist
Virginia Polytechnic Institute and State University
Blacksburg, VA

John E. Bloor
University of Tennessee
Knoxville, Tennessee

George Bodner
Purdue University
West Lafayette, Indiana

Paul W. Bohn
University of Illinois
Urbana, Illinois

Valerie Borek
Boothwyn, Pennsylvania

Erwin Boschmann
Indiana University
Indianapolis, Indiana

Wayne B. Bosma
Bradley University
Peoria, Illinois

Ole Bostrup
The Danish Society of The History of Chemistry
Fredensborg, Denmark

F. Javier Botana
University of Cadiz
Cadiz, Spain

Lawrence H. Brannigan
Southwestern Illinois College
Bellville, Illinois

Ronald Brecher
GlobalTox International Consultants Inc.
Guelph, Ontario, Canada

Jeffrey C. Bryan
University of Wisconsin-La Crosse
La Crosse, Wisconsin

Robert A. Bulman
National Radiological Protection Board
Oxfordshire, England

Jean-Claude Bunzli
Swiss Federal Institute of Technology
Lausanne, Switzerland

Douglas Cameron
Montana Tech of the University of Montana
Butte, Montana

Michael C. Cann
University of Scranton
Scranton, Pennsylvania

Charles E. Carraher Jr.
Florida Atlantic University
Boca Raton, Florida

Henry A. Carter
Augustana University College
Camrose, Alberta, Canada

John Castro
Huntington Station, New York

Lyman R. Caswell
Texas Woman's University
Denton, Texas

Raymond Chang
Williams College
Williamstown, Massachusetts

Philip J. Chenier
University of Wisconsin
Eau Claire, Wisconsin

Peter E. Childs
University of Limerick
Limerick, Ireland

Gregory R. Choppin
Florida State University
Tallahassee, Florida

Abraham Clearfield
Texas A&M University
College Station, Texas

Brian P. Coppola
University of Michigan
Ann Arbor, Michigan

Eugene C. Cordero
San José State University
San José, California

Paul A. Craig
Rochester Institute of Technology
Rochester, New York

M. J. Crawford
Ludwig-Maximilians-Universtät-München
Munich, Germany

Mary R. S. Creese
University of Kansas
Lawrence, Kansas
(retired)

Bartow Culp
Purdue University
West Lafayette, Indiana

Carlos Eduardo da Silva Cortes
Universidade Federal Fluminense
Rio de Janeiro, Brazil

G. Brent Dawson
University of North Carolina at Greensboro
Greensboro, North Carolina

Stephanie E. Dew
Centre College
Danville, Kentucky

Anthony Diaz
Central Washington University
Ellensburg, Washington

Diego J. Diaz
University of Illinois at Urbana-Champaign
Urbana, Illinois

James S. Dickson
Iowa State University
Ames, Iowa

David A. Dobberpuhl
Creighton University
Omaha, Nebraska

Daryl J. Doyle
Kettering University
Flint, Michigan

Michael Eastman
Northern Arizona University
Flagstaff, Arizona

Andrew Ede
University of Alberta
Edmonton, Canada

Edward M. Eyring
University of Utah
Salt Lake City, Utah

Scott E. Feller
Wabash College
Crawfordsville, Indiana

Kurt W. Field
Bradley University
Peoria, Illinois

Matthew A. Fisher
St. Vincent College
Latrobe, Pennsylvania

George Fleck
Smith College
Northampton, Massachusetts

Michael J. Fosmire
Purdue University
West Lafayette, Indiana

Mark Freilich
The University of Memphis
Memphis, Tennessee

Ken Geremia
Copper Development Association, Inc.
New York, New York

Gregory S. Girolami
University of Illinois, Champaign-Urbana
Urbana, Illinois

Carmen Giunta
Le Moyne College
Syracuse, New York

Harold Goldwhite
California State University, Los Angeles
Los Angeles, California

Barbara E. Goodman
University of South Dakota
Vermillion, South Dakota

Gordon W. Gribble
Dartmouth College
Hanover, New Hampshire

Robert K. Griffith
West Virginia University
Morgantown, West Virginia

Neena Grover
Colorado College
Colorado Springs, Colorado

William G. Gutheil
University of Missouri-Kansas City
Kansas City, Missouri

David G. Haase
North Carolina State University
Raleigh, North Carolina

Arthur M. Halpern
Indiana State University
Terre Haute, Indiana

R. Adron Harris
The University of Texas at Austin
Austin, Texas

John Harwood
Tennessee Technological University
Cookeville, Tennessee

Ian S. Haworth
University of Southern California
Los Angeles, California

James M. Helt
CUNY- College of Staten Island
Staten Island, New York

J. Carver Hill
Cary, North Carolina

Darleane C. Hoffman
Lawrence Berkeley National Laboratory, University of California
Berkeley, California

Thomas A. Holme
University of Wisconsin-Milwaukee
Milwaukee, Wisconsin

Chaim T. Horovitz
Emeritus Professor
Rehovot , Israel

John D. Hostettler
San José State University
San José, California

Brian R. James
University of British Columbia
Vancouver, British Columbia, Canada

Frank A. J. L. James
The Royal Institution
London, England

Mark Jensen
Argonne National Laboratory
Argonne, Illinois

Melvin D. Joesten
Vanderbilt University
Nashville, Tennessee

W. L. Jolly
University of California, Berkeley
Berkeley, California

Loretta L. Jones
University of Northern Colorado
Greeley, Colorado

David A. Juckett
Michigan State University
Lansing, Michigan
Barros Research Institute
Holt, Michigan

Paul J. Karol
Carnegie Mellon University
Pittsburgh, Pennsylvania

Vladimir Karpenko
Charles University
Prague, Czech Republic

George B. Kauffman
California State University, Fresno
Fresno, California

W. Frank Kinard
College of Charleston
Charleston, SC

Thomas M. Klapotke
Ludwig-Maximillians-Universität München
Munich, Germany

Kyle Knight
University of Tennessee at Chattanooga
Chattanooga, Tennessee

Paul E. Koch
Penn State, Behrend College
Erie, Pennsylvania

Doris K. Kolb
Bradley University
Peoria, Illinois

Kenneth E. Kolb
Bradley University
Peoria, Illinois

Daniel P. Kramer
Kramer & Asscociates, Ltd.
Centerville, Ohio

Ágúst Kvaran
University of Iceland
Reykjavik, Iceland

J. J. Lagowski
The University of Texas
Austin, Texas

Michael Laing
Professor Emeritus
University of Natal
Durban, South Africa

Rattan Lal
The Ohio State University
Columbus, Ohio

Thomas H. Lane
Dow Corning Coporation
Midland, Michigan

Alan B. G. Lansdown
Imperial College School of Medicine
London, United Kingdom

Amanda Lawrence
University of Illinois, Champaign-Urbana
Urbana, Illinois

Peter B. Leavens
University of Delaware
Newark, Delaware

G. J. Leigh
University of Sussex
Brighton, United Kingdom

H. Eugene LeMay, Jr.
University of Nevada
Reno, Nevada

Kenneth B. Lipkowitz
North Dakota State University
Fargo, North Dakota

Ingrid A. Lobo
University of Texas at Austin
Austin, Texas

Walter Loveland
Oregon State University
Corvallis, Oregon

Phyllis A. Lyday
U.S. Geological Survey
Reston, Virginia

Paul A. Maggard Jr.
North Carolina State University
Raleigh, North Carolina

Michael E. Maguire
Case Western Reserve University
Cleveland, Ohio

Keith L. Manchester
University of the Witwatersrand, Johannesburg
South Africa

Dean F. Martin
University of South Florida
Tampa, Florida

Barbara B. Martin
University of South Florida
Tampa, Florida

James D. Martin
North Carolina State University
Raleigh, North Carolina

Anthony F. Masters
University of Sydney
New South Wales, Australia

Seymour Mauskopf
Duke University
Durham, North Carolina

Sean McMaughan
Texas A&M University
College Station, Texas

Paul F. McMillan
University College London, Christopher Ingold Laboratories
London, United Kingdom

Edward A. Meighen
McGill University
Montreal, Canada

Louis Messerle
University of Iowa
Iowa City, Iowa

Donald E. Moreland
Professor Emeritus
North Carolina State University
Raleigh, North Carolina

Matthew J. Morra
University of Idaho, Soil & Land Resources Division
Moscow, Idaho

Richard Mowat
North Carolina State University
Raleigh, North Carolina

Frank Mucciardi
McGill University
Montreal, Quebec, Canada

Kathleen L. Neeley
University of Kansas
Lawrence, Kansas

John Michael Nicovich
Georgia Institute of Technology
Atlanta, Georgia

Robert Noiva
University of South Dakota School of Medicine
Vermillion, South Dakota

Charles E. Ophardt
Elmhurst College
Elmhurst, Illinois

Richard Pagni
University of Tennessee
Knoxville, Tennessee

Gus J. Palenik
University of Florida
Gainsville, Florida

Robert P. Patterson
North Carolina State University, Crop Science Department
Raleigh, North Carolina

Harry E. Pence
State University of New York, College at Oneonta
Oneonta, New York

Mark A. Pichaj
Biola University
La Marada, California

John Pickering
San José State University
Nuclear Science Facility
San José, California

Tanja Pietraß
New Mexico Tech
Socorro, New Mexico

A. G. Pinkus
Baylor University
Waco, Texas

Jennifer L. Powers
Kennesaw State University
Kennesaw, Georgia

Suzanne T. Purrington
North Carolina State University
Raleigh, North Carolina

Ray Radebaugh
National Institute of Standards and Technology, Cryogenics Technologies Group
Boulder, Colorada

Thomas B. Rauchfuss
University of Illinois, Champaign-Urbana
Urbana, Illinois

Geoffrey W. Rayner-Canham
Sir Wilfred Grenfell College

Corner Brook, Newfoundland, Canada

Marelene Rayner-Canham
Sir Wilfred Grenfell College
Corner Brook, Newfoundland, Canada

Jeffrey C. Reid
North Carolina Geological Survey
Raleigh, North Carolina

Richard E. Rice
Florence, Montana

D. Paul Rillema
Wichita State University
Wichita, Kansas

Frank Rioux
St. John's University
College of St. Benedict
St. Joseph, Minnesota

Robert Rittenhouse
Walla Walla College
College Place, Washington

Elizabeth S. Roberts-Kirchhoff
University of Detroit Mercy
Detroit, Michigan

Brett Robinson
Horticultural and Food Research Institute of New Zealand
Palmerston North, New Zealand

Herbert W. Roesky
University of Göttingen
Göttingen, Germany

Hiranya S. Roychowdhury
New Mexico State University
Las Cruces, New Mexico

Gordon Rutter
Royal Botanic Garden Edinburgh
Edinburgh, United Kingdom

Martin D. Saltzman
Providence College
Providence, Rhode Island

Jerry L. Sarquis
Miami University
Oxford, Ohio

Nancy N. Sauer
Los Alamos National Laboratory
Los Alamos, New Mexico

Wolf-Dieter Schubert
Vienna University of Technology
Vienna, Austria

Joachim Schummer
University of South Carolina
Columbia, South Carolina

Alan Schwabacher
University of Wisconsin-Milwaukee
Milwaukee, Wisconsin

A. Truman Schwartz
Macalester College
Saint Paul, Minnesota

William M. Scovell
Bowling Green State University
Bowling Green, Ohio

Lydia S. Scratch
Ottawa, Ontario, Canada

Peter Rudolf Seidl
Universidade Federal do Rio de Janeiro
Rio de Janeiro, Brazil

N. M. Senozan
California State University
Long Beach, California

James F. Shackelford
University of California at Davis
Davis, California

Stephanie Dionne Sherk
Westland, Michigan

Ben Shoulders
The University of Texas at Austin
Austin, Texas

David B. Sicilia
University of Maryland
College Park, Maryland

Herbert B. Silber
San José State University
San José, California

Robert J. Silva
Lawrence Berkeley National Laboratory, University of California
Berkeley, California

Sharron W. Smith
Hood College
Frederick, Maryland

C. Graham Smith
Penicuik, United Kingdom

Mary L. Sohn
Florida Institute of Technology
Melbourne, Florida

David Speckhard
Loras College
Dubuque, Iowa

Jonathan N. Stack
North Dakota State University
Fargo, North Dakota

Conrad L. Stanitski
University of Central Arkansas
Conway, Arkansas

Donald H. Stedman
University of Denver
Denver, Colorado

Anthony N. Stranges
Texas A&M University
College Station, Texas

Dan M. Sullivan
University of Nebraska at Omaha
Omaha, Nebraska

Tammy P. Taylor
Los Alamos National Laboratory
Los Alamos, New Mexico

Ann T. S. Taylor
Wabash College
Crawfordsville, Indiana

Sidney Toby
Rutgers University
Piscataway, New Jersey

Reginald P. T. Tomkins
New Jersey Institute of Technology
Newark, New Jersey

Anthony S. Travis
The Hebrew University
Jerusalem, Israel

Donald G. Truhlar
University of Minnesota
Minneapolis, Minnesota

Georgios Tsaparlis
University of Ioannina
Ioannina, Greece

Dennis G. Tuck
University of Windsor
Windsor, Ontario, Canada
(deceased)

Geraldo Vicentini
Instituto de Quimíca, University of São Paulo
São Paulo, Brazil

Nanette M. Wachter
Hofstra University
Hempstead, New York

George H. Wahl Jr.
North Carolina State University
Raleigh, North Carolina

Helen R. Webb
Monash University
Clayton, Victoria, Australia

Laurence E. Welch
Knox College
Galesburg, Illinois

J. Reed Welker
University of Arkansas
Fayetteville, Arkansas
(retired)

Guang Wen
Tottori University
Tottori, Japan

Todd W. Whitcombe
University of Northern British Columbia
Prince George, British Columbia, Canada

Vivienne A. Whitworth
Ryerson University
Toronto, Ontario, Canada

David A. Williams
University College London
London, United Kingdom

Robert K. Wismer
Millersville University
Millersville, Pennsylvania

Ronni Wolf
Kaplan Medical Center
Rechovot, Israel

Adele J. Wolfson
Wellesley College
Wellesley, Massachusetts

Burghard Zeiler
Wolfram Bergbau und Hüttenges m.b.H.
Austria, Germany

Lea B. Zinner
Departamento de Química
São Paulo, Brazil

Cynthia G. Zoski
Georgia State University
Atlanta, Georgia

Thomas M. Zydowsky
FMC Corporation
Princeton, New York

Kekulé, Friedrich August

GERMAN CHEMIST
1829–1896

Friedrich August Kekulé was born on September 7, 1829, in Darmstadt, Hesse (later part of Germany). He showed an early aptitude for both languages and drawing and wanted to be an architect. He began his architecture studies at the University of Geissen in 1847, but after attending the lectures of the famous chemist Justus von Liebig he switched to chemistry. Kekulé had great interest in the theoretical aspects of chemistry, and less in the more practical applications that so interested von Liebig. On von Liebig's advice Kekulé went to Paris in 1851 to further his chemical studies.

Paris during the 1850s was an ideal place for a young scientist, as there was a great deal of interest in that city in theoretical chemistry, particularly in the structure of molecules. Preexisting ideas (such as the dualism of Swedish chemist Jöns J. Berzelius) stressed that molecules formed because of the inherent electrical charges that individual elements possessed (which were sometimes opposing and therefore attractive). Organic molecules were not in keeping with the dualism concept, but some scientists proposed that they could be derived from a number of simple inorganic molecules.

German chemist Friedrich Kekulé, known for his work and theories in molecular structure, such as the tetravalent structure of carbon.

Kekulé returned to Germany in 1852, obtained his doctoral degree at Geissen in that year, and then spent a year working in Switzerland. This was followed by two years in London (1853–1855), where he met the chemist Alexander Williamson. Williamson had extended Charles Gerhardt's "type theories" to explain how ethers could be derived from the water type. Kekulé, with Williamson's encouragement, extended type theory further and introduced a new type—the methane or marsh gas type. This led to the development of the tetravalent model of carbon and the understanding that carbon forms rings and chains.

Kekulé's principal insight was to realize that type theory did not take into account the specific combining power (or valences) of specific atoms. In 1857 Kekulé suggested that carbon was tetravalent, his suggestion based on the specific chemistries of the compounds that carbon formed with elements such as hydrogen (CH_4) and chlorine (CCl_4). Kekulé extended his ideas in the following year by suggesting that two carbon atoms bonded together in the formation of hydrocarbons such as ethane (C_2H_6). Similarly, additional carbon and hydrogen units could be added, extending the carbon

JOSEF LOSCHMIDT (1821–1895)

Before Kekulé dreamed of his structure for benzene, Josef Loschmidt proposed the ring structure of the molecule. Unfortunately, Loschmidt did not publish his theory in a widely read scientific journal. As a result, the credit for this revolutionary theory is hotly debated today.

—Valerie Borek

atom chain and forming an ordered series. A similar structural theory was developed independently at around the same time by the Scottish chemist Archibald Scott Couper, working in the laboratory of Adolphe Wurtz in Paris. Publication of Couper's paper was delayed by Wurtz (until Kekulé's had appeared) and the structural theory of organic chemistry is really a culmination of the efforts of Kekulé and Couper.

Kekulé was offered the position of professor of chemistry at the University of Ghent in Belgium in 1858. There his linguistic abilities stood him in good stead, as he had to lecture in French. In 1867 Kekulé was called to the University of Bonn and remained there until his death, on July 13, 1896.

In 1859 Kekulé started to use graphical representations of organic molecules, in part to emphasize the tetravalent nature of carbon atoms and their ability to form chains. He then turned his attention to the structure of benzene (C_6H_6), a compound with unusual properties that could not be explained by any theories of the day.

Kekulé proposed in 1865 that benzene had a structure in which six carbon atoms formed a ring, with alternating single and double bonds. However, the chemistry of benzene was not always consistent with this structural formula. (All of the carbon atoms in a benzene molecule were equal and equivalent in terms of the reactions of benzene.) To overcome this problem, Kekulé suggested in 1872 that there were two forms of benzene, in dynamic **equilibrium**. Kekulé's dynamical theory proved to be only partially correct. In 1933 Linus Pauling used quantum mechanics to explain more fully the nature of benzene.

equilibrium: condition in which two opposite reactions are occurring at the same speed, so that concentrations of products and reactants do not change

On March 11, 1890, on the occasion of the twenty-fifth anniversary of his announcement of his benzene theory, Kekulé gave a speech in Berlin in which he revealed that both his structural theories and the structure of benzene were revealed to him in dreams. Scholars have tended to dismiss his account of his own creative processes and have placed more stock in the early training that Kekulé received in architecture as the key to his inspiration. SEE ALSO BERZELIUS, JÖNS JAKOB; LIEBIG, JUSTUS VON; ORGANIC CHEMISTRY; PAULING, LINUS.

Martin D. Saltzman

Bibliography

Gillis, Jean (1972). "Friedrich August Kekule." In *The Dictionary of Scientific Biography,* Volume 7. New York: Scribners.

Wotiz, John H. (1993). *The Kekule Riddle: A Challenge for Chemists and Psychologists.* Clearwater, FL: Cache River Press.

Kelsey, Frances Kathleen Oldham

CANADIAN-AMERICAN PHARMACOLOGIST AND PHYSICIAN
1914–

Born Frances Kathleen Oldham in Cobble Hill, Vancouver Island, Canada, on July 24, 1914, Oldham grew up in the country and always wanted to be a scientist. She earned B.S., M.S., and Ph.D. degrees in pharmacology. During the 1930s Oldham and fellow pharmacologist Fremont Ellis Kelsey studied the effects of the drug quinine on pregnant rabbits and their embryos at the University of Chicago. They discovered that the adult rabbits could

Pharmacologist Frances Oldham Kelsey, who rejected the use of the sedative thalidomide in the United States.

metabolize the drug, due to the presence of an enzyme in their bodies, but that the embryos died, because they had not yet acquired this enzyme. They were among the first scientists to verify that some drugs that are safe for adult humans are dangerous to human embryos. Oldham and Kelsey married in 1943. Frances Kelsey earned an M.D. degree in 1950; the Kelseys then moved to South Dakota, where Frances practiced medicine part time, taught pharmacology, and reviewed scientific articles.

metabolize: performing metabolism—the processes that produce complex substances from simpler components, with a consequent use of energy (anabolism) and those that break down complex food molecules, thus liberating energy (catabolism)

In 1960 the Kelseys moved to Washington, D.C., where Frances was offered a job at the Food and Drug Administration (FDA). Her job was to evaluate applications from drug companies that wished to market new drugs. She had sixty days to evaluate each application. The first drug she was asked to evaluate was thalidomide, a sedative prescribed to pregnant women for relief from morning sickness. The drug had been marketed to the masses in West Germany, where it had first been developed, and in many other countries, including Canada.

Frances Kelsey and the pharmacologist and chemist who assisted her in the review of thalidomide were immediately concerned that the information that had been provided by the drug firm did not prove the drug's safety. The company was asked to resubmit their application. At around this time, Kelsey began to read reports that some patients who had taken thalidomide had developed peripheral neuritis, a disease whose symptoms included a painful tingling in the arms and feet. Kelsey rejected a second application. Remembering her earlier research with quinine, Kelsey wondered what thalidomide might do to the developing fetuses of pregnant women. The drug company continued to pressure Kelsey to approve its application, even contacting her superiors at the FDA. She continued to reject their applications. In 1961 European pediatricians began reporting an epidemic of phocomelia among newborns, a developmental anomaly characterized by short limbs, toes attached to hips, and flipperlike arms. By November 1961 a German pediatrician had determined that the teratogen (a substance causing one or more developmental abnormalities in fetuses) was thalidomide. From 1957 to 1962, more than 10,000 children in forty-six countries were estimated to have deformities owing to their mothers' use of thalidomide during early pregnancy. By her insistence on getting thorough, all-sided information on drug safety before drug application approval, Kelsey had prevented the use of thalidomide in the United States.

On August 17, 1962, President John F. Kennedy presented Kelsey with the President's Award for Distinguished Federal Civilian Service, in recognition of her having prevented (almost single-handedly) a national tragedy. In late 1962 the U.S. Congress passed the Kefauver–Harris Amendments to the Food, Drug, and Cosmetics Act (named after the senator and congressman who sponsored the bills in Congress), which gave medical officers at the FDA more power. The amendments outlawed drug testing on humans without informed consent, and mandated that a new drug could not be approved until scientific evidence had established that the drug was safe and effective. As of 2001 Kelsey still worked at the FDA as one of their leading compliance officers. In 2000 she was inducted into the National Women's Hall of Fame, whose Web site states: "Kelsey is both a woman of courage and one of reason—demanding of herself and others in her profession high standards of science and integrity." SEE ALSO ENZYMES; TERATOGEN.

Kathleen L. Neeley

Bibliography

Bren, Linda (2001). "Frances Oldham Kelsey: FDA Medical Reviewer Leaves Her Mark on History." *FDA Consumer* 35(2):24–29. Also available from <http://www.fda.gov>.

Truman, Margaret (1976). *Women of Courage.* New York: William Morrow.

Internet Resources

National Women's Hall of Fame. "Frances Kathleen Oldham Kelsey, Ph.D., M.D. (1914–)." Available from <http://www.greatwomen.org>.

Khorana, Har Gobind

AMERICAN BIOCHEMIST
1922–

Har Gobind Khorana was born in a small village in British India, in which his family was among the few literate residents. He received his M.S. from the University of Lahore, and in 1945 he was awarded a grant to study for a Ph.D. at the University of Liverpool. He then went on to complete post-doctoral work in Switzerland and at Cambridge. It was there, while working with Alexander Todd, that Khorana became interested in both nucleic acids and proteins, the study of which became his life's work.

In 1952 G. M. Shrum, head of the British Research Council on the campus of the University of British Columbia, Vancouver, Canada, offered Khorana the opportunity to form his own research group on whatever topic he wished. His group became very successful in developing methods for synthesizing phosphate **ester** derivatives of nucleic acids, and in 1959 he and John G. Moffatt announced the **synthesis** of acetyl coenzyme A (acetyl CoA), a molecule essential to the biochemical processing of proteins fats and carbohydrates. Prior to this work, the coenzyme had to be extracted from yeast by a very laborious and expensive process, so this discovery led to Khorana's international recognition within the scientific community and he received many job offers as a result. He accepted the position of codirector of the Institute for Enzyme Research at the University of Wisconsin.

American chemist Har Gobind Khorana, corecipient, with Robert W. Holley and Marshall W. Nirenberg, of the 1968 Nobel Prize in physiology or medicine, "for their interpretation of the genetic code and its function in protein synthesis."

In the early 1960s it had been recognized that **DNA** and **RNA** (in the form of messenger RNA [mRNA]) were somehow involved in the synthesis of proteins in living cells. Whereas the basic building blocks of DNA are the four nucleotides adenosine (A), **cytosine** (C), **guanine** (G), and **thymine** (T)—in RNA, **uracil** (U) is substituted for thiamine—the basic building blocks of all proteins are twenty amino acids strung together in different sequences to produce individual proteins. In 1961, Marshall W. Nirenberg, and Heinrich J. Matthaei announced that they had created a synthetic mRNA, which, when inserted into *E. coli* bacteria, always caused the addition of one amino acid phenylalanine to a growing strand of linked amino acid. They also determined that if they synthesized RNA with three units of uracil joined together, it caused an amino acid chain consisting entirely of phenylalanine to be produced.

These experiments, which proved that mRNA transmits the genetic information from DNA, thus directing the creation of specific complex proteins, stimulated Khorana to use his expertise in **polynucleotide synthesis** to uncover the exact mechanisms involved. The results were spectacular. Within a few short years his research group was able to establish which serial combinations of nucleotides form which specific amino acids; that nucleotide instructions (genetic **code**) are always transmitted to the cell in groups of three called codons; and that some of the codons direct the cell to start or stop the manufacture of proteins. For this work Khorana, along with Nirenberg and biochemist Robert W. Holley, was awarded the Nobel Prize in physiology in 1968.

In 1970 Khorana announced the creation of the first artificial DNA gene of yeast. At the same time, he and most of his research team moved to the Massachusetts Institute of Technology (MIT) because, as he explained, "You stay intellectually alive longer if you change your environment every so often" (McMurray, p. 1089). Since going to MIT, Khorana has reported major advances concerning how rhodopsin, the photo**receptor** in the human eye, functions. SEE ALSO CODON; NUCLEIC ACIDS; TODD, ALEXANDER.

John E. Bloor

ester: organic species containing a carbon atom attached to three moieties: an O via a double bond, an O attached to another carbon atom or chain, and an H atom or C chain; the R(C=O)OR functional group

synthesis: combination of starting materials to form a desired product

DNA: deoxyribonucleic acid—the natural polymer that stores genetic information in the nucleus of a cell

RNA: ribonucleic acid—a natural polymer used to translate genetic information in the nucleus into a template for the construction of proteins

cytosine: heterocyclic, pyrimidine, amine base found in DNA

guanine: heterocyclic, purine, amine base found in DNA

thymine: one of the four bases that make up a DNA molecule

uracil: heterocyclic, pyrimidine, amine base found in RNA

polynucleotide synthesis: formation of DNA or RNA

code: mechanism to convey information on genes and genetic sequence

receptor: area on or near a cell wall that accepts another molecule to allow a change in the cell

Bibliography

McMurray, Emily J., ed. (1995). *Notable Twentieth-Century Scientists.* Detroit, MI: Gale Research.

Kinase

adenosine triphosphate (ATP): molecule formed by the condensation of adenine, ribose, and triphosphoric acid, HOP(O)OH–O–(O)OH–OP(O)OH–OH; it is a key compound in the mediation of energy in both plants and animals

phosphorylation: the process of addition of phosphates into biological molecules

glucose: common hexose monosaccharide; monomer of starch and cellulose; also called grape sugar, blood sugar, or dextrose

metabolism: the complete range of biochemical processes that take place within living organisms; comprises processes that produce complex substances from simpler components, with a consequent use of energy (anabolism), and those that break down complex food molecules, thus liberating energy (catabolism)

code: mechanism to convey information on genes and genetic sequence

Kinases are enzymes that transfer a phosphate group from **adenosine triphosphate (ATP)**, or other trinucleotide, to a number of biological substrates, such as sugars or proteins. They are part of a larger family of enzymes known as group transferases, but are limited to phosphate transfers. A typical reaction catalyzed by a kinase (e.g., hexokinase) is the **phosphorylation** of **glucose** upon its entry into a cell

$$\text{Glucose} + \text{ATP} \rightarrow \text{Glucose-6-phosphate} + \text{ADP}$$

This reaction sets the stage for the subsequent **metabolism** of glucose via a number of metabolic pathways. Creatine kinase (CK), which transfers a phosphate from ATP to creatine, acts to store some of the energy of ATP in the muscle molecule creatine. CK levels in blood are measured in blood tests to diagnose heart attacks, as damaged heart muscle cells release CK into the bloodstream. Recently there has been great interest in the phosphorylation of specific amino acids in proteins. This modification acts as a regulation of the protein's activity. Hormonal signals outside a cell initiate a cascade of biochemical events inside the cell that include activation of a number of protein kinases. Of further interest in kinase regulation is the fact that several cancer-causing genes (oncogenes) **code** for kinases. The lack of regulation of these genes may be important in the etiology of cancer. SEE ALSO Amino Acid; Enzymes.

C. Larry Bering

Bibliography

Moran, Laurence E.; Scrimgeour, K. Gray; Horton, H. Robert; et al. (1994). *Biochemistry*, 2nd edition. Englewood Cliffs, NJ: Prentice Hall.

Internet Resources

"The Protein Kinase Resource Center." San Diego Supercomputer Center. Available from <http://www.sdsc.edu>.

Kinetics

Chemical kinetics is the study of the rates of chemical reactions. Such reaction rates range from the almost instantaneous, as in an explosion, to the almost unnoticeably slow, as in corrosion. The aim of chemical kinetics is to make predictions about the composition of reaction mixtures as a function of time, to understand the processes that occur during a reaction, and to identify what controls its rate.

Rates and Rate Laws

spectroscopy: use of electromagnetic radiation to analyze the chemical composition of materials

The rate of a chemical reaction is defined as the rate of change of the concentration of one of its components, either a reactant or a product. The experimental investigation of reaction rates therefore depends on being able to monitor the change of concentration with time. Classical procedures for reactions that take place in hours or minutes make use of a variety of techniques for determining concentration, such as **spectroscopy** and electrochemistry. Very fast reactions are studied spectroscopically. Spectroscopic procedures are available for monitoring reactions that are initiated by a rapid pulse of electromagnetic radiation and are over in a few femtoseconds (1 fs $= 10^{-15}$ s).

The analysis of kinetic data commonly proceeds by establishing a *rate law*, a mathematical expression for the rate in terms of the concentrations of the reactants (and sometimes products) at each stage of the reaction. For instance, it may be found that the rate of consumption of a reactant is proportional to the concentration of the reactant, in which case the rate law is

$$\text{Rate} = k[\text{Reactant}]$$

where [Reactant] denotes the concentration of the reactant and k is called the rate constant. The rate constant is independent of the concentrations of any species in the reaction mixture but depends on the temperature. A reaction with a rate law of this form is classified as a *first-order rate law*. More generally, a reaction with a rate law of the form

$$\text{Rate} = k[\text{Reactant A}]^a[\text{Reactant B}]^b\ldots$$

is said to be of order a in A, of order b in B, and to have an overall order of a + b + ... Some rate laws are far more complex than these two simple examples and many involve the concentrations of the products.

The advantage of identifying the reaction order is that all reactions with the same rate law (but different characteristic rate constants) behave similarly. For example, the concentration of a reactant in a first-order reaction decays exponentially with time at a rate determined by the rate constant

$$[\text{Reactant}] = [\text{Reactant}]_0 e^{-kt}$$

where $[\text{Reactant}]_0$ is the initial concentration of the reactant. On the other hand, all second-order reactions lead to the following time-dependence of the concentration:

$$[\text{Reactant}] = \frac{[\text{Reactant}]_0}{1 + kt[\text{Reactant}]_0}$$

Figure 1 shows the time-dependence predicted by these expressions. It is common to report the time-dependence of first-order reactions in terms of the *half-life*, $t_{1/2}$, of the reactant, the time needed for its concentration to fall to half its initial value. For a first-order reaction (but not for other orders)

$$t_{1/2} = \frac{\ln 2}{k}$$

Thus, reactions with large rate constants have short half-lives.

Reaction Mechanisms

The identification of a rate law provides valuable insight into the reaction mechanism, the sequence of elementary steps by which a reaction takes place. The aim is to identify the reaction mechanism by constructing the rate law that it implies. This procedure may be simplified by identifying the *rate-determining step* of a reaction, the slowest step in a sequence that determines the overall rate. Thus, if the proposed mechanism is A → B followed by B → C, and the former is much faster than the latter, then the overall rate of the reaction will be equal to the rate of A → B, for once B is formed, it immediately converts into C.

In general, for a mechanism of many steps (including their reverse), the construction of the overall rate law is quite difficult, requiring an approximation or a computer for a numerical analysis. One common approximation

Figure 1. The time-dependence of the concentration of the reactant in first-order and second-order reactions. The two reactions have the same initial rate. Note how the reactant takes longer to disappear in a second-order reaction.

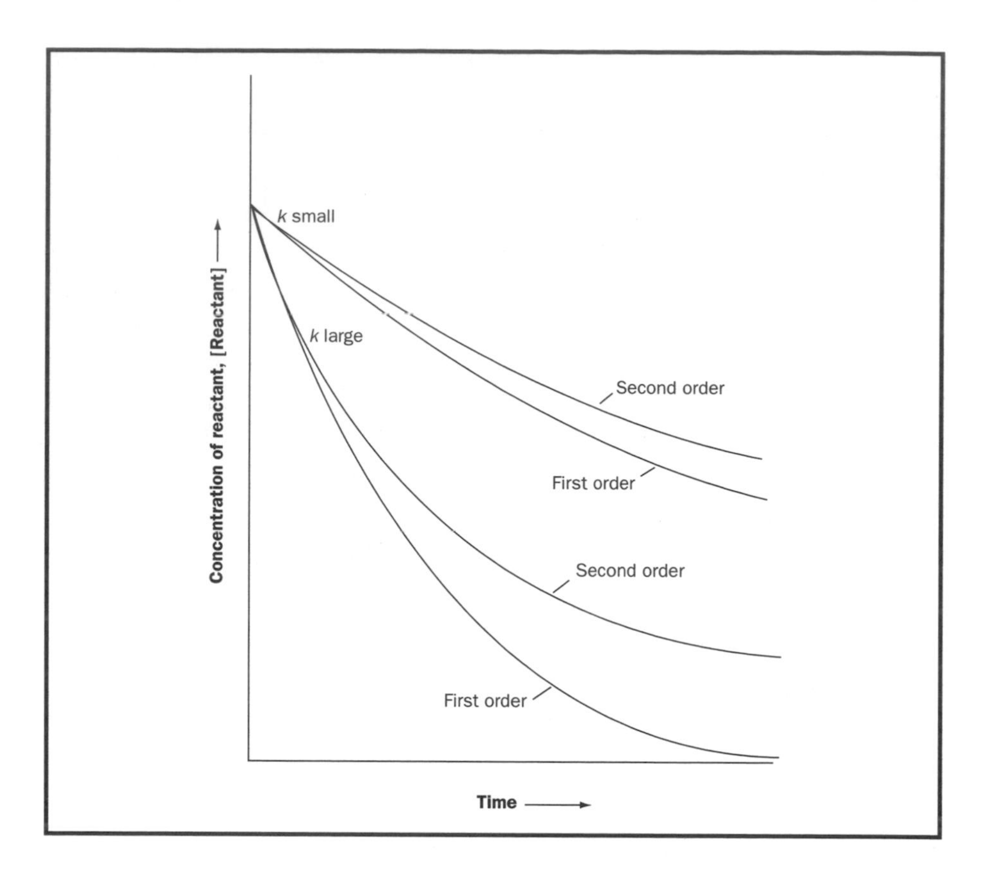

intermediate: molecule, often short-lived, that occurs while a chemical reaction progresses but is not present when the reaction is complete

phase: homogeneous state of matter

Kinetics is the study of the rate of chemical reactions, such as the acid dehydration of sugar, shown in this photograph.

is the *steady-state assumption*, in which the net rate of formation of any **intermediate** (B in the present example) is set equal to zero. A hazard of using kinetic information to identify a reaction mechanism, however, is that more than one mechanism might result in the same rate law, especially when approximate solutions are derived. For this reason, proposed reaction mechanism must be supported by additional evidence.

The Origin of Reaction Rates

Once a reaction mechanism has been identified, attention turns to the molecular properties that govern the values of the rate constants that occur in the individual elementary steps. A clue to the factors involved is provided by the experimental observation that the rate constants of many reactions depend on temperature according to the Arrhenius expression

$$\ln k = A - \frac{E_a}{RT}$$

where E_a is called the *activation energy.*

The simplest model that accounts for the Arrhenius expression is the *collision theory* of gas-**phase** reaction rates, in which it is supposed that reaction occurs when two reactant molecules collide with at least a minimum kinetic energy (which is identified with the activation energy, Figure 2). A more sophisticated theory is the *activated complex theory* (also known as the *transition state theory*), in which it is supposed that the reactants encounter each other, form a loosened cluster of atoms, then decompose into products.

Reactions in solution require more detailed consideration than reactions in gases. It is necessary to distinguish between "diffusion-controlled" and

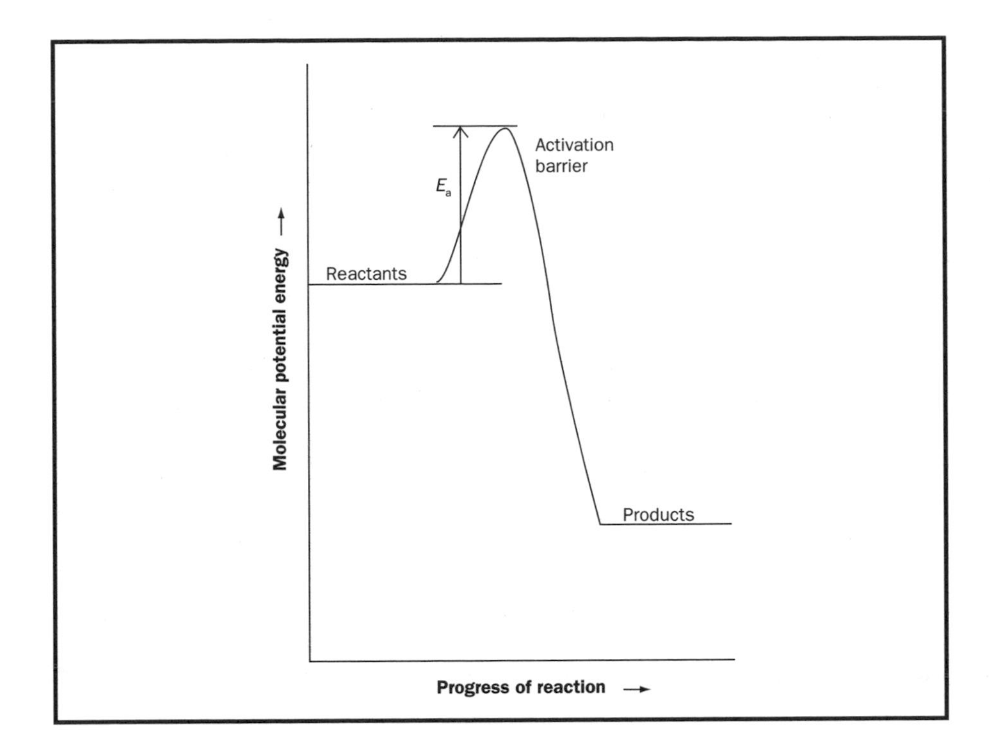

Figure 2. A "reaction profile" showing how molecular potential energy varies as reactants change into products. Only reactants that encounter each other with a kinetic energy equal to the height of the activation barrier can go on to form products. The height of the activation barrier is the activation energy of the reaction.

"activation-controlled" reactions. In a diffusion-controlled reaction, the rate is controlled by the ability of the reactants to migrate through the solvent and encounter each other. In an activation-controlled reaction, the rate is controlled by the ability of the reactants that have met each other to acquire enough energy to react.

The rate of a reaction may also be increased by finding a **catalyst**, a substance that takes part in a reaction by providing an alternative pathway with a lower activation energy but is regenerated in the process and is therefore not consumed. **Catalysis** is the foundation of the chemical industry and a great effort is made to discover or fabricate efficient, economical catalysts. It is also the foundation of life, because the biological catalysts known as enzymes (elaborate protein molecules) control almost every aspect of an organism's function. SEE ALSO CATALYSIS AND CATALYSTS; ENZYMES; PHYSICAL CHEMISTRY.

catalyst: substance that aids in a reaction while retaining its own chemical identity

catalysis: the action or effect of a substance in increasing the rate of a reaction without itself being converted

Peter Atkins

Bibliography

Atkins, Peter, and de Paula, Julio (2002). *Physical Chemistry.* New York: W.H. Freeman.

Atkins, Peter, and Jones, Loretta (2001). *Chemical Principles.* New York: W.H. Freeman.

Laidler, K. J. (1987). *Chemical Kinetics.* New York: Harper and Row.

Pilling, M. J., and Seakins, P. W. (1996). *Reaction Kinetics.* London: Oxford University Press.

Krebs, Hans Adolf

BRITISH BIOCHEMIST
1900–1981

Hans Krebs was born into a prosperous and well-educated family in Hildesheim, Germany. His father was a physician who specialized in otolaryngology, and it was Hans's intention to follow in his father's footsteps and

British biochemist Sir Hans Adolf Krebs, corecipient of the 1953 Nobel Prize in physiology or medicine, "for his discovery of co-enzyme A and its importance for intermediary metabolism."

become a physician. Krebs was educated at the Gymnasium Andreanum, and after World War I, he went on to study medicine at the Universities of Göttingen, Freiburg, and Berlin. In 1925 he earned an M.D. degree at the University of Hamburg. He was at this point passionately attracted to medical research, and he did not enter medical practice. In 1926, Krebs became an assistant to Professor Otto Warburg at the prestigious Kaiser Wilhelm Institute for Biology in Berlin, a post he held until 1930. Warburg (who later won the 1931 Nobel Prize in medicine) encouraged Krebs to pursue a career in research.

In 1931 Krebs moved to Freiburg to teach medicine. It was there that he authored (with Kurt Henseleit) his first important paper, which examined liver function in mammals and described how ammonia was converted to urea in liver cells. Krebs also studied the syntheses of uric acid and purines in birds. However, Krebs's research was cut short when the Nazis came to power in 1933. Krebs was Jewish, and he was therefore summarily fired from his post. He left Germany for England, taking a position at the School of Biochemistry at Cambridge University at the invitation of Sir Frederick Gowland Hopkins (who had won the 1929 Nobel Prize in medicine). In 1935 Krebs moved to the University of Sheffield to become a lecturer in pharmacology.

At Sheffield Krebs embarked upon the work that would elucidate some of the complex reactions of cell **metabolism** (the processes that extract energy from food). This extraction of energy is achieved via a series of chemical transformations that remove energy-rich electrons from molecules obtained from food. These electrons pass along a chain of molecular carriers in a way that ultimately gives rise to water and **adenosine triphosphate (ATP)**, which is the primary source of chemical energy that powers cellular activity.

metabolism: the complete range of biochemical processes that take place within living organisms; comprises processes that produce complex substances from simpler components, with a consequent use of energy (anabolism), and those that break down complex food molecules, thus liberating energy (catabolism)

adenosine triphosphate (ATP): molecule formed by the condensation of adenine, ribose, and triphosphoric acid, HOP(O)OH–O–(O)OH–OP(O)OH–OH; it is a key compound in the mediation of energy in both plants and animals

Krebs found that the pivotal mechanism of cell metabolism was a cycle. The cycle starts with glycolysis, which produces acetyl coenzyme A (acetyl CoA) from food molecules—carbohydrates, fats, and certain amino acids. The acetyl CoA reacts with oxaloacetate to form citric acid. The citric acid then goes through seven reactions that reconvert it back to oxaloacetate, and the cycle repeats. There is a net gain of twelve molecules of ATP per cycle. Not only does this cycle (known as the Krebs cycle, and also as the tricarboxylic acid cycle and the citric acid cycle) generate the chemical energy to run the cell, it is also a central component of the syntheses of other biomolecules.

carboxylic acid: one of the characteristic groups of atoms in organic compounds that undergoes characteristic reactions, generally irrespective of where it occurs in the molecule; the $-CO_2H$ functional group

Krebs published his groundbreaking paper on this cyclic component of cell metabolism in the journal *Enzymologia* in 1937, and it quickly became a foundational concept in biochemistry and cell biology. It was for this research that Krebs won the Nobel Prize in medicine in 1953. (He remains one of the most often cited scientists in cell biology, with his work being noted more than 11,000 times since 1961, when the citation records of original articles in cell biology began being counted.)

Krebs worked in both research and applied science in the area of cell metabolism and nutrition. During World War II he developed a bread that helped to keep the British people nourished at a time of food shortages. He developed new analytical techniques for research in cell biology and investigated other metabolic reactions, such as the **synthesis** of glutamic acid.

synthesis: combination of starting materials to form a desired product

He was also an energetic instructor. His students went on to become directors of laboratories and to win many prizes.

In 1954 Krebs was appointed the Whitley Chair of Biochemistry at Oxford University. That same year he received the Royal Medal of the **Royal Society** of London. In 1958, for his scientific work and his contributions to the lives of British people, Krebs was knighted. Even after his retirement in 1967, he continued to do research on liver disease, the genetic bases of metabolic diseases, and the link between poor nutrition and juvenile delinquency. In addition to his Nobel Prize and Royal Medal, he received honorary degrees from nine universities. SEE ALSO GLYCOLYSIS; KREBS CYCLE.

Andrew Ede

Royal Society: The U.K. National Academy of Science, founded in 1660

Bibliography

Holmes, Frederic Lawrence (1991–1993). *Hans Krebs,* Vols. I and II. New York: Oxford University Press.

Krebs, Hans, and Johnson, W. A. (1937). "The Role of Citric Acid in Intermediate Metabolism in Animal Tissues." *Enzymologia* 4:148–156.

Krebs, Hans, and Martin, Anne (1981). *Reminiscences and Reflections.* Oxford, UK: Clarendon Press.

Internet Resources

More information available from <www.nobel.se/medicine/laureates/1953/>.

Krebs Cycle

The Krebs cycle is a series of enzymatic reactions that catalyzes the aerobic **metabolism** of fuel molecules to carbon dioxide and water, thereby generating energy for the production of **adenosine triphosphate (ATP)** molecules. The Krebs cycle is so named because much of its elucidation was the work of the British biochemist Hans Krebs. Many types of fuel molecules can be drawn into and utilized by the cycle, including acetyl coenzyme A (acetyl CoA), derived from glycolysis or fatty acid **oxidation**. Some amino acids are metabolized via the enzymatic reactions of the Krebs cycle. In **eukaryotic cells**, all but one of the enzymes catalyzing the reactions of the Krebs cycle are found in the mitochondrial matrixes.

The sequence of events known as the Krebs cycle is indeed a cycle; oxaloacetate is both the first reactant and the final product of the metabolic pathway (creating a loop). Because the Krebs cycle is responsible for the ultimate oxidation of metabolic **intermediates** produced during the metabolism of fats, proteins, and carbohydrates, it is the central mechanism for metabolism in the cell. In the first reaction of the cycle, acetyl CoA condenses with oxaloacetate to form citric acid. Acetyl CoA utilized in this way by the cycle has been produced either via the oxidation of fatty acids, the breakdown of certain amino acids, or the oxidative decarboxylation of **pyruvate** (a product of glycolysis). The citric acid produced by the condensation of acetyl CoA and oxaloacetate is a tri**carboxylic acid** containing three **carboxylate** groups. (Hence, the Krebs cycle is also referred to as the citric acid cycle or tricarboxylic acid cycle.)

After citrate has been formed, the cycle machinery continues through seven distinct enzyme-catalyzed reactions that produce, in order, isocitrate, α-ketoglutarate, succinyl coenzyme A, succinate, fumarate, malate, and

metabolism: the complete range of biochemical processes that take place within living organisms; comprises processes that produce complex substances from simpler components, with a consequent use of energy (anabolism), and those that break down complex food molecules, thus liberating energy (catabolism)

adenosine triphosphate (ATP): molecule formed by the condensation of adenine, ribose, and triphosphoric acid, HOP(O)OH–O–(O)OH–OP(O)OH–OH; it is a key compound in the mediation of energy in both plants and animals

oxidation: process that involves the loss of electrons (or the addition of an oxygen atom)

eukaryotic cell: cell characterized by membrane-bound organelles, most notably the nucleus, and that possesses chromosomes whose DNA is associated with proteins

intermediate: molecule, often short-lived, that occurs while a chemical reaction progresses but is not present when the reaction is complete

pyruvate: anion of pyruvic acid produced by the reaction of oxygen with lactic acid after strenuous exercise

carboxylic acid: one of the characteristic groups of atoms in organic compounds that undergoes characteristic reactions, generally irrespective of where it occurs in the molecule; the $–CO_2H$ functional group

carboxylate: structure incorporating the –COO– group

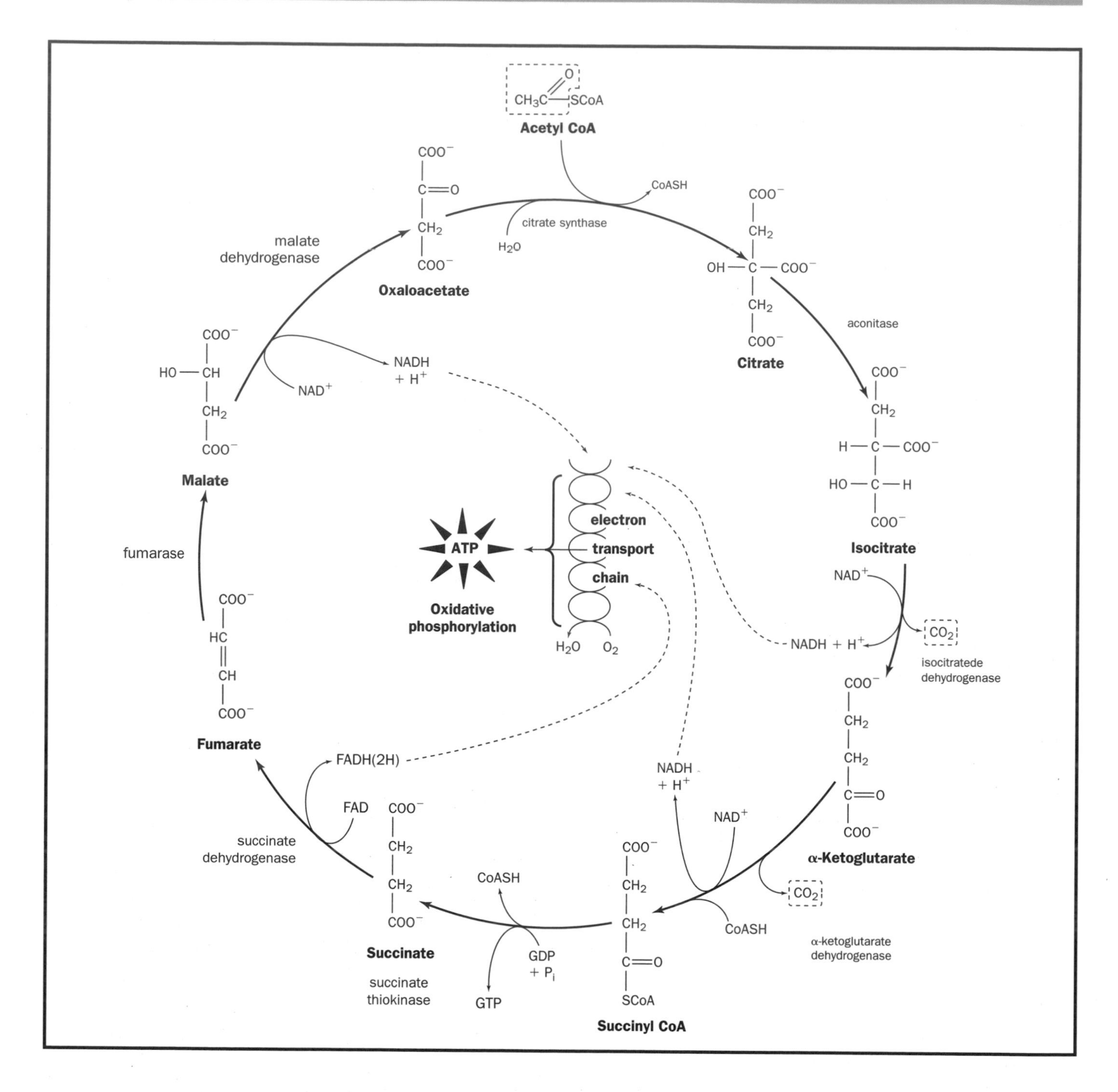

Figure 1. Krebs Cycle.

oxaloacetate. The freshly produced oxaloacetate, in turn, reacts with yet another molecule of acetyl CoA, and the cycle begins again. Each turn of the Krebs cycle produces two molecules of carbon dioxide, one guanosine triphosphate molecule (GTP), and enough electrons to generate three molecules of NADH and one molecule of $FADH_2$.

The Krebs cycle is present in virtually all eukaryotic cells that contain mitochondria, but functions only as part of aerobic metabolism (when oxygen is available). This oxygen requirement is owing to the close relationship between the mitochondrial electron transport chain and the Krebs cycle. In the Krebs cycle, four **oxidation–reduction reactions** occur. A high energy phosphate bond in the form of GTP is also generated. (This high

oxidation–reduction reaction: reaction, sometimes called redox, that involves the movement of electrons between reactants to form products

energy phosphate bond is later transferred to adenosine diphosphate [ADP] to form adenosine triphosphate [ATP].) As the enzymes of the Krebs cycle oxidize fuel molecules to carbon dioxide, the coenzymes NAD^+, FAD, and coenzyme Q (also known as ubiquinone) are reduced. In order for the cycle to continue, these reduced coenzymes must become reoxidized by transferring their electrons to oxygen, thus producing water. Therefore, the final acceptor of the electrons produced by the oxidation of fuel molecules as part of the Krebs cycle is oxygen. In the absence of oxygen, the Krebs cycle is inhibited.

The citric acid cycle is an amphibolic pathway, meaning that it can be used for both the **synthesis** and degradation of biomolecules. Besides acetyl CoA (generated from **glucose**, fatty acids, or ketogenic amino acids), other biomolecules are metabolized by the cycle. Several amino acids are degraded to become what are intermediates of the cycle. Likewise, odd-chain fatty acids are metabolized to form succinyl coenzyme A, another intermediate of the cycle. Krebs cycle intermediates are also used by many organisms for the synthesis of other important biomolecules. Some organisms use the Krebs cycle intermediates α-ketoglutarate and oxaloacetate in the synthesis of several amino acids. Succinyl coenzyme A is utilized in the synthesis of **porphyrin** rings, used in heme manufacture and **chlorophyll biosynthesis**. Oxaloacetate and malate are utilized in the synthesis of glucose, in a process known as gluconeogenesis. SEE ALSO GLYCOLYSIS; KREBS, HANS ADOLF.

Robert Noiva

synthesis: combination of starting materials to form a desired product

glucose: common hexose monosaccharide; monomer of starch and cellulose; also called grape sugar, blood sugar, or dextrose

porphyrin: common type of heterocyclic ligand that has four five-member rings with a nitrogen, all of which lie in a plane; nitrogen atom lone pairs are available for coordinate covalent bonds

chlorophyll: active molecules in plants undergoing photosynthesis

biosynthesis: formation of a chemical substance by a living organism

Bibliography

Berg, Jeremy M.; Tymoczko, John L.; and Stryer, Lubert (2002). *Biochemistry*, 5th edition. New York: W. H. Freeman.

Voet, Donald; Voet, Judith G.; and Pratt, Charlotte W. (2002). *Fundamentals of Biochemistry*, updated edition. New York: Wiley.

Krypton

MELTING POINT: −157.36°C
BOILING POINT: −153.22°C
DENSITY: 2.818 g/cm³
MOST COMMON IONS: None

Krypton (from the Greek word *kryptos*, meaning "hidden"), is the second heaviest of the **noble gases**. It was discovered in 1898 by Sir William Ramsay and Morris Travers during their experiments with liquid air, air that has been liquefied by cooling. It has a concentration of 1.14 ppm by volume in Earth's atmosphere. It is present in the Sun and in the atmosphere of Mars.

noble gas: element characterized by inert nature; located in the rightmost column in the Periodic Table

At room temperature krypton is a colorless, odorless gas. Upon freezing it forms a white crystal with a face-centered cubic structure. In a vacuum discharge tube, it emits primarily a mixture of green and yellow light. During the late twentieth century the wavelength of light corresponding to krypton's 605.78-nanometer (2.4×10^{-5}-inch) **spectral line** was the internationally adopted definition of the meter. Krypton gas is used in the manufacture of fluorescence lights and flashlamps used in high-speed photography.

spectral line: line in a spectrum representing radiation of a single wavelength

Krypton is produced deep within stars during **nucleosynthesis**. It has six naturally occurring (i.e., stable) **isotopes**, the most abundant of which is krypton-84 (57%). Some long-lived radioactive isotopes exist as well.

nucleosynthesis: creation of heavier elements from lighter elements via fusion reactions in stars

isotope: form of an atom that differs by the number of neutrons in the nucleus

Two of them, krypton-85 (half-life = 10.7 y) and krypton-81 (half-life = 210,000 y) have been used to date well water. Radioactive krypton is produced in **fission** reactions of heavy elements. Thus, radioactive isotopes of krypton have always formed part of the natural radiation background of Earth's atmosphere.

fission: process of splitting an atom into smaller pieces

Although a noble gas, krypton is not entirely unreactive. One krypton compound, krypton difluoride (KrF_2), is commercially available in small quantities. SEE ALSO GASES; NOBLE GASES; RAMSAY, WILLIAM; TRAVERS, MORRIS.

Richard Mowat

Bibliography

Almqvist, Ebbe (2003). *History of Industrial Gases.* New York: Kluwer Academic/Plenum Publishers.

Lide, David R., ed. (2003). *The CRC Handbook of Chemistry and Physics,* 84th edition. Boca Raton, FL: CRC Press.

Lanthanides

The lanthanide or **rare earth elements** (atomic numbers 57 through 71) typically add electrons to the 4f orbitals as the **atomic number** increases, but lanthanum ($4f^0$) is usually considered a lanthanide. Scandium and yttrium are also chemically similar to lanthanides. Lanthanide chemistry is typically that of +3 cations, and as the atomic number increases, there is a decrease in radius for each lanthanide, known as the "lanthanide contraction." Because bonding within the lanthanide series is usually predominantly ionic, the lanthanide contraction often determines the differences in properties of lanthanide compounds and ions. Lanthanide compounds often have high coordination numbers between 6 and 12. SEE ALSO CERIUM; DYSPROSIUM; ERBIUM; EUROPIUM; GADOLINIUM; HOLMIUM; LANTHANUM; LUTETIUM; PRASEODYMIUM; PROMETHIUM; SAMARIUM; TERBIUM; THULIUM; YTTERBIUM.

rare earth elements: older name for the lanthanide series of elements, from lanthanum to lutetium

atomic number: the number assigned to an atom of an element that indicates the number of protons in the nucleus of that atom

Herbert B. Silber

Bibliography

Bünzli, J.-C. G., and Choppin, G. R. (1989). *Lanthanide Probes in Life, Chemical, and Earth Sciences.* New York: Elsevier.

Cotton, Simon (1991). *Lanthanides and Actinides.* New York: Oxford University Press.

Lanthanum

MELTING POINT: 920°C
BOILING POINT: 3,469°C
DENSITY: 6.16 g/cm^3
MOST COMMON IONS: La^{3+}

Elemental lanthanum has a ground state (electronic configuration) of $[Xe]5d6s^2$. Naturally occurring lanthanum is a mixture of two stable **isotopes**, ^{138}La and ^{139}La. The element was discovered in 1839 by Carl Gustaf Mosander in the form of the lanthanum oxide, at that time called "lanthana." The name is derived from the Greek *lanthanein* ("to lie hidden"), as the element had been overlooked due to its similarity to the earlier discovered cerium.

isotope: form of an atom that differs by the number of neutrons in the nucleus

rare earth elements: older name for the lanthanide series of elements, from lanthanum to lutetium

lanthanides: a family of elements (atomic number 57 through 70) from lanthanum to lutetium having from 1 to 14 4f electrons

Monazite and bastnasite are the principal lanthanum ores, in which lanthanum occurs together with other members of the **rare earth elements** or the **lanthanides**. It can be separated from the other rare earths by ion exchange or solvent extraction techniques. Lanthanum is a silver-white, malleable, and **ductile metal**. It is soft enough to be cut with a knife. The metal is rapidly oxidized when exposed to air. Cold water attacks lanthanum only slowly, but reaction with hot water is fast.

ductile: property of a substance that permits it to be drawn into wires

metal: element or other substance the solid phase of which is characterized by high thermal and electrical conductivities

Lanthanum chemistry is dominated by the trivalent lanthanum(III) ion, La^{3+}. This ion forms ionic bonds with **ligands** containing an oxygen or nitrogen donor atom. The ground state electronic configuration of La^{3+} is $[Xe]4f^0$. Due to the absence of unpaired 4f electrons, lanthanum(III) compounds are colorless, both in solution and in the solid state. Lanthanum(III) oxide is added to optical glass to increase its refractive index and alkali resistance. Lanthanum-nickel **alloys** are being used in the storage of hydrogen gas. Thulium(III)-**doped** lanthanum oxybromide ($LaOBr{:}Tm^{3+}$) is a blue-emitting phosphor used in x-ray intensifying screens. SEE ALSO Cerium; Dysprosium; Erbium; Europium; Gadolinium; Holmium; Lanthanides; Lutetium; Neodymium; Praseodymium; Promethium; Samarium; Terbium; Ytterbium.

ligand: molecule or ion capable of donating one or more electron pairs to a Lewis acid

alloy: mixture of two or more elements, at least one of which is a metal

dope: to add a controlled amount of an impurity to a very pure sample of a substance, which can radically change the properties of a substance

Koen Binnemans

Bibliography

Cotton, Simon (1991). *Lanthanides and Actinides.* New York: Oxford University Press.

Kaltsoyannis, Nikolas, and Scott, Peter (1999). *The f Elements.* New York: Oxford University Press.

Lavoisier, Antoine

FRENCH CHEMIST
1743–1794

Antoine-Laurent Lavoisier, born in Paris, France, is considered the father of modern chemistry. During the course of his career, Lavoisier managed to transform just about every aspect of chemistry. But Lavoisier was not just a scientist. He was involved in French taxation politics during a turbulent time in the country's history—the French Revolution (the first major social revolution proclaiming the liberty of the individual [ca. 1789–1799]). Because of his involvement with the ruling class, he was executed during the revolutionary days known as the Terror, at the height of his scientific career.

Just before and during the French Revolution, another revolution was taking place. In any study of the history of chemistry, the period between 1770 and 1790 is commonly regarded as the "Chemical Revolution." This revolution, which marked the beginnings of modern chemistry, occurred in large part as a result of Lavoisier's scientific excellence and brilliant experimental capabilities. He played a role in many aspects of the Chemical Revolution, including the abandonment of the phlogiston theory of **combustion**, the evolution of the concept of an element, and the development of a new chemical nomenclature.

combustion: burning, the reaction with oxygen

Perhaps Lavoisier's most important accomplishment was his role in the dismantling of the phlogiston theory of combustion. Phlogiston was a substance believed to be emitted during combustion and the **calcination** of **metals**. Earlier chemists, such as the Germans Johann Becher (1635–1682) and George Stahl (1660–1734), supposed that a metal was composed of calx and phlogiston, and that burning resulted from the loss of phlogiston. The fact that metals actually gained weight during combustion was usually

calcine: to heat or roast to produce an oxide (e.g., CaO from calcite)

metal: element or other substance the solid phase of which is characterized by high thermal and electrical conductivities

French chemist Antoine-Laurent Lavoisier, considered to be the founder of modern chemistry.

explained away by the theory that phlogiston had negative weight. Lavoisier, like some others, saw that it was illogical for anything to have negative weight.

To prove his supposition that phlogiston did not exist, Lavoisier introduced quantitative measurement to the laboratory. Using precise weighing, he showed that in all cases of combustion where an increase in weight was observed, air was absorbed, and that when a calx was burned with charcoal, air was liberated. In addition to showing by precise measurement that phlogiston did not exist, Lavoisier's findings also implied that the total weight of the substances taking part in a chemical reaction remains the same before and after the reaction—an early statement of the law of conservation of mass. By ridding the chemical world of the phlogiston theory of combustion using quantitative analysis, Lavoisier was able to push chemistry toward its modern state. No longer would counterintuitive notions such as a substance having negative weight occupy the minds of chemists.

Likewise, Lavoisier's work was also able to refute the theory that the world was composed of either one, two, three, or four elements. Lavoisier defined an element as the "last point which analysis is capable of reaching," or in modern terms, a substance that cannot be broken down any further into its components. This break from the theories of the ancient world allowed chemists to pursue the study of chemistry with a different outlook of the world. By defining elements as the last points of analysis, Lavoisier opened up new investigative possibilities. In his classic textbook *Elements of Chemistry* (generally acknowledged to be the first modern chemistry textbook), he compiled a list of all the substances he could not break down into simpler substances, that is, he created the first table of elements (although not the Periodic Table of later years). By acknowledging that there could be more elements than his preliminary list provided, Lavoisier left the search for more elements to his successors.

Lavoisier's dismantling of the phlogiston theory and his systematic definition of an element caused many chemists to view basic concepts differently and to embrace the principles of Lavoisier's new chemistry. One of the methods Lavoisier used to spread his ideas was to construct a new and logical system for naming chemicals. Working with Claude Berthollet and Antoine Fourcroy, Lavoisier developed a new nomenclature based on three general principles: (1) Substances should have one fixed name, (2) names ought to reflect composition when known, and (3) names should generally be chosen from Greek or Latin roots. This new nomenclature was published in 1787, and it swayed even more chemists to adopt the new chemistry.

Nevertheless, Lavoisier did not always hit on the right theories for the right reasons. For example, he believed that acidity was caused by the presence of oxygen in a compound. Lavoisier concluded in 1776 that oxygen was the part of a compound that was responsible for the property of acidity because he had isolated it from so many acids. In fact, oxygen means "acid former." According to Lavoisier, the other portion of the compound combined with the oxygen was called an "acidifiable base" and it was responsible for the specific properties of the compound. Although these concepts turned out to be wrong, the thinking behind them is important since it represented the first systematic attempt to chemically characterize acids and bases.

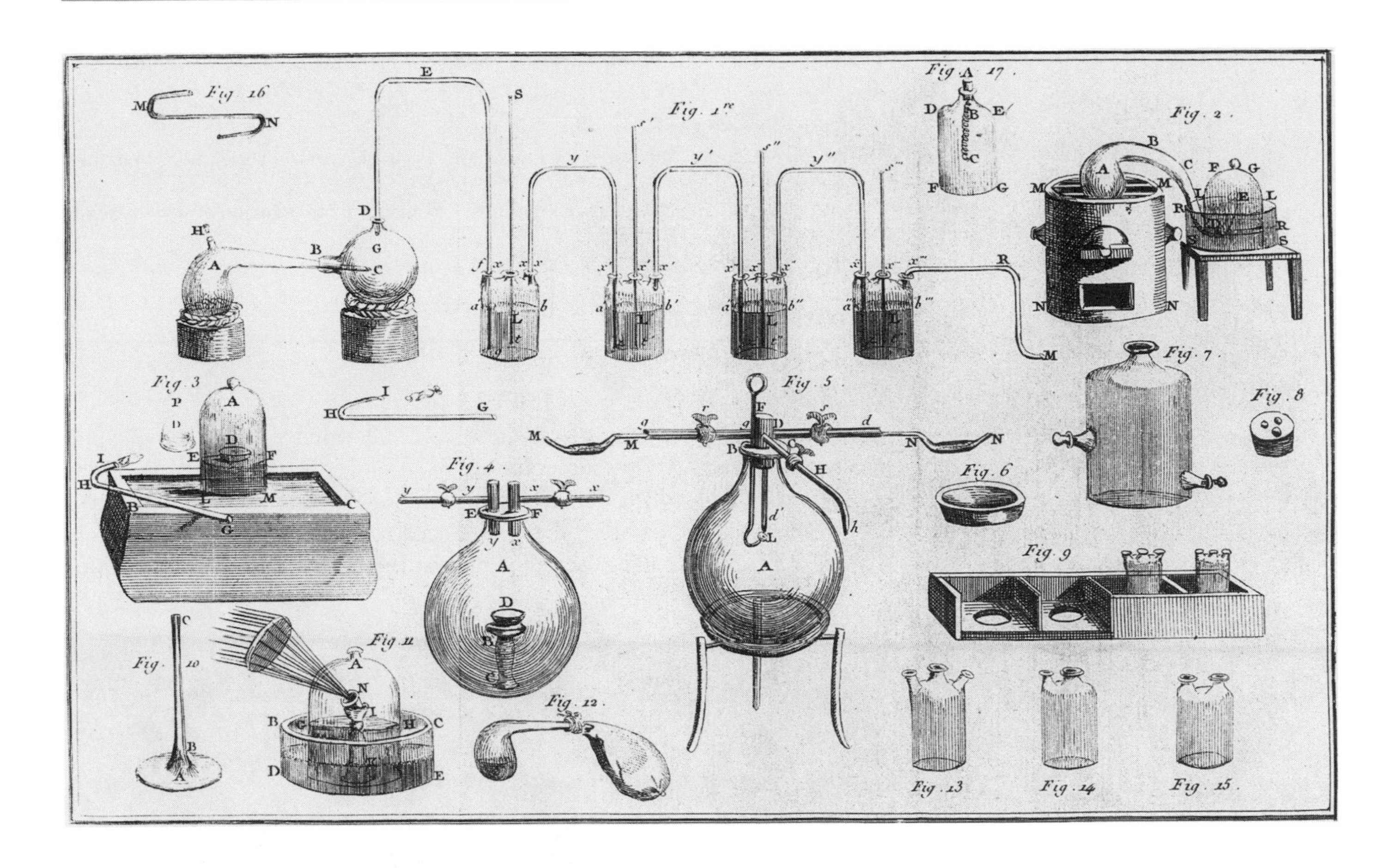

A rendering of instruments in Lavoisier's laboratory.

Lavoisier was not only interested in the theoretical aspects of chemistry. He also devoted much of his time to studying more practical topics, such as the best ways of lighting streets in a large town. In addition, Lavoisier took part in the development of what was to become the metric system and he was involved in improving the manufacture of gunpowder.

Although Lavoisier was independently wealthy, thanks to a considerable fortune inherited from his mother, he sought to increase his wealth in order to pursue his scientific career on a larger scale. For this reason, he entered the Ferme, a private company whose members purchased the privilege of collecting national taxes. During the French Revolution, the tax collectors of the Ferme were the subject of popular hatred. Although he carried out his duties honestly, Lavoisier was associated with the perceived corruption of the tax collection system. At the height of the Revolution, Lavoisier was arrested and executed by beheading in 1794.

Lavoisier's untimely death ended an era in the history of chemistry. With his contributions to chemistry ranging from developing the modern concept of combustion to establishing the language of chemistry, Lavoisier provided the foundation for the study of chemistry as a modern science. SEE ALSO BERTHOLLET, CLAUDE-LOUIS.

Lydia S. Scratch

Bibliography

Donovan, Arthur (1993). *Antoine Lavoisier: Science, Administration and Revolution.* Oxford, U.K.: Blackwell.

Jaffe, Bernard (1976). *Crucibles: The Story of Chemistry from Ancient Alchemy to Nuclear Fission.* New York: Dover.

Yount, Lisa (1997). *Antoine Lavoisier: Founder of Modern Chemistry.* Springfield, NJ: Enslow Publishers.

Internet Resources

Beretta, Marco, ed. "Panopticon Lavoisier." Available from <http://moro.imss.fi.it/lavoisier>.

Poirier, Jean-Pierre. "Lavoisier's Friends." Available from <http://historyofscience.free.fr>.

Lawrence, Ernest

AMERICAN PHYSICIST
1901–1958

isotope: form of an atom that differs by the number of neutrons in the nucleus

Ernest Orlando Lawrence was a pioneer of "big science," the use of complicated and expensive instrumentation by large teams of researchers. He is best known for inventing the cyclotron, one of the first and most successful "atom smashers." With this particle accelerator, Lawrence and his colleagues were able to make new radioactive **isotopes**, synthesize transuranium

American physicist Ernest Orlando Lawrence, recipient of the 1939 Nobel Prize in physics, "for the invention and development of the cyclotron and for results obtained with it, especially with regard to artificial radioactive elements."

elements that do not occur in nature, and advance knowledge of the atomic nucleus.

Youth and Education

According to his mother, Lawrence was "born grown up" on August 8, 1901, in Canton, South Dakota, a rural town of less than one thousand inhabitants. Ernest's father, Carl, was superintendent of the Canton public schools when his eldest son was born. A second son, John, was born in 1904. Both boys demonstrated early interest in science and technology and they eventually worked together on medical applications of the cyclotron.

Lawrence graduated from the University of South Dakota in 1922 with a bachelor of arts in chemistry. His original intention had been to become a physician, but as a sophomore he met Lewis Akeley, the university's lone physics professor. Akeley soon recognized the young man's talent and transformed him into a physicist. Although it was at the time impossible to major in that discipline at the university, Akeley provided special tutorials for his prize student and thus prepared him for graduate study.

In 1922 Lawrence entered the University of Minnesota, where he became a research student of W. F. G. Swann and was awarded a master of science in physics. His thesis became the basis for his first publication. When Swann moved to the University of Chicago and then to Yale, Lawrence followed him, earning his Ph.D. from the latter institution in 1925. After his degree he remained at Yale, first as a research fellow and soon thereafter as a faculty member.

The Atomic Nucleus and the Cyclotron

Ernest Lawrence's experimental skill, hard work, and professional ambition were soon common knowledge among many physicists, and offers of employment came from a number of universities. The most attractive was from the University of California in Berkeley, an institution that was eager to build a reputation as a world-class center for scientific research and education. In 1928 Lawrence moved across the country to assume an associate professorship at Berkeley. Two years later he was a full professor, the youngest in the history of the university.

The first quarter of the twentieth century was a time of great intellectual ferment in the physical sciences. Experimental discoveries such as x rays, cathode rays (electrons), and radioactivity demanded drastic revisions in the prevailing concept of atomic structure. There was convincing evidence, largely obtained in European laboratories, that atoms consisted of minuscule, incredibly dense, positively charged nuclei surrounded by negative electrons. The new quantum or wave mechanics, developed by German physicist Max Planck, Danish physicist Niels Bohr, German physicist Werner Heisenberg, Austrian physicist Erwin Schrödinger, and others, appeared to provide the theoretical and mathematical tools to explain this structure. But what was the nature of the nucleus and the forces that held together the positively charged protons that, according to classical electrostatic arguments, should be repelling each other?

It was to such problems that Lawrence soon applied his experimental genius. He was confident that nuclei could be probed by bombarding atoms with protons and other subatomic particles. What was needed was a machine to accelerate these tiny projectiles to high velocities and energies. The design for such a device came to him in the spring of 1929, while reading a paper by Norwegian engineer Rolf Wideröe. The prototype for the "magnetic-resonance accelerator" was built by Niels Edlefsen, Lawrence's first Ph.D. student, in early 1930.

The first "cyclotron," as it came to be known, consisted of a flat glass cylinder containing two semicircular D-shaped electrodes. These electrodes (called "dees") were attached to a radio-frequency oscillator that would cause them to alternate polarity rapidly between positive and negative charges. A strong electromagnet, with 4-inch pole faces above and below the apparatus, created a magnetic field **perpendicular** to the plane of the electrodes. In operation, the glass chamber was pumped down to a near vacuum and protons (hydrogen ions) were injected into the center of the device. As the dees changed polarity, the positive protons would be alternatively pulled and pushed in a circular orbit of increasing diameter. As the orbit increased, so did the velocity and energy of the particles. When the stream of protons reached the desired energy, they were deflected and directed at the intended target.

perpendicular: condition in which two lines (or linear entities like chemical bonds) intersect at a 90-degree angle

The first model cyclotron, a leaky gadget coated with sealing wax that cost about $25, was soon replaced with a brass box. Before long, the size of the device began to grow. The 4-inch version was succeeded by a 9-inch model, followed by cyclotrons 11, 27, 34, 60, and finally 184 inches in diameter. Each increase in size required larger electromagnets, more electric power, and more money. The energy of the accelerated particles also increased, and that was the point of the enterprise—to obtain more powerful probes. At the 11-inch stage, Lawrence and his team "split" their first atom, and the 184-inch cyclotron attained his goal of particles with energies of 100 million electron volts. In all this research, significant contributions were made by Lawrence's students and collaborators at the Radiation Laboratory, especially M. Stanley Livingston and Edwin M. McMillan.

Applications, Issues, and Analysis

In 1939 Ernest Lawrence was awarded the Nobel Prize in physics for his invention of the cyclotron. That same year Austrian physicists Lise Meitner and Otto Frisch correctly concluded that the experimental results of German chemists Otto Hahn and Fritz Strassmann indicated that neutrons cause uranium atoms to split (undergo **fission**) into smaller fragments, with the release of great quantities of energy. It was also in 1939 that the German army invaded Poland, launching World War II (1939–1945). With the latter two events, the study of the atomic nucleus and atomic energy literally became a matter of life and death.

fission: process of splitting an atom into smaller pieces

Lawrence's 184-inch cyclotron was soon modified to separate uranium isotopes and became the prototype for the **calutrons** used for similar purposes in the **Manhattan Project**. Both during and after the war, Lawrence was one of the most politically influential American scientists; he served on many key committees, including the Scientific Panel that advised on the first use of the atomic bomb.

calutron: electromagnetic device that separates isotopes based on their masses

Manhattan Project: government project dedicated to creation of an atomic weapon; directed by General Leslie Groves

Ernest Lawrence died on August 27, 1958, in Palo Alto, California. Never a gifted mathematician or theoretician, he was a brilliant experimentalist with the entrepreneurial skills to enthusiastically promote his vision of "big science." Thanks to the cyclotron and other accelerators there are hundreds of radioactive isotopes that have applications in medicine and elsewhere, dozens of subatomic particles, and at least seventeen artificial elements that stretch beyond uranium in the Periodic Table. Lawrence's legacy is commemorated in the Lawrence Berkeley Laboratory, the Lawrence Hall of Science, the Lawrence Livermore Laboratory, and, most appropriately, in element number 103 (which concludes the actinide series). Lawrencium (Lr) was the first element named for an American. SEE ALSO Bohr, Niels; Heisenberg, Werner; Meitner, Lise; Planck, Max; Radioactivity; Schrödinger, Erwin; Transactinides.

A. Truman Schwartz

Bibliography

Childs, Herbert (1968). *An American Genius: The Life of Ernest Orlando Lawrence, Father of the Cyclotron.* New York: Dutton.

Davis, Nuel Pharr (1968). *Lawrence and Oppenheimer.* New York: Simon & Schuster.

Heilbron, J. L., and Seidel, Robert W. (1989). *Lawrence and His Laboratory: A History of the Lawrence Berkeley Laboratory*, Vol. 1. Berkeley: University of California Press.

Herkin, Gregg (2002). *Brotherhood of the Bomb: The Tangled Lives and Loyalties of Robert Oppenheimer, Ernest Lawrence, and Edward Teller.* New York: Henry Holt.

Lawrencium

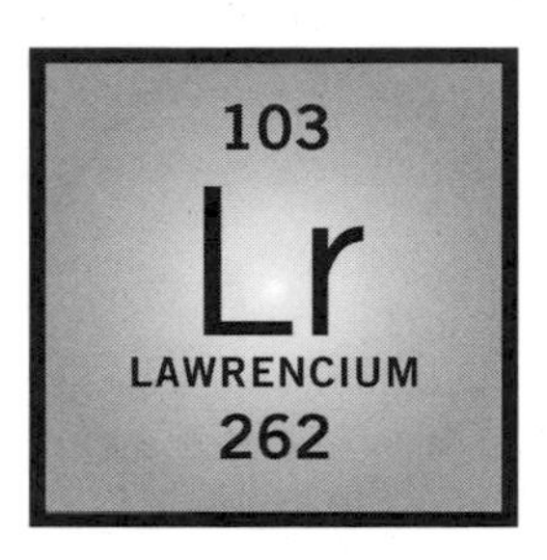

MELTING POINT: Unknown
BOILING POINT: Unknown
DENSITY: Unknown
MOST COMMON IONS: Lw^{3+}

Lawrencium is a synthetic radioactive element and the last member of the actinide series. It was discovered in 1961 by Albert Ghiorso and his coworkers, who bombarded a target of **isotopes** of californium (^{249}Cf–^{252}Cf) with boron projectiles (either ^{10}B or ^{11}B) using the Heavy Ion Linear Accelerator (HILAC) at the University of California, Berkeley, producing isotopes of unknown element 103 of masses 257 and 258. The atoms recoiling after the **nuclear** reaction were caught on metalized Mylar tape, which was then moved past a series of α-detectors, and the decays of a few new atoms of element 103 ($^{257}103$ and $^{258}103$), having half-lives of seconds or less, were recorded. The element is named after Ernest O. Lawrence, the inventor of the cyclotron and the founder of the Berkeley Radiation Laboratory. "Lawrencium" was suggested and accepted by the International Union of Pure and Applied Chemistry, but the originally suggested symbol of Lw was changed to Lr. In 1965 Evgeni D. Donets at the Joint Institutes of Nuclear Research in Dubna, Russia, using a double-recoil technique, confirmed the **atomic number** of element 103 by linking its decay, via either electron capture followed by α-decay, or α-decay followed by electron capture, to its known granddaughter, an isotope of fermium (^{252}Fm).

All Lr isotopes are radioactive, decaying by **α-particle** emission, electron capture, and/or spontaneous **fission**, and have half-lives ranging from a few tenths of a second for the isotope of mass 252, the lightest isotope, to 3.6 hours for the isotope of mass 262, the heaviest. The ground state electronic configuration of the gaseous Lr atom is believed to be $[Rn]5f^{14}6d7s^2$, by analogy to its **lanthanide** homologue, lutetium (element 71). Its most stable ion in **aqueous solution** is Lr^{3+}, and attempts to reduce it to Lr^{2+}

isotope: form of an atom that differs by the number of neutrons in the nucleus

nuclear: having to do with the nucleus of an atom

atomic number: the number assigned to an atom of an element that indicates the number of protons in the nucleus of that atom

α-particle: subatomic particle with 2+ charge and mass of 4; a He nucleus

fission: process of splitting an atom into smaller pieces

lanthanides: a family of elements (atomic number 57 through 70) from lanthanum to lutetium having from 1 to 14 4f electrons

aqueous solution: homogenous mixture in which water is the solvent (primary component)

or Lr^{1+} in aqueous solution have been unsuccessful. SEE ALSO ACTINIUM; BERKELIUM; EINSTEINIUM; FERMIUM; LAWRENCE, ERNEST; MENDELEVIUM; NEPTUNIUM; NOBELIUM; PLUTONIUM; PROTACTINIUM; RADIOACTIVITY; RUTHERFORDIUM; THORIUM; TRANSMUTATION; URANIUM.

Darleane C. Hoffman

Bibliography

Ghiorso, A.; Sikkeland, A.; Larsh, A. E.; and Latimer, R. M. (1961). "New Element, Lawrencium, Atomic Number 103." *Physical Review Letters* 6:473–475.

Hoffman, Darleane C.; Ghiorso, Albert; and Seaborg, Glenn T. (2000). *The Transuranium People: The Inside Story*. Singapore: World Scientific Publishing.

Hoffman, Darleane C., and Lee, Diana M. (1999). "Chemistry of the Heaviest Elements—One Atom at a Time." *Journal of Chemical Education* 76(3):331–347.

Seaborg, Glenn T., and Loveland, Walter D. (1990). *The Elements Beyond Uranium*. New York: Wiley.

Lead

MELTING POINT: 327.5°C
BOILING POINT: 1,740.0°C
DENSITY: 11.34 g/cm^3
MOST COMMON IONS: Pb^{2+}, Pb^{4+}

Lead makes up only about 0.0013 percent of Earth's crust but was well known in the ancient world and was even mentioned in the Book of Exodus. The word "lead" is derived from the Anglo-Saxon word *laedan*. Lead's symbol, Pb, comes from the Latin word for lead, *plumbum*. Because of lead's long use in piping, the word "plumber" comes from that same root. Lead is an extremely dense but malleable **metal** that is very resistant to corrosion.

metal: element or other substance the solid phase of which is characterized by high thermal and electrical conductivities

Lead is sometimes found free in nature but is usually obtained from ores such as galena (PbS) or cerussite ($PbCO_3$), from which it is easily mined and refined. Most lead is obtained by simply roasting galena in hot air. About one-third of the lead used in the United States is obtained through recycling efforts.

Lead has seen many uses over the ages. As a constituent of pewter (an **alloy** of tin and lead), lead was a component of Roman eating and drinking utensils. It has been suggested that the decline of the Roman Empire may have been tied to this use, since acidic foodstuffs extract small amounts of lead, a cumulative human poison. Lead's use as a pottery glaze has been banned for the same reason—the danger of lead ingestion via the extraction of the lead by food and drink. During the twentieth century, a **volatile** form of lead—tetraethyl lead [$Pb(CH_2CH_3)_4$]—was developed and widely used to improve the octane level of gasoline. That use has also been banned for health and environmental reasons.

alloy: mixture of two or more elements, at least one of which is a metal

volatile: low boiling, readily vaporized

Lead remains in wide use in electrical cable sheathing, automobile batteries, lead crystal, radiation protection, and some solders. SEE ALSO INORGANIC CHEMISTRY; RADIOACTIVITY.

George H. Wahl Jr.

Bibliography

Internet Resources

Jefferson Lab. "It's Elemental: The Element Lead." Available from <http://education.jlab.org/itselemental/ele082.html>.

Le Bel, Joseph-Achille

FRENCH CHEMIST
1847–1930

French chemist Joseph-Achille Le Bel, who, with Jacobus van't Hoff, was the founder of modern stereochemistry.

Joseph-Achille Le Bel, born in Pechelbronn, France, was, with Dutch physical chemist Jacobus Hendricus van't Hoff, the cofounder of modern stereochemistry. They independently established the relation between optical activity and asymmetric carbon compounds.

Le Bel was born into a wealthy family that controlled the petroleum industry in Pechelbronn, Alsace. In 1865 he was sent to the École Polytechnique in Paris to obtain a chemical education and spent most of his time there doing chemical research. After graduation, he worked with the French chemists Antoine Balard and Adolphe Wurtz in Paris, in between intermediate periods of refinery construction at home. Finally in 1889, he sold his shares in the family business and established a private laboratory in Paris where he devoted himself to organic chemistry and, in his later years, paleontology, botany, and philosophy. An independent thinker who never held an academic appointment, Le Bel did manage to achieve general recognition as a chemist and even became president of the French Chemical Society in 1892.

In 1874, at the age of twenty seven, Le Bel presented a brief paper to the Paris Chemical Society that led to his scientific fame, although it may be regarded as his only outstanding contribution to the field of chemistry. By the late 1840s the great chemist and microbiologist Louis Pasteur had separated two sorts of tartrate crystals of the same composition, each crystal shape being the mirror image of the other. These crystals in solution not only rotated the plane of polarized light to a certain angle (optical activity), the rotation also occurred in opposite directions. Pasteur called such pairs of substances optical **isomers**, and because they showed no difference in chemical properties, they were represented by the same constitutional structural formula in the new chemical structure theory. Le Bel then extended the structure theory, from constitutional structural to configurational representations in three-dimensional space, to account for the difference in optical isomers. He argued that if a tetravalent carbon atom combined with four different groups, as in tartrate, the carbon must be asymmetric in three-dimensional space (i.e., without a symmetry plane or center). Furthermore, for each such asymmetric carbon there were exactly two different structures (stereoisomers), each being the mirror image of the other, just like the crystal shapes of Pasteur. Le Bel's structure theory could not explain optical activity, but it explained and predicted which compounds had stereoisomers and which did not, an approach that he also extended to nitrogen compounds.

isomer: molecules with identical compositions but different structural formulas

Oddly enough, van't Hoff, with whom Le Bel had worked in Wurtz's laboratory shortly beforehand, independently arrived at the same theory from a different starting point at virtually the same time. SEE ALSO Chirality; Pasteur, Louis; van't Hoff, Jacobus.

Joachim Schummer

Bibliography

Leicester, Henry M. (1973). "Le Bel, Joseph Achille." In *Dictionary of Scientific Biography*, Vol. VIII, ed. Charles C. Gillispie. New York: Scribners.

French chemist Nicolas Leblanc, inventor of the process for transforming sodium chloride (NaCL, or common salt) into soda ash.

Leblanc, Nicolas

FRENCH CHEMIST AND SURGEON
1742–1806

Nicolas Leblanc invented a method of making alkali soda from salt that became one of the most important chemical processes of the nineteenth century. Leblanc was born in Issoudun, France; his father managed an ironworks. After completing his medical education in about 1780, Leblanc became a private physician in the house of Philippe Égalité, duc d'Orléans (1747–1793). France had been suffering from an acute shortage of alkali from traditional vegetable sources. Alkali was critical in the manufacture of glass, textiles, paper, soap, and other products. In 1775 the French Royal Academy offered a prize to anyone who could develop a process for transforming common salt (sodium chloride) into soda ash. With Égalité's support, Leblanc achieved the goal by 1789 and opened a small factory at Saint Denis that began production in 1791.

What became known as the Leblanc process was actually several interrelated processes. Salt was first reacted with sulphuric acid in a cast-iron pan, then in a **reverberator furnace** (in which heat was applied from a flame blown from a separate chamber, not in direct contact with the salt), to produce saltcake (sodium sulphate), with hydrochloric acid released as a waste gas. Saltcake was used to make sodium carbonate, or roasted with limestone (calcium carbonate) and coal or coke to produce "black ash." This mixture of sodium carbonate, calcium sulphide, sodium sulphide, lime, salt, carbon, and ash could be treated further with hot water to produce impure sodium carbonate in solution, evaporated into soda crystals (washing soda), or heated to yield anhydrous sodium carbonate. The latter, in turn, could be reacted with lime to made caustic soda (sodium hydroxide), the strongest commercial alkali then available.

reverberator furnace: furnace or kiln used in smelting that heats material indirectly by deflecting a nearby flame downward from the roof

Leblanc personally benefited little from his innovation. The National Assembly granted him a fifteen-year patent in September 1791, but three years later the revolutionary government sequestered his factory and made his patents public, giving Leblanc only meager compensation for his assets. Napoléon Bonaparte returned the plant to him in 1802, but by then Leblanc was too poor to resume production and, in 1806 he took his own life. (In 1855 Napoléon III gave Leblanc's heirs a payment in lieu of the 1775 prize.)

Leblanc's process—by greatly reducing the cost and boosting the efficiency of alkali for the key industries that depended on it—boosted European industrialization for two generations. The year of Leblanc's suicide, the Saint Gobain Company opened a soda ash factory; by 1818 French producers were turning about roughly 10,000 to 15,000 tons of Leblanc soda ash per year. British producers, discouraged until a prohibitive tax on salt was repealed in 1823, embraced the new process and surpassed the French by the middle of the century. U.S. alkali makers remained wedded to **potash** but imported Leblanc soda ash after 1850. Germany took the lead in Leblanc soda production in the 1870s.

potash: the compound potassium oxide, K_2O

By that time the Leblanc process was facing competition from the newer Solvay (ammonia soda) alkali. The dominance of Leblanc soda was extended by improvements, most notably the Deacon process (1868), which converted the wasteful and harmful hydrochloric acid gases into chlorine, and the

Chance process (1882), which recovered waste sulphur. By the turn of the twentieth century, however, the Leblanc and Solvay processes were eclipsed by new electrolytic methods for making chlorine and caustic soda. SEE ALSO Alkali Metals; Industrial Chemistry, Inorganic; Sodium.

David B. Sicilia

Bibliography

Haber, L. F. (1958). *The Chemical Industry during the Nineteenth Century: A Study of the Economic Aspect of Applied Chemistry in Europe and North America.* Oxford, U.K.: Clarendon Press.

Landes, David S. (1972) *The Unbound Prometheus: Technological Change and Industrial Development in Western Europe from 1750 to the Present.* Cambridge, MA: Harvard University Press.

Morgan, Sir Gilbert T., and Pratt, David D. (1938). *British Chemical Industry: Its Rise and Development.* London: Edward Arnold & Co.

Le Châtelier, Henri

FRENCH CHEMIST
1850–1936

Henri-Louis Le Châtelier was born into a family of architects, engineers, and scientists in Paris. His family home was like a drop-in center for France's leading chemists. Le Châtelier became a well-known industrial chemist himself, interested in **metallurgy**, cements, glasses, fuels, explosives, and, most famously, chemical **equilibrium**. By mixing theoretical work with practical applications, Le Châtelier became one of France's most valuable and productive chemists.

metallurgy: the science and technology of metals

equilibrium: condition in which two opposite reactions are occurring at the same speed, so that concentrations of products and reactants do not change

Le Châtelier served as an army lieutenant during the Franco–Prussian War in 1870, after which he returned to school to finish his degree. His intention was to become a mining engineer, but he changed his mind when he was offered a professorship in chemistry at the École des Mines in Paris.

Le Châtelier's first area of investigation was the chemistry of cements. By repeating the experiments of Antoine Lavoisier on the preparation of plaster of paris, Le Châtelier learned that when a cement comes into contact with water, a solution is formed that yields an interlaced, **coherent mass** of minute crystals.

coherent mass: mass of particles that stick together

Through his studies on cements, Le Châtelier became interested in the applications of thermodynamics to chemistry. It was while working with thermodynamics that Le Châtelier devised what became known as Le Châtelier's principle in 1884. This principle states that if a system is in a state of equilibrium and one of the conditions is changed, such as the pressure or temperature, the equilibrium will shift in such a way as to try to restore the original equilibrium condition. Using Le Châtelier's principle as a guide the efficiency of chemical processes can be improved by shifting a system to yield more of the desired product. For example, the German chemist Fritz Haber employed the principle in his development of a process for synthesizing ammonia. Le Châtelier considered missing this practical application of his principle to be the greatest blunder in his scientific career.

In addition to his work on equilibrium, Le Châtelier was involved in many other areas of chemistry and science in general. For instance, he investigated the cause and prevention of mining disasters. He helped produce safer explosives and conducted research that was used to develop the torch now used in cutting and welding steel. He also designed instruments when none existed to suit his purposes, such as the thermocouple to measure very high temperatures.

Le Châtelier was additionally involved in other areas of research. He wrote numerous biographies and spent much time working on articles on social welfare, the relationship of science to economics, and the relationship between pure and applied science. Le Châtelier was also devoted to teaching, working at the École des Mines until he retired at the age of sixty-nine. He advised his students to be content with adding to the structure of science, but to pay attention to anything unusual they observed.

Le Châtelier contributed much to chemistry. He bridged the gaps between theory and practice by choosing research problems that looked as if they would have industrial applications. For this reason, he became one of the most successful researchers in combining theoretical science with practical applications. SEE ALSO EQUILIBRIUM; HABER, FRITZ; LAVOISIER, ANTOINE; THERMODYNAMICS.

Lydia S. Scratch

Bibliography

Segal, Bernice C. (1985). *Chemistry: Experiment and Theory.* New York: Wiley.

Internet Resources

Kurtz, Jim, and Oliver, John. "Henri Louis Le Chatelier: A Man of Principle." Woodrow Wilson National Fellowship Foundation. Available from <http://www.woodrow.org>.

Le Châtelier, Henri. "A General Statement of the Laws of Chemical Equilibrium." Available from <http://webserver.lemoyne.edu>.

Leclanché, Georges

FRENCH ENGINEER AND INVENTOR
1839–1882

In 1866 Georges-Lionel Leclanché was granted French patent no. 71,865, which described a remarkable advancement in the technology of the **primary electrochemical cell**. Typically called a battery, an electrochemical cell generates electrical current by chemical reactions at the two electrodes of the cell, the cathode and the anode. In early batteries the current was originally carried through an electrically conductive liquid; later improvements substituted a conductive paste for the liquid. Although Leclanché did not invent this type of battery, commonly called a dry cell, his version, only slightly modified since then, is used today to power millions of devices from toys to portable computers.

primary electrochemical cell: voltaic cell based on an irreversible chemical reaction

Leclanché was born on October 9, 1839, in Paris, France. His father was a cultured and politically active lawyer in the French government during a tempestuous time in that country's history. Because of shifting political winds, young Georges and his father spent the better part of eighteen years away from Paris. Upon his return in 1856, Leclanché enrolled in the École Centrale Imperiale des Arts et Manufactures, where he majored in **metallurgy**. He was far more interested in analytical and industrial chemistry, however, and after graduating in 1860, he became a laboratory manager in a company that manufactured lead salts.

metallurgy: the science and technology of metals

Politically active like his father, Leclanché was forced to flee to Belgium in 1863 because of his opposition to France's involvement in Mexico. While there, he became interested in electrochemical research. He returned to Paris the following year and became a chemist in the materials laboratory of a railroad company, where he further developed his cell. After his cell was patented, he started a company for its manufacture and continued to perfect its design until his untimely death from throat cancer in 1882, at the age of forty-three.

Although electrochemical cells had been developed and used since the beginning of the nineteenth century, they suffered several practical disadvantages: They generally contained costly or dangerous ingredients (such as platinum or mercury), and they were composed of delicate components in liquid solutions, which limited their portability. Leclanché's original cell used a manganese dioxide/carbon cathode, a zinc anode, and ammonium chloride as the **electrolyte solution**. Although this was technically a "wet" cell, later advances in design replaced the electrolyte solution with a conductive paste. The main advantages of the Leclanché cell were the low cost of its components and the cell's robust construction, which allowed it to be manufactured cheaply and utilized widely at a time when batteries were the only source of electricity. SEE ALSO ELECTROCHEMISTRY; NEW BATTERY TECHNOLOGY.

electrolyte solution: a liquid mixture containing dissolved ions

Bartow Culp

Bibliography

Davis, Jack (1967). "Georges-Lionel Leclanché." *Electrochemical Technology* 5:487–490.

Heise, George W., and Cahoon, N. Corey, eds. (1971). *The Primary Battery*, Vol. 1. New York: Wiley.

Leloir, Luis

ARGENTINE CHEMIST
1906–1987

Luis Leloir was born in Paris in 1906. His parents were Argentine, and he resided in Buenos Aires from the age of two and for most of his career until his death in 1987. Leloir received his M.D. from the University of Buenos Aires in 1932, after which he worked at the Institute of Physiology with Professor Bernardo Houssay on the action of **adrenalin** in carbohydrate **metabolism**. In 1936 he collaborated with Sir F. G. Hopkins at the Biochemical Laboratory in Cambridge, England. He returned to Buenos Aires to study the **oxidation** of fatty acids in the liver and on the formation of **angiotensin**. He was forced to leave Argentina in 1943 when Houssay's laboratory was closed by the government of Juan Perón in response to Houssay's public criticism of the dictatorship. During the next few years Leloir worked in the United States in the laboratory of Carl and Gerty Cori at Washington University in St. Louis, Missouri, and D. E. Green at Columbia University in New York City.

adrenalin: chemical secreted in the body in response to stress

metabolism: the complete range of biochemical processes that take place within living organisms; comprises processes that produce complex substances from simpler components, with a consequent use of energy (anabolism), and those that break down complex food molecules, thus liberating energy (catabolism)

oxidation: process that involves the loss of electrons (or the addition of an oxygen atom)

angiotensin: chemical that causes a narrowing of blood vessels

Leloir returned to Argentina in 1945 to take part in the founding of the Instituto de Investigaciones Bioquimicas supported by the Jaime Campomar Foundation. He became the institute's first director. Most of the research at the institute was conducted on a shoestring budget, and some of the complex equipment there was ingeniously built by Leloir or his colleagues

Argentine chemist Luis Leloir, recipient of the 1970 Nobel Prize in chemistry, "for his discovery of sugar nucleotides and their role in the biosynthesis of carbohydrates."

chromatography: the separation of the components of a mixture in one phase (the mobile phase) by passing it through another phase (the stationary phase) making use of the extent to which the components are absorbed by the stationary phase

galactose: six-carbon sugar

glucose: common hexose monosaccharide; monomer of starch and cellulose; also called grape sugar, blood sugar, or dextrose

biosynthesis: formation of a chemical substance by a living organism

R. Caputto, C. E. Cardini, R. Trucco, and A. C. Paladini from everyday objects. For instance, a fraction-collecting device for column **chromatography** was built with a Meccano set (a toy from Leloir's childhood that was similar to an erector set); its collection bottles moved about with a toy railroad train.

In spite of inadequate funding and the resulting limitations, work began at the new institute on the metabolism of the sugar **galactose**; it led to the discovery of **glucose** 1,6-diphosphate and uridine diphosphate glucose (UDPG). The latter substance was the first sugar nucleotide discovered. Leloir continued his work in this sector of research and discovered several other sugar nucleotides. He and his coworkers determined that UDPG is the nucleotide that provides the glucose units in the **biosynthesis** of glycogen. Glycogen is the form of polymeric carbohydrate in which energy is stored inside an animal cell, while starch is the form in which energy is stored in plants. Further work demonstrated that the donor of the glucose units for the biosynthesis of starch is adenosine diphosphate glucose (ADPG.) This work was quite important in improving scientists' understanding of the processes by which carbohydrates are converted to energy and the glycogen storage diseases. Leloir received the Nobel Prize in chemistry in 1970; his was the first Nobel Prize in chemistry ever awarded to an Argentine.

Leloir is best remembered for producing Nobel Prize–quality research under very difficult conditions and for contributing significantly to the development of scientific analysis and discovery in Argentina. SEE ALSO CARBOHYDRATES; CORI, CARL AND GERTY; NUCLEOTIDE.

Lawrence H. Brannigan

Bibliography

Scientific American, ed. (1955). *The Physics and Chemistry of Life.* (1955). New York: Simon & Schuster.

Internet Resources

The Nobel Foundation. "Nobel e-Museum." Available from <http://www.nobel.se>.

Southwest Baptist University Web site. Available from <http://www.sbuniv.edu>.

Lewis, Gilbert N.

AMERICAN PHYSICAL CHEMIST
1875–1946

Gilbert Newton Lewis was born on October 25, 1875, in West Newton, Massachusetts. A precocious child, he received his early education at home and learned to read by the age of three. When Lewis was nine, his family moved to Lincoln, Nebraska. He attended the University of Nebraska for two years and in 1893 transferred to Harvard University, from which he received his B.S. in 1896.

After a brief stint as a teacher at Phillips Academy in Andover, Massachusetts, Lewis returned to Harvard, where he obtained his M.A. in 1898 and Ph.D. in 1899. He subsequently studied at the universities at Göttingen and Leipzig in Germany (1900–1901) and then returned to Harvard as an instructor (1901–1906). In 1907 Lewis became an assistant professor at the Massachusetts Institute of Technology, where he soon rose to the rank of full professor.

In 1912 Lewis accepted a position as dean and chairman of the College of Chemistry at the University of California, Berkeley. He remained at Berkeley for the rest of his life and transformed the chemistry department there into a world-class center for research and teaching. His reforms in the way chemistry was taught, a **catalyst** for the modernization of chemical education, were widely adopted throughout the United States. Lewis introduced thermodynamics to the curriculum, and his book on the same subject became a classic. He also brought to the study of physical chemistry such concepts as fugacity, activity and the activity coefficient, and ionic strength.

American chemist Gilbert N. Lewis, a theorist of chemical bonds, or valence.

catalyst: substance that aids in a reaction while retaining its own chemical identity

attraction: force that brings two bodies together, such as two oppositely charged bodies

valence: combining capacity

noble gas: element characterized by inert nature; located in the rightmost column in the Periodic Table

At the beginning of the twentieth century physicists tried to relate the electronic structure of atoms to two basic chemical phenomena: the chemical bond (the **attraction** between atoms in a molecule) and **valence** (the quality that determines the number of atoms and groups with which any single atom or group will unite chemically and also expresses this ability to combine relative to the hydrogen atom). German chemist Richard Abegg was the first to recognize in print the stability of the group of eight electrons, the arrangement of outer electrons that occurs in **noble gases** and is often attained when atoms lose or gain electrons to form ions. Lewis called this the "group of eight," and American chemist and physicist Irving Langmuir labeled it an "octet."

In 1902, while explaining the laws of valence to his students at Harvard, Lewis conceived a concrete model for this process, something Abegg had not done. He proposed that atoms were composed of a concentric series of cubes with electrons at each of the resulting eight corners. This "cubic atom" explained the cycle of eight elements in the Periodic Table and corresponded to the idea that chemical bonds were formed by the transfer of electrons so each atom had a complete set of eight electrons. Lewis did not publish his theory, but fourteen years later it became an important part of his theory on the shared electron-pair bond.

In 1913 Lewis and Berkeley colleague William C. Bray proposed a theory of valence that differentiated two different types of bond: a polar bond formed by the transfer of electrons and a **nonpolar** bond not involving electron transfer. In 1916 Lewis published his seminal article suggesting that the chemical bond is a pair of electrons shared or held jointly by two atoms. He depicted a single bond by two cubes sharing an edge, or more simply by double dots in what has become known as Lewis dot structure.

nonpolar: molecule, or portion of a molecule, that does not have a permanent, electric dipole

According to Lewis's octet rule, each atom should be surrounded by four pairs of electrons, either shared or free pairs. Lewis derived structures for **halogen** molecules, the ammonium ion, and oxy acids, inexplicable according to previous valence theories. He viewed polar bonds as unequally shared electron pairs. Because the complete transfer of electrons was only an extreme case of polarity, he abandoned his earlier dualistic view; the polar theory was just a special case of his more general theory.

halogen: element in the periodic family numbered VIIA (or 17 in the modern nomenclature) that includes fluorine, chlorine, bromine, iodine, and astatine

Lewis's shared electron-pair theory languished until Langmuir revived and elaborated it beginning in 1919. It was soon accepted as the Lewis–Langmuir theory, one of the most fundamental concepts in the history of chemistry.

Lewis's acid-base concept is also well known to introductory-level chemistry students. A Lewis acid, for example, BF_3, $AlCl_3$, or SO_3, is a substance

coordination chemistry: chemistry involving complexes of metal ions surrounded by covalently bonded ligands

ligand: molecule or ion capable of donating one or more electron pairs to a Lewis acid

coordinate covalent bond: covalent bond in which both of the shared electrons originate on only one of the bonding atoms

isotope: form of an atom that differs by the number of neutrons in the nucleus

that can accept a pair of electrons from a Lewis base, for example, NH_3 or OH^-, which is a substance that can donate a pair of electrons. It can be applied to various areas, for example, **coordination chemistry**: The metal ion is a Lewis acid, the **ligand** is a Lewis base, and the resulting formation of a **coordinate covalent bond** corresponds to a Lewis acid–base reaction.

Lewis made additional valuable contributions to the theory of colored substances, radiation, relativity, the separation of **isotopes**, heavy water, photochemistry, phosphorescence, and fluorescence. As a major in the U.S. Army Chemical Warfare Service during World War I, he worked on defense systems against poison gases. From 1922 to 1935 he was nominated numerous times for the Nobel Prize in chemistry. Lewis's death, while measuring the dielectric constant of hydrogen cyanide on March 23, 1946, precluded his receiving the prize, which is not awarded posthumously. SEE ALSO Acid-Base Chemistry; Lewis Structures.

George B. Kauffman

Bibliography

"Gilbert Newton Lewis: 1875–1946." Papers presented at the 183rd National Meeting of the American Chemical Society, Las Vegas, NV. *Journal of Chemical Education* 61: (January 1984) 3–21, (February 1984) 93–116, (March 1984) 185–215.

Hildebrand, Joel H. (1958). "Gilbert N. Lewis." *Biographical Memoirs, National Academy of Sciences* 31:209–235.

Leicester, Henry M., ed. (1968). *Source Book in Chemistry 1900–1950,* pp. 100–106. Cambridge, MA: Harvard University Press.

Lewis, Edward S. (1998). *A Biography of Distinguished Scientist Gilbert Newton Lewis.* Lewiston, NY: Edwin Mellen Press.

Lewis, Gilbert N. (1916). "The Atom and the Molecule." *Journal of the American Chemical Society* 38:762–785.

Lewis, Gilbert Newton (1923). *Valence and the Structure of Atoms and Molecules.* New York: Chemical Catalog Co. Reprinted, New York: Dover, 1966.

Lewis, Gilbert Newton, and Randall, Merle F. (1923). *Thermodynamics and the Free Energy of Chemical Substances.* New York: McGraw-Hill.

Lewis Acids ***See Acid-Base Chemistry.***

Lewis Structures

In 1902, while trying to find a way to explain the Periodic Table to his students, the chemist Gilbert Newton Lewis discovered that the chemistry of the main-group elements could be explained using a model in which electrons arranged around atoms are conceived as occupying the faces of concentric cubes. This model was based on four assumptions.

1. The number of electrons in the outermost cube of an atom is equal to the number of electrons lost when the atom forms positive ions;
2. each neutral atom has one more electron in its outermost cube than the atom that precedes it in the Periodic Table;
3. it takes eight electrons—an octet—to complete a cube;
4. once an atom has an octet of electrons in its outermost cube, the cube becomes part of the cote of electrons about which the next cube is built.

Lewis determined the formulas of simple ionic compounds (such as NaCI) by theorizing that atoms gain electrons if the outermost cube is more than half full, and lose electrons if the cube is less that half-full (until the cube is either full or empty). Sodium, for example, loses the one electron in its outermost cube at the same time that chlorine gains the electron it needs to fill its outermost cube.

As understanding of the structure of the atom developed, it became apparent why the magic number of electrons for each of the main-group elements was eight. The outermost **atomic orbitals** for these elements are the *s* and *p* orbitals in a given shell, and it takes eight electrons in its outermost shell: [He] $2s^2$ $2p^3$. It therefore has to gain three electrons to fill this shell.

atomic orbital: mathematical description of the probability of finding an electron around an atom

The electrons in an outermost shell are known as **valence** electrons. The number of bonds an element can form is called valence (from the Latin *valens*, "to be strong"). An atom's valence electrons are those electrons that can be gained or lost in a chemical reaction. Because the electrons that occupy filled *d* or *f* subshells are seldom disturbed in a chemical reaction, we can also define an atom's valence electrons as the electrons that are not present in an atom of the preceding rare gas (ignoring filled *d* or *f* subshells). Gallium, for example, has the following electron configuration: [Ar] $4s^2$ $3d^{10}$ $4p^1$. The 4*s* and 4*p* electrons can be lost in a chemical reaction, but the electrons in the filled 3*d* **subshell** cannot. Gallium therefore has three valence electrons.

valence: combining capacity

subshell: electron energy sublevel, of which there are four: s, p, d, and f

By 1916 Lewis realized that there was another way that atoms can combine to achieve an octet of valence electrons: They can share electrons and form a **covalent bond**. Two fluorine atoms, for example, by sharing a pair of electrons can form a stable F_2 molecule in which each atom has an octet of valence electrons. A pair of oxygen atoms, by sharing two pairs of electrons, can form an O_2 molecule in which each atom has a total of eight valence electrons.

covalent bond: bond formed between two atoms that mutually share a pair of electrons

Whenever Lewis applied his model to covalent compounds, he noted that the atoms seemed to share pairs of electrons. He also noted that most compounds contained even numbers of electrons, which suggested that electrons exist in pairs. He therefore replaced his cubic model of the atom, in which eight electrons were oriented toward the surfaces of a cube, with a model based on pairs of electrons. In this notation, each atom is surrounded by up to four pairs of dots, corresponding to the eight possible valence electrons. This symbolism is still in use today. The only significant modification is the use of lines to indicate covalent bonds formed by the sharing of a pair of electrons. The Lewis structures for F_2 and O_2 are written as follows:

:F̤̈—F̤̈:

:Ö=Ö:

The prefix "co-" is used to indicate that two or more entities are joined or have equal standing (as in, for example, *coexist*, *cooperate*, and *coordinate*). It is therefore appropriate that the term "covalent bond" is used to describe molecular bonds that result from the sharing of one or more pairs of electrons.

The Lewis structure of a compound can be arrived at by trial and error. We start by notating symbols that contain the correct number of valence electrons for the atoms in the molecule. We then pair electrons to indicate covalent bonds until we come up with a Lewis structure in which each atom (with the exception of hydrogen atoms) has an octet of valence electrons. The trial-and-error method for writing Lewis structures can be time-consuming, however. For all but the simplest molecules, the step-by-step process is faster.

Step 1: Determine the total number of valence electrons.

Step 2: Write the skeleton structure of the molecule.

Step 3: Assign two valence electrons to each covalent bond in the skeleton structure.

Step 4: Try to complete the octets of the atoms by distributing the remaining valence electrons as nonbonding electrons.

The first step involves calculating the number of valence electrons in the molecule or ion. For a neutral molecule it is the sum of the valence electrons of each atom. If the molecule carries an electric charge, we add one electron for each negative charge or subtract one electron for each positive charge.

Consider the chlorate (ClO_3^-) ion. A chlorine atom (Group VIIa) has seven valence electrons, and each oxygen atom (Group VIa) has six valence electrons. Because the chlorate ion has a charge of −1, it contains one more electron than a neutral ClO_3 molecule. Thus, the ClO_3^- ion has a total of twenty-six valence electrons.

$$ClO_3^-: \ 7 + 3(6) + 1 = 26$$

electronegative: capable of attracting electrons

The second step in this process involves deciding which atoms in the molecule are connected by covalent bonds. This can be the most difficult step in the process. As a rule, the least **electronegative** element is at the center of the molecule. It is also useful to note that the formula of the compound often provides a hint to the skeleton structure. The formula for the chlorate ion, for example, suggests the following skeleton structure.

```
O—Cl—O
   |
   O
```

The third step assumes that the skeleton structure of the molecule is based on covalent bonds. The valence electrons are therefore divided into two categories: bonding electrons and nonbonding electrons. Because it takes two electrons to form a covalent bond, we can calculate the number of nonbonding electrons in the molecule by subtracting two electrons for each bond in the skeleton structure from the total number of valence electrons.

There are three covalent bonds in the skeleton structure of the chlorate ion. As a result, six of the twenty-six valence electrons must be used as bonding electrons. This leaves twenty nonbonding electrons in the valence shell.

26 valence electrons
−6 bonding electrons
20 nonbonding electrons

The fourth step in the process by which Lewis structures are generated involves using the nonbonding valence electrons to complete the octets of the atoms in the molecule. Each oxygen atom in the ClO_3^- ion already has two electrons, the electrons in each Cl–O bond. Each oxygen atom therefore needs six nonbonding electrons to complete the octet. Thus, it takes eighteen nonbonding electrons to satisfy the octets of the three oxygen atoms. This leaves one pair of nonbonding electrons, which can be used to fill the octet of the central atom.

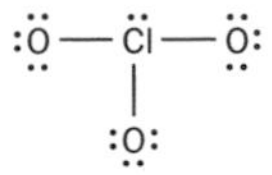

```
:Ö—C̈l—Ö:
   |
  :O:
```

Occasionally, we encounter a molecule that does not seem to have enough valence electrons. When this happens, we have to remember why atoms share electrons in the first place. If we cannot achieve a satisfactory Lewis structure by having two atoms share a single pair of electrons, it may be possible to achieve this goal by having them share two or even three pairs of electrons. Consider **formaldehyde** (H_2CO), for example, which contains twelve valence electrons.

formaldehyde: name given to the simplest aldehyde HC(O)H, incorporating the –C(O)H functional group

$$H_2CO: \ 2(1) + 4 + 6 = 12$$

The formula of this molecule suggests the following skeleton structure.

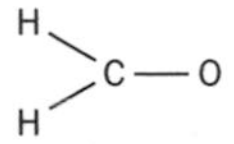

```
H
 \
  C—O
 /
H
```

There are three covalent bonds in this skeleton structure, which means that six valence electrons must be used as bonding electrons. This leaves six nonbonding electrons. It is impossible, however, to complete the octets of the atoms in this molecule with only six nonbonding electrons. When the nonbonding electrons are used to complete the octet of the oxygen atom, the carbon atom has a total of only six valence electrons.

```
H
 \
  C—Ö:
 /
H
```

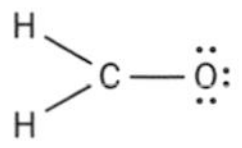

We therefore assume that the carbon and oxygen atoms share two pairs of electrons. There are now four bonds in the skeleton structure, which leaves only four nonbonding electrons. This is enough, however, to satisfy the octets of the carbon and oxygen atoms.

```
H
 \
  C=Ö:
 /
H
```

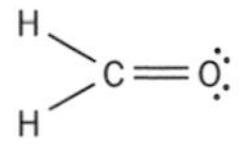

Every once in a while, we encounter a molecule for which it is impossible to write a satisfactory Lewis structure. Consider boron trifluoride (BF_3), for example, which contains twenty-four valence electrons.

$$BF_3: \ 3 + 3(7) = 24$$

There are three covalent bonds in what is the most reasonable skeleton structure for the molecule. Because it takes six electrons to form the skeleton structure, there are eighteen nonbonding valence electrons. Each fluorine atom needs six nonbonding electrons to complete its octet. Thus, all nonbonding electrons are consumed by the three fluorine atoms. As a result, we run out of electrons, and the boron atom still has only six valence electrons.

```
:F̈
   \
    B—F̈:
   /
:F̈
```

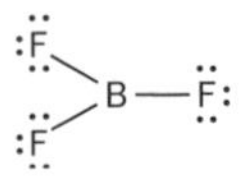

For reasons that are not discussed here, the elements that form strong double or triple bonds are C, N, O, P, and S. Because neither boron nor fluorine falls in this group, we have to stop short with what appears to be an unsatisfactory Lewis structure.

It is also possible to encounter a molecule that seems to have too many valence electrons. When that happens, we expand the valence shell of the central atom. Consider the Lewis structure for sulfur tetrafluoride SF_4, for example, which contains thirty-four valence electrons.

$$SF_4\text{: } 6 + 4(7) = 34$$

There are four covalent bonds in the skeleton structure for SF_4. Because this structure uses eight valence electrons to form the covalent bonds that hold the molecule together, there are twenty-six nonbonding valence electrons.

Each fluorine atom needs six nonbonding electrons to complete its octet. Because there are four of these atoms, we need twenty-four nonbonding electrons for this purpose. But there are twenty-six nonbonding electrons in this molecule. We have already completed the octets for all five atoms, and we still have one pair of valence electrons. We therefore expand the valence shell of the sulfur atom to hold more than eight electrons.

SEE ALSO Bonding; Lewis, Gilbert N.; Molecules.

George Bodner

Liebig, Justus von

GERMAN CHEMIST
1803–1873

elemental analysis: determination of the percent of each atom in a specific molecule

Justus von Liebig, one of the founders of modern chemistry, was born on May 12, 1803, in Darmstadt, Hesse, Germany. His father was a manufacturer of drugs and paints. As an adolescent Liebig performed many experiments using materials from his father's business while neglecting other studies. He was apprenticed to an apothecary at age fifteen; however, his real interest was chemistry. He enrolled at the University of Bonn in 1820 to attend the lectures of Wilhelm Kastner. When Kastner left for the University of Erlangen, Liebig followed him there and received his doctoral degree in 1822 after only two years of study.

German chemist Justus von Liebig.

Liebig soon realized that his knowledge of chemistry was deficient and using the patronage of the grand duke of Hesse was able to study in Paris from 1822 to 1824. Paris was the leading center for the study of chemistry at this time, and here Liebig was able to attend the lectures of such famous chemists as Joseph Gay-Lussac, Louis Thénard, and Pierre Dulong. Parisian chemistry stressed a rigorous and quantitative approach that was lacking in European chemistry. In Paris Liebig had the opportunity to work in the laboratory of Gay-Lussac, where he acquired skills in the **elemental analysis** of inorganic and organic compounds, as well as in the systematic methodology of chemical research.

In 1825 Liebig returned to Germany and was offered the professorship of chemistry at the University of Geissen. He stayed in Geissen until 1851, at which time he was called to the chair of chemistry at the University of Munich where he remained until his death on April 18, 1873.

In Geissen Liebig drawing from his studies in Paris established the model for chemical education that was soon copied by other German educators. His students learned by working in the laboratory with their mentor, starting with simple procedures, working their way through more complex exercises, and finally graduating to their own independent research.

One of Liebig's many achievements at Geissen was the development of more efficient **combustion** techniques for the elemental analysis of organic compounds. The impetus for this was his inability to get a good result in his analysis of a compound he had isolated from urine (which he had named hippuric acid) using the conventional methods that were available to him. These methods, which were tedious and time-consuming, included the use of very small amounts of a given sample, which amplified the experimental error. In 1830 Liebig devised a technique that allowed the use of larger samples and he was able to quantify the amounts of carbon and hydrogen in organic compounds. The trapping of the gaseous combustion products water vapor and carbon dioxide on pre-weighed absorbents was part of his technique. This procedure greatly reduced error and was simple enough so that Liebig's students were able to analyze all types of organic compounds almost routinely, which greatly enhanced the existing knowledge of the variety of organic compounds in nature.

combustion: burning, the reaction with oxygen

Liebig pioneered methods for the analysis of nitrogen, sulfur, and **halogens** in organic compounds, in addition to his contributions to the analysis of carbon and hydrogen in these compounds. Liebig was the founding editor of one of the first chemical journals, *Annalen der chemie* in 1832. SEE ALSO GAY-LUSSAC JOSEPH-LOUIS; ORGANIC CHEMISTRY; PROTEINS.

Martin D. Saltzman

halogen: element in the periodic family numbered VIIA (or 17 in the modern nomenclature) that includes fluorine, chlorine, bromine, iodine, and astatine

Bibliography

Brock, W. H. (1997). *Justus von Liebig: The Chemical Gatekeeper.* New York: Cambridge University Press.

Light *See Radiation.*

Lipid Bilayers

Lipid bilayers form the fundamental structures of cell membranes and thus provide a semipermeable interface between the interior and exterior of a cell and between compartments within the cell. Bilayer-forming lipids are amphipathic molecules (containing both **hydrophilic** and **hydrophobic** components). The hydrophilic fragment, typically termed the lipid headgroup, is charged, or polar, whereas the hydrophobic section consists of a pair of alkyl chains (typically between 14 and 20 carbon atoms in length). A typical class of bilayer-forming lipid is the diacyl phosphatidylcholines, in which phosphate and tetramethyl ammonium moieties comprise the polar head groups and two fatty acids chains constitute the hydrophobic portions.

lipid: a nonpolar organic molecule; fatlike; one of a large variety of nonpolar hydrophobic (water-hating) molecules that are insoluble in water

hydrophilic: a part of a molecule having an affinity for water

hydrophobic: a part of a molecule that repels water

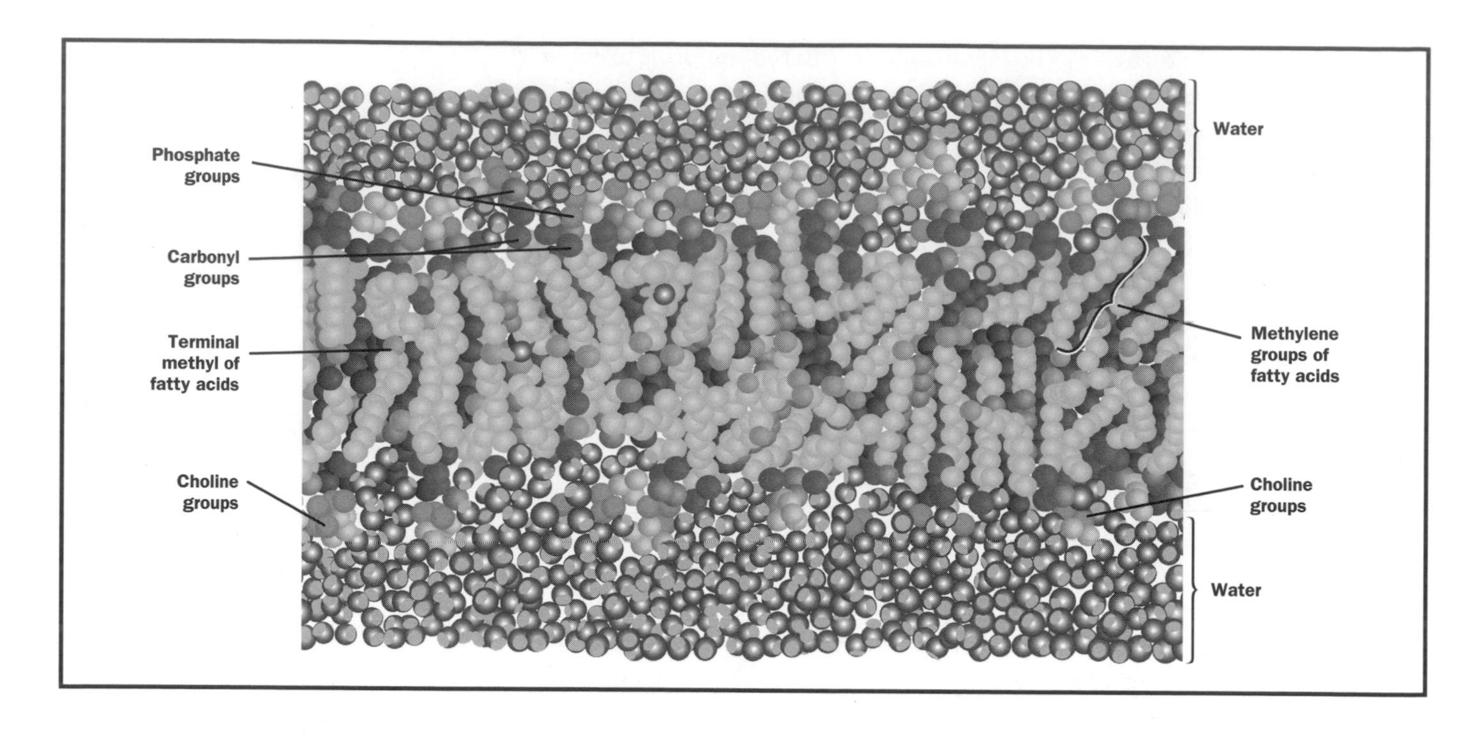

Figure 1. An image of a dipalmitolyphosphatidylcholine lipid bilayer from a molecular dynamics computer simulation.

In an aqueous environment the lipids self-assemble into structures that minimize contact between water molecules and the hydrophobic components of the lipids by forming two leaflets (monolayers); this arrangement brings the hydrophobic tails of each leaflet in direct contact with each other, and leaves the head groups in contact with water (see Figure 1). Lipid bilayers are classified as lyotropic and thermotropic liquid crystals, meaning that their structures are a function of both water content and temperature, respectively. Under conditions of low hydration and/or low temperature, the lipid bilayer is ordered with straight alkyl chains and a regular arrangement of head groups, and is in a gel state. At high water content and/or elevated temperatures the bilayer exists in the (biologically relevant) fluid state, characterized by a high degree of disorder in both chain and head group conformations (Figure 1 represents the fluid bilayer state).

melting point: temperature at which a substance in the solid state undergoes a phase change to the liquid state

phase: homogeneous state of matter

The alkyl chain **melting point**, the gel to fluid transition temperature, is a function of chain length (shorter chains will dissociate at lower temperatures) and the extent of unsaturation within the chains (greater double bond content lowers the melting temperature). The fluid **phase** bilayer has a hydrophobic thickness on the order of 30 angstroms, and (because of the hydrophobic chains) is a significant barrier to the passage of charged or polar solutes. The bilayer is fluid in the sense that lipid molecules and other membrane constituents readily undergo lateral diffusion within the plane of the membrane. SEE ALSO LIPIDS; MEMBRANE.

Scott E. Feller
Ann T. S. Taylor

Bibliography

Berg, Jeremy M.; Tymoczko, John L.; and Stryer, Lubert (2002). *Biochemistry*, 5th edition. New York: W. H. Freeman.

Voet, Donald; Voet, Judith G.; and Pratt, Charlotte (1999). *Fundamentals of Biochemistry*. New York: Wiley.

Lipids

Lipids are a class of biomolecules that is defined by their solubility in organic solvents, such as chloroform, and their relative insolubility in water. Interactions among lipids and of lipids with other biomolecules arise largely from their **hydrophobic** ("water-hating") nature. Lipids can be divided into two main categories according to their structures: those that are based on fatty acids, and those that are based on **isoprene**, a branched, five-carbon chain.

lipid: a nonpolar organic molecule; fatlike; one of a large variety of nonpolar hydrophobic (water-hating) molecules that are insoluble in water

hydrophobic: a part of a molecule that repels water

isoprene: common name for 2-methyl-1,3 butadiene, the monomer of the natural rubber polymer

Fatty Acid–Based Lipids

Fatty acids are unbranched carboxylic acids, usually containing an even number of carbon atoms (between 12 and 24, inclusive). If there are no double bonds between carbon atoms, the fatty acid is saturated; if there are double bonds between carbon atoms, the fatty acid is unsaturated. Naturally occurring unsaturated fatty acids have one to six double bonds, with the double bonds separated by at least two single bonds; the double bonds have the *cis* configuration. These double bonds inhibit "packing" of the molecules (in solids), which lowers the fatty acid **melting point**. Many physical properties of lipid substances are determined by the extent of unsaturation. Polyunsaturated omega-3 (ω-3) fatty acids, so named because the double bond between the third to last (ω-3) and fourth to last (ω-4) carbons, are commonly found in cold-water fish and are thought to play an important role in many **neurological** functions.

melting point: temperature at which a substance in the solid state undergoes a phase change to the liquid state

neurologic: of or pertaining to the nervous system

In response to stress conditions, various tissues convert polyunsaturated fatty acids having twenty carbons to a family of compounds called eicosanoids. Eicosanoids include prostaglandins, thromboxanes, prostacyclins, and leukotrienes, and are generally involved in inflammation and pain sensation. Aspirin, acetaminophen, and other **analgesics** work by inhibiting the initial reactions required for the conversion of fatty acids to eicosanoids.

analgesic: compound that relieves pain, e.g., aspirin

The **carboxylic acid** group of a fatty acid molecule provides a convenient place for linking the fatty acid to an alcohol, via an **ester** linkage. If the fatty acid becomes attached to an alcohol with a long carbon chain, the resultant substance is called a wax. Waxes are very hydrophobic, and thus repel water. Glycerol, a three-carbon compound with an alcohol group at each carbon, very commonly forms esters with fatty acids. When glycerol and a fatty acid molecule are combined, the fatty acid portion of the resultant compound is called an "acyl" group, and the glycerol portion is referred to as a "glyceride." Using this nomenclature system, a triacylglyceride has three fatty acids attached to a single glycerol molecule; sometimes this name is shortened to "triglyceride." Triglyceride substances are commonly referred to as

carboxylic acid: one of the characteristic groups of atoms in organic compounds that undergoes characteristic reactions, generally irrespective of where it occurs in the molecule; the $-CO_2H$ functional group

ester: organic species containing a carbon atom attached to three moieties: an O via a double bond, an O attached to another carbon atom or chain, and a H atom or C chain; the R(C=O)OR functional group

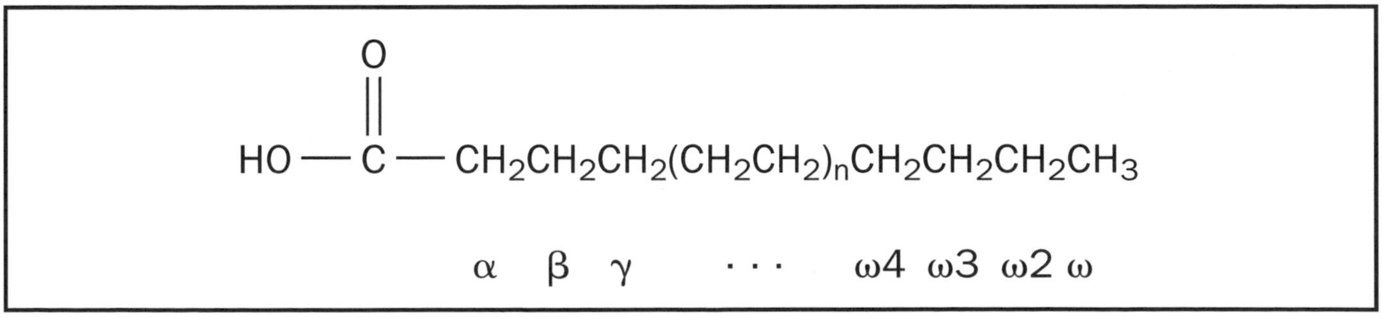

Figure 1. Generic fatty acid.

$$HO-\overset{\overset{\displaystyle O}{\|}}{C}-CH_2CH_2CH_2(CH_2CH_2)_nCH_2CH_2CH_2CH_3$$

$$\alpha \quad \beta \quad \gamma \quad \cdots \quad \omega 4 \quad \omega 3 \quad \omega 2 \quad \omega$$

fats or oils, depending on whether they are solid or liquid at room temperature. Triglycerides are an energy reserve in biological systems. Diacylglycerides are commonly found in nature with acyl chains occurring at two adjacent carbons, and are the basis of phospholipid chemistry.

Isoprene-Based Lipids

The other class of lipid molecules, based on a branched five-carbon structure called isoprene, was first identified via steam distillation of plant materials. The extracts are called "essential oils." They are often fragrant, and are used as medicines, spices, and perfumes. A wide variety of structures is obtained by fusing isoprene monomer units, leading to a very diverse set of compounds, including terpenes, such as β-carotene, pinene (turpentine), and carvone (oil of spearmint); and steroids, such as testosterone, cholesterol, and **estrogen**.

estrogen: female sex hormone

Lipid Organization

"Like oil and water" is a saying based on the minimal interaction of lipids with water. Although this saying is apt for isoprene-based lipids and bulky fatty acid–based lipids such as waxes and triglycerides, it is not apt for all lipids (e.g., it does not apply to substances composed of fatty acids or diacylglycerides).

Fatty acids and diacylglycerides are often amphipathic; that is, the carboxylic acid "head" is **hydrophilic** and the hydrocarbon "tail" is hydrophobic. When a fatty acid or triglyceride substance is placed in water, structures that maximize the interactions of the hydrophilic heads with water and minimize the interactions of the hydrophobic tails with water are formed. At low lipid concentrations a monolayer is formed, with hydrophilic heads associating with water molecules and hydrophobic tails "pointing" straight into the air (see Figure 2).

hydrophilic: a portion of a molecule having an affinity for water

As the concentration of lipid is increased, the surface area available for monolayer formation is reduced, leading to the formation of alternative structures (depending on the particular lipid and condition). Compounds that have a relatively large head group and small tail group, such as fatty acids and detergents, form spherical structures known as micelles. The concentration of lipid required for micelle formation is referred to as the critical micelle concentration (CMC). Other hydrophobic molecules, such as molecules within dirt, triacylglycerides, and other large organic molecules, associate with the hydrophobic tail portion of a micelle.

Compounds that have approximately equal-sized heads and tails tend to form bilayers instead of micelles. In these structures, two monolayers of lipid molecules associate tail to tail, thus minimizing the contact of the hydrophobic portions with water and maximizing hydrophilic interactions. Lipid molecules can move laterally (within a single layer of the bilayer, called a leaflet), but movement from one leaflet to the opposing leaflet is much more difficult.

$$H_2CO\text{: } 2(l) + 4 + 6 = 12$$

Often these bilayer sheets can wrap around in such a way as to form spherical structures, called **vesicles** or **liposomes** (depending on their size). Several new anticancer treatments are based upon the packaging of chemo-

vesicle: small compartment in a cell that contains a minimum of one lipid bilayer

liposome: sac formed from one or more lipid layers that can be used for drug transport to cells in the body

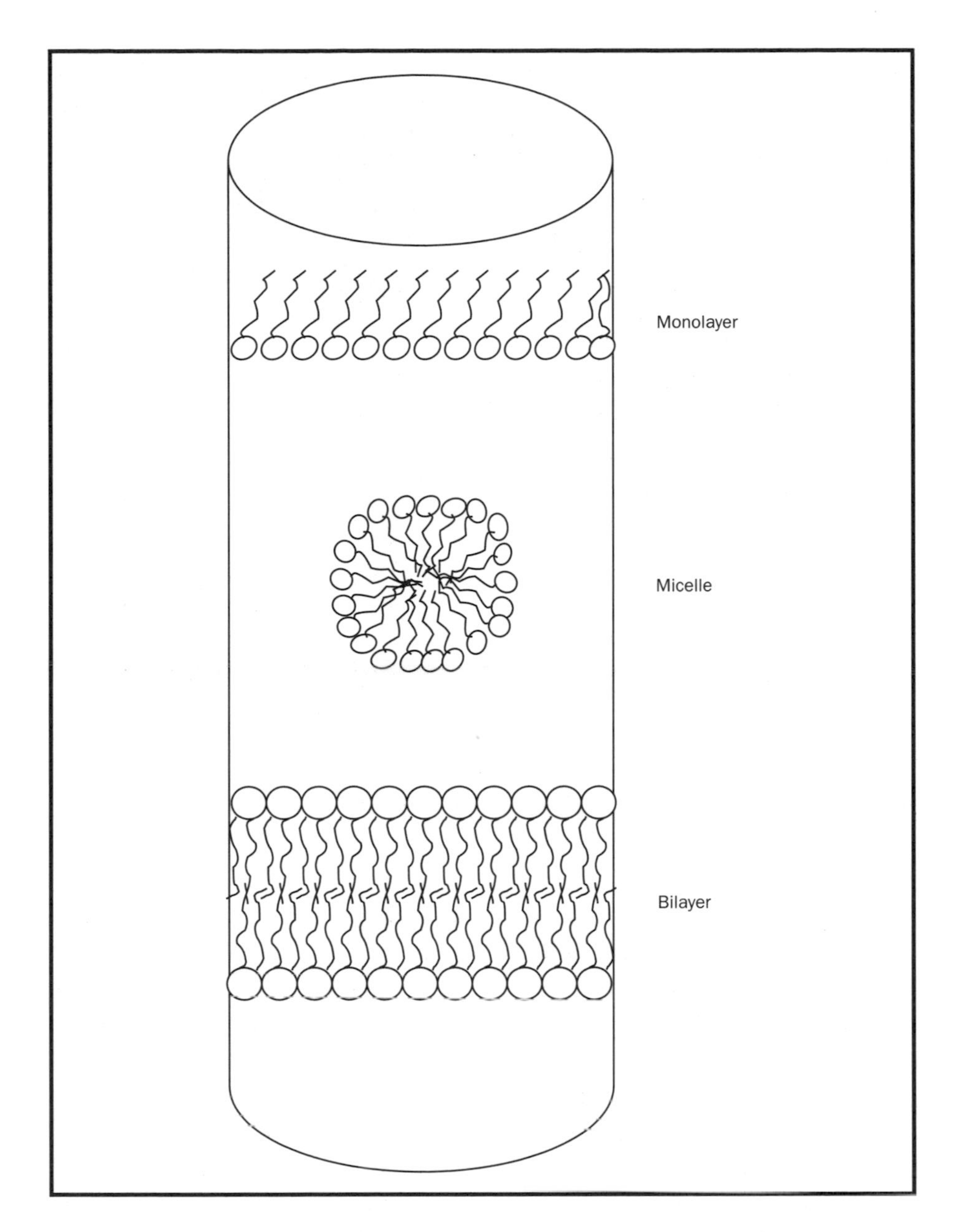

Figure 2. Assembly of lipids into more complex structures. At low concentrations, lipids form monolayers, with the polar head group (represented as a circle) associating with the water, while the hydrophobic tails (represented as lines) associate with the air. As the concentration of lipid increases, either miscelles or bilayers form, depending upon the lipid and conditions.

therapeutic agents inside liposomes and then directing the liposomes to a specific target tissue.

Lipids can also form structures in conjunction with various proteins. A cell membrane consists of a lipid bilayer that holds within it a variety of proteins that either transverse the bilayer or are associated more loosely with the bilayer. Cholesterol can insert into the bilayer, and this helps to regulate the fluidity of the membrane.

A variety of lipid-protein complexes are used in the body to transport relatively water-insoluble lipids, such as triglycerides and cholesterol, in circulating blood. These complexes are commonly called lipoproteins; they contain both proteins and lipids in varying concentrations. The density of these lipoproteins depends on the relative amounts of protein, because lipids are less dense than protein. Low density lipoproteins, or LDLs, have a relatively higher ratio of lipid to protein. LDLs are used to transport cholesterol and triglycerides from the liver to the tissues. In contrast, high density

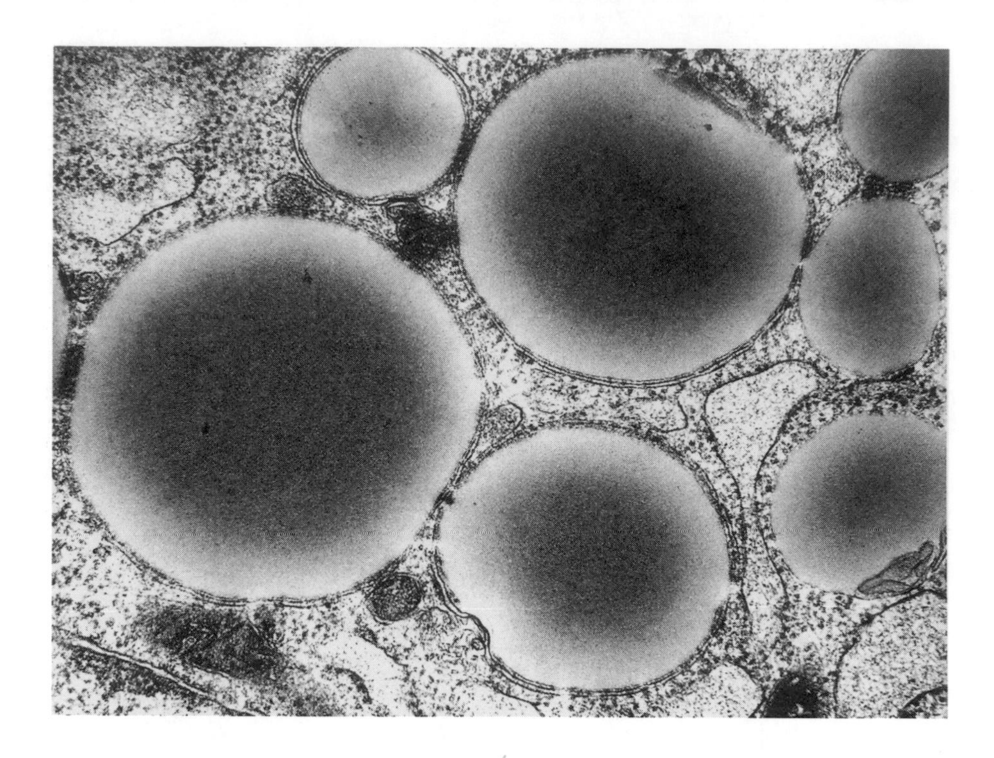

An electron micrograph of lipid droplets in the fat cell of a rat.

lipoproteins, or HDLs, have a relatively lower ratio of lipid to protein and are used in the removal of cholesterol and fats from tissues.

Functions of Lipids

Lipids perform a variety of tasks in biological systems. Terpenes, steroids, and eicosanoids act as communication molecules, either with other organisms or with other cells within the same organism. The highly reduced carbon atoms in triglycerides help to make fats an ideal energy storage compound.

Some of the functions of lipids are related to the structures they form. The micelle formation characteristic of fatty acids, detergents, and soaps in **aqueous solution** helps to dissolve dirt and other hydrophobic materials. Lipid bilayers play many vital roles. Liposomes are used to deliver drugs to desired tissues. A cell membrane, because of its hydrophobic core, is a substantial barrier to the passage of ions, allowing the cell interior to have concentrations of ions different from those of the extracellular environment. Bilayers are good electrical insulators, and aid in the transmission of nerve impulses along the conducting portions of nerve fibers. The importance of lipids in neural function is seen in diseases in which these insulators are lost, such as multiple sclerosis, or not properly maintained, such as Tay-Sachs disease.

aqueous solution: homogenous mixture in which water is the solvent (primary component)

Although they are a chemically diverse assortment of compounds, lipids share a number of properties. The amphipathic nature of lipid molecules encourages the formation of more complex structures such as micelles, bilayers, and liposomes. These structures, as well as the actual lipid substances themselves, affect all aspects of cell biology. SEE ALSO FATS AND FATTY ACIDS; LIPID BILAYERS; MEMBRANE; PHOSPHOLIPIDS.

Ann T. S. Taylor
Scott E. Feller

Bibliography

Voet, Donald; Voet, Judith G.; and Pratt, Charlotte (1999). *Fundamentals of Biochemistry.* New York: Wiley.

Internet Resources

King, Michael W. Lipoproteins. Indiana University School of Medicine. Available from <http://www.indstate.edu/thcme/>.

Liquid Crystals

In 1888 the Austrian botanist and chemist Friedrich Reinitzer, interested in the chemical function of cholesterol in plants, noticed that the cholesterol derivative cholesteryl benzoate had two distinct melting points. At 145.5°C (293.9°F) the solid compound melted to form a turbid fluid, and this fluid stayed turbid until 178.5°C (353.3°F), at which temperature the turbidity disappeared and the liquid became clear. On cooling the liquid, he found that this sequence was reversed. He concluded that he had discovered a new state of matter occupying a niche between the crystalline solid and liquid states: the liquid crystalline state. More than a century after Reinitzer's discovery, liquid crystals are an important class of advanced materials, being used for applications ranging from clock and calculator displays to temperature sensors.

Mesophases

In a crystalline solid, the molecules are well ordered in a **crystal lattice**. When a crystal is heated, the thermal motions of the molecules within the **lattice** become more vigorous, and eventually the vibrations become so strong that the crystal lattice breaks down and the molecules assume a disordered liquid state. The temperature at which this process occurs is the **melting point**. Although the transition from a fully ordered structure to a fully disordered one takes place in one step for most compounds, this

crystal lattice: three-dimensional structure of a crystaline solid

lattice: systematic geometrical arrangement of atomic-sized units that constitute the structure of a solid

melting point: temperature at which a substance in the solid state undergoes a phase change to the liquid state

A digital fish finder, the display of which is made with the use of liquid crystal diodes.

transition is not a universal behavior. For some compounds, this process of diminishing order as temperature is increased occurs via one or more intermediate steps. The intermediate phases are called *mesophases* (from the Greek word *mesos*, meaning "between"), or liquid crystalline phases. Liquid crystalline phases have properties intermediate between those of fully ordered crystalline solids and liquids. Liquid crystals are fluid and can flow like liquids, but the magnitudes of some electrical and mechanical properties of individual liquid crystals depend on the direction of the measurement (either along the main crystal axis or in another direction not along the main axis). Typical liquid crystals have rodlike or disklike shapes. Classes of liquid crystalline states or mesophases can be distinguished according to degrees of internal order.

phase: homogeneous state of matter

The least ordered liquid crystalline **phase** for rodlike molecules is the nematic phase (N), in which the long axes of individual molecules have an approximate direction (which is called the director, n). A nematic phase material has a low viscosity and is therefore very fluid. The term "nematic" is derived from the Greek word for thread (after the threadlike microscopic textures exhibited by nematic phase substances). In the smectic phases, the molecules have more order than molecules existing in the nematic phase. Just as in the nematic phase, the molecules have their long axes more or less parallel to the director. Additionally, the molecules are more or less confined to layers. The term "smectic" is derived from the Greek word for soap (owing to the fact that smectic liquid crystals have mechanical properties similar to those of concentrated aqueous soap solutions). The smectic phases are divided into classes based on degree of molecular order; the smectic A phase (SmA) and the smectic C phase (SmC) are the most studied ones. In the SmA phase, the molecules are **perpendicular** to the smectic layer planes, whereas in the SmC phase they are tilted. Substances assuming these phases have some fluidity, but their viscosities are much higher than that of a nematic phase substance. In Figure 1, the arrangements of molecules in the nematic, smectic A, and smectic C phases are shown schematically. Chiral molecules (molecules lacking a center of symmetry) can assume a *cholesteric* phase, also called a chiral nematic phase. In this mesophase, the molecules have helical arrangements. The pitch is the distance along a longitudinal axis corresponding to a full turn of the **helix**.

perpendicular: condition in which two lines (or linear entities like chemical bonds) intersect at a 90-degree angle

helix: form of a spiral or coil such as a corkscrew

Typical mesophases formed by disklike molecules are columnar, wherein the molecules are "stacked" into columns. The columns are in turn packed together to form two-dimensional arrays. In addition to the columnar arrangements, the molecules can become ordered in a way that is comparable to a heap of coins spread on a flat surface—the *discotic nematic* phase.

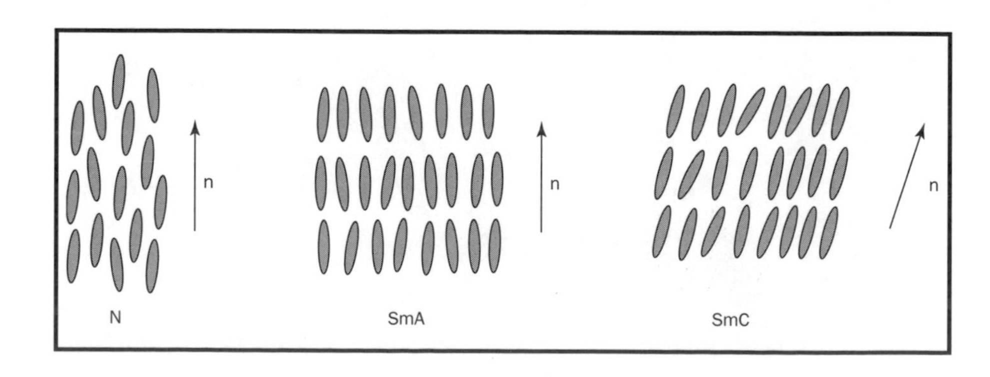

Figure 1. Rodlike molecules in the nematic phase (N), the smectic A phase (SmA) and in the smectic C phase (SmC). The director is denoted as n.

Displays

By far the most important application of liquid crystals is display devices. Liquid crystal displays (LCDs) are used in watches, calculators, and laptop computer screens, and for instrumentation in cars, ships, and airplanes. Several types of LCDs exist. In general their value is due to the fact that the orientation of the molecules in a nematic phase substance can be altered by the application of an external electric field, and that liquid crystals are anisotropic fluids, that is, fluids whose physical properties depend on the direction of measurement. It is not pure liquid crystalline compounds that are used in LCDs, but liquid crystal mixtures having optimized properties.

The simplest LCDs that display letters and numbers have no internal light source. They make use of surrounding light, which is selectively reflected or absorbed. An LCD is analogous to a mirror that is made nonreflective at distinct places on its surface for a certain period. The main advantage of an LCD is low energy consumption. More ad-vanced LCDs need back light, color filters, and advanced electronics to display complex figures. The best-known LCD is the so-called twisted nematic display.

Liquid Crystal Thermometers

The use of liquid crystals as temperature sensors is possible because of the selective reflection of light by chiral nematic (cholesteric) liquid crystals. A chiral nematic liquid crystal reflects light having a characteristic wavelength determined by its pitch and by the viewing angle (the angle between the eye of the observer and the surface of the liquid crystal). Because the pitch of a chiral nematic compound is temperature-dependent, observed color is a function of temperature. Liquid crystals can therefore serve as thermometers. By mixing chiral nematic compounds, thermometers can be customized to be effective in a desired temperature range. The color variation of some liquid crystal thermometers extends across the entire visible light spectrum within changes of a few tenths of a degree centigrade. For use in devices, microcapsules containing chiral nematic mixtures are mixed with binder materials. Liquid crystal thermometers find application in medicine (medical thermography). A liquid crystal thermometer attached to the skin can measure temperature variations of the skin. This can be useful in the detection of skin cancer, as tumors have different temperatures than surrounding tissues. In electronics, liquid crystal temperature sensors can pinpoint bad connections within a circuit board by detecting the characteristic local heating. The color changes of gadgets such as "mood rings" are a manifestation of chiral nematic mixtures. SEE ALSO INORGANIC CHEMISTRY.

Koen Binnemans

Bibliography

Collings, Peter J. (1990). *Liquid Crystals: Nature's Delicate Phase of Matter.* Princeton, NJ: Princeton University Press.

Collings, Peter J., and Hird, Michael (1997). *Introduction to Liquid Crystals: Chemistry and Physics.* London: Taylor and Francis.

Demus, Dietrich; Goodby, John; Gray, George W.; et al., eds. (1998). *Handbook of Liquid Crystals,* Vols. 1–3. Weinheim, Germany: Wiley-VCH.

Liquids

A liquid is one state in which matter can exist. A liquid can take the shape of any container it is placed in (unlike a solid), but the volume of the liquid will always remain constant (unlike a gas).

Liquid-Particle Movement

On a molecular level the molecules of a liquid are arranged, or ordered somewhere between the order of a solid and the randomness of a gas. The particles comprising a solid occur in an ordered fashion, producing a characteristic three-dimensional configuration that is present throughout the entire structure. The forces between particles in a solid are strong, holding the particles in a rigid form. A solid is therefore noncompressible and cannot flow to take the shape of its container. Conversely, in a gas the particles have no regularity in their arrangement, have essentially unrestricted movement, and are widely separated. Because the forces between particles are small, the particles may move apart and fill the available space. The shape and volume of a gas may therefore be changed.

A molecule of a liquid experiences an environment similar to that of a solid—it is in close proximity to its neighbors and has similar packing density. In a liquid, however, particles have no long-range order (i.e., only on a localized basis). The **intermolecular forces** between particles of a liquid are stronger than the kinetic energies of the molecules, which are thus held close together. These forces, however, do not hold the molecules in a rigid structure. Subsequently the molecules can move with respect to each other, allowing a liquid to flow. A liquid is minimally compressible and much denser than a gas, and maintains a constant volume.

intermolecular forces: force that arises between molecules; generally it is at least one order of magnitude weaker than the chemical bonding force

The particles comprising liquids can be molecules or atoms depending on the chemical nature of the substance. The general characteristics of a liquid are the same irrespective of its composition (molecules versus atoms) but **hydrogen bonding** can increase the attractive forces between molecules making a liquid flow less easily.

hydrogen bonding: intermolecular force between the H of an N–H, O–H, F–H bond and a lone pair on O, N, or F of an adjacent molecule

Decreasing the kinetic energy of particles of a liquid by cooling will eventually result in the change of state from liquid to solid. Similarly, increasing the kinetic energy by heating will result in the state change from liquid to gas. For example, at a pressure of one atmosphere, pure water changes from liquid to solid at 0°C (32°F) and from liquid to gas at 100°C (212°F).

Surface Tension

Surface tension is the appearance of a film over the top of a liquid, making the liquid behave as if it had a skin. It is because of surface tension that a small object like a pin or an insect can be supported on the surface of a liquid. This phenomenon is caused by the attraction between molecules of a liquid. In the bulk of a liquid any individual molecule is attracted equally in all directions by all adjacent molecules. At the surface of the liquid, however, there are no molecules attracting the surface molecules other than those in the liquid itself. This leads to a net attraction for the surface molecules into the liquid, in a direction parallel to the surface.

The most noticeable effect of surface tension is to reduce the surface of the liquid to the smallest possible size. Surface tension on water, for exam-

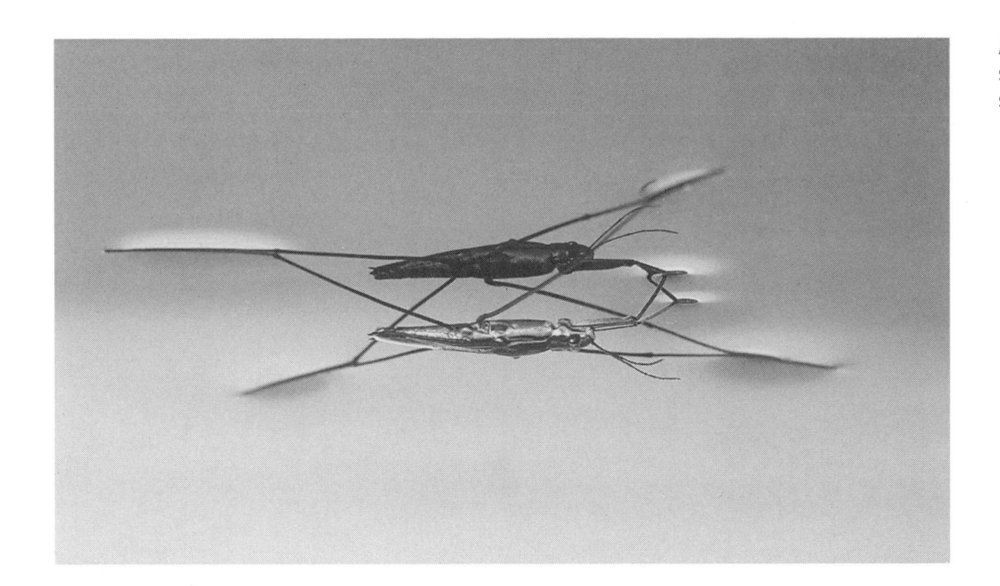

A waterstrider bug standing on the surface of water, a demonstration of the surface tension of water.

ple, causes water droplets to form the familiar bead shape (particularly noticeable on a wax surface). This surface tension can be destroyed by the addition of a detergent, causing the water droplet to spread out to a thin film covering the surface of the container.

To expand the surface of a liquid, the forces pulling the particles inward must be overcome. Surface tension is a measure of the amount of energy required to increase the surface area of a liquid by a given amount. A liquid displaying hydrogen bonding will have a higher surface tension than one that does not. For example, at 20°C (68°F) water has a surface tension of 7.29×10^{-2} joules per meter squared (the amount of energy that must be applied to water to increase the surface area by 1 meter squared). For mercury the surface tension is even higher (4.6×10^{-1} joules per meter squared at 20°C); this is due to the stronger metallic bonds between the atoms of mercury.

Viscosity

Viscosity is a measure of how thick (viscous) and sticky a liquid is. Viscosity reduces the ability of a liquid to flow. Any liquid that can flow readily (such as water) will have a low viscosity. Liquids with a high viscosity (such as molasses and motor oil) will flow more slowly and with greater difficulty.

Viscosity can be measured by timing how long it takes for the liquid to flow through a capillary tube or how long it takes for a steel ball to fall through the liquid. At a molecular level viscosity is a function of the attractive forces of the molecules of the liquid and, to a lesser extent, the presence of structural components (such as long-chain molecules) that can become entangled (a form of steric hindrance). Temperature also greatly affects viscosity: as temperature increases, viscosity decreases. This is because higher levels of kinetic energy are more able to overcome the intermolecular attractive forces.

Liquid Crystals

The term "liquid crystal" describes an intermediate state between a solid and a liquid. On heating to a specific temperature, some solids become

The liquids in this photograph have different levels of viscosity or fluidity.

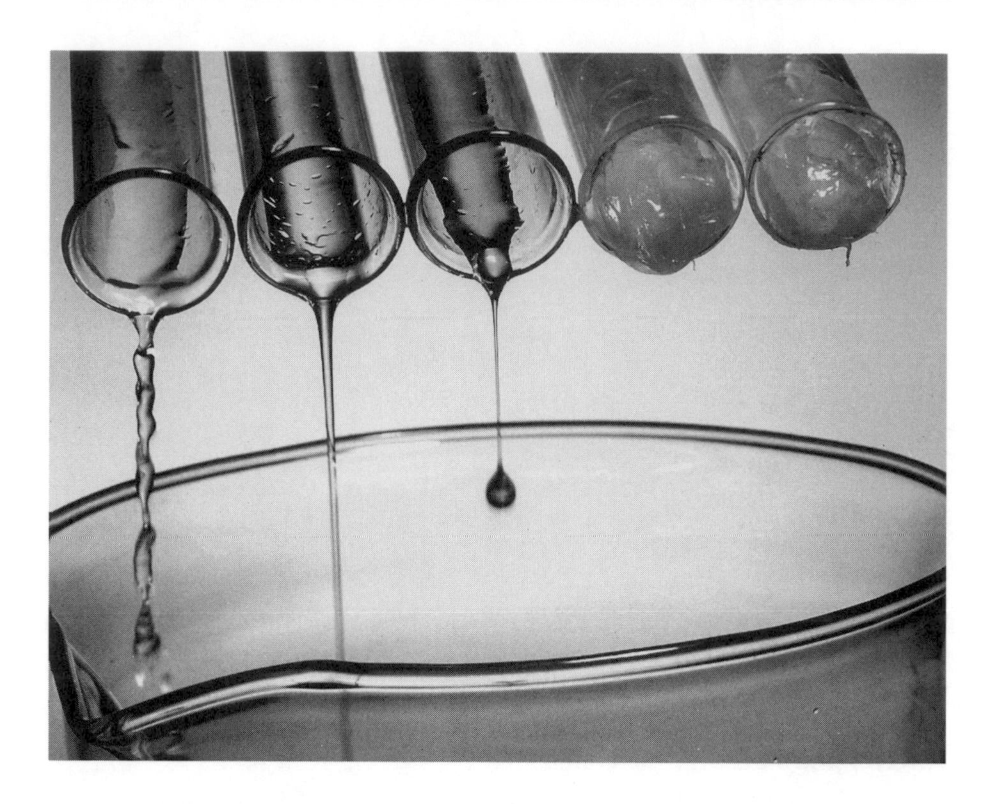

cloudy liquids, then become clear when the temperature is raised even further. For such substances, the temperature range at which this intermediate state—called the liquid crystal state—exists is always the same. In 1888 Austrian botanist and chemist Frederich Reinitzer was the first to discover a substance (cholesterol benzoate) that exhibited this behavior. Cholesterol benzoate melts at 145°C (293°F) to form a cloudy, milklike liquid, but at 179°C (354°F) the liquid becomes clear.

Liquid crystals are composed of long, rodlike molecules that can be ordered in a number of ways. There are three groups of liquid crystals: cholesteric, nematic, and smectic. In the cholesteric form all of the molecules in one layer are aligned in the same manner, but adjacent layers have the molecules twisted with respect to each layer (commonly encountered in crystals of cholesterol from which the name originates). In the case of a nematic liquid crystal, the axes of the molecules are aligned but the ends of the molecules are not aligned or adjacent to each other. The smectic liquid crystal is characterized by the axes and the ends of the molecules being aligned. The ordering of the molecules is altered by changes in pressure, temperature, and electric and magnetic fields.

Glasses

Glasses are supercooled liquids that form a noncrystalline solid. The most frequently encountered glass is the supercooled form of liquid silicon dioxide—the glass used in windows, for example. When silicon dioxide is heated to 1,600°C (2,912°F), it forms a viscous liquid. Upon rapid cooling, silicon-oxygen bonds are formed before the atoms are able to arrange themselves in a regular pattern characteristic of a solid.

Glasses share a number of common characteristics regardless of their origins. All glasses are transparent or translucent, hard, brittle, and resistant to chemical attack. Glasses can be made from a range of acidic oxides including lead, boron, and phosphorus. When a glass is heated to its softening point, the material begins to crystallize and becomes more brittle and opaque. Toughening of glass can be brought about by rapid cooling during its production or by chemical treatment of the surface. To produce colored glass, metallic oxides or other compounds are added. For window glass other compounds can be added to make the characteristics more desirable; these include sodium carbonate, calcium oxide (lime), or calcium carbonate. SEE ALSO GLASS; MOLECULAR STRUCTURE; PHYSICAL CHEMISTRY.

Gordon Rutter

Bibliography

Brady, James E., and Holum, John R. (1993). *Chemistry: The Study of Matter and Its Changes*. New York: John Wiley and Sons.

Collings, Peter J., and Hird, Mike (1997). *Introduction to Liquid Crystals: Chemistry and Physics*. London: Taylor and Francis.

Hansen, Jean Pierre, and McDonald, Ian R. (1990). *Theory of Simple Liquids*. London: Academic Press.

Marcus, Yitzhak (1977). *Introduction to Liquid State Chemistry*. New York: John Wiley and Sons.

Lister, Joseph

BRITISH SURGEON
1827–1912

Joseph Lister is known as the founder of antiseptic surgery, a significant advance in medicine developed in the nineteenth century. Infection of wounds and surgical incisions was a major cause of hospital deaths before Lister developed a way of preventing these infections with chemical antiseptics. His discovery made surgery much safer and permitted surgeons to perform operations not previously attempted because of the high risk of fatal infections.

Lister attended University College in London, England, for both his undergraduate and medical education, graduating with honors in 1852. His first position following graduation was as a staff surgeon at University College Hospital. In 1854 he accepted an appointment in Edinburgh, Scotland, as an assistant to Dr. James Syme, a prominent surgeon and noted professor of surgery. In 1856 Lister married Dr. Syme's daughter, Agnes, and accepted a position as an assistant surgeon at the Royal Infirmary in Edinburgh. Lister developed a reputation as both a skillful surgeon and excellent teacher. In 1861 he was appointed professor of surgery at the University of Glasgow where he began the experiments that led to the practice of antiseptic surgery.

When Lister became a practicing surgeon in 1852, conditions in surgical wards were truly appalling. Most surgeons operated with unwashed hands and dirty instruments while wearing bloodstained operating coats that were never washed. The patients then rested in beds with dirty linens that were often not even changed between patients. Consequently many patients survived the operation only to die from gangrene or blood poisoning. Nearly all surgical patients experienced infections and the smell of **putrefaction** permeated surgical wards. The cause of infection was generally attributed to "bad air" and was considered an unavoidable aspect of all hospitals.

putrefaction: decomposition of organic matter

In 1864 Lister discussed the possible causes of putrefaction with a chemistry colleague who suggested that Lister read the publications of a French

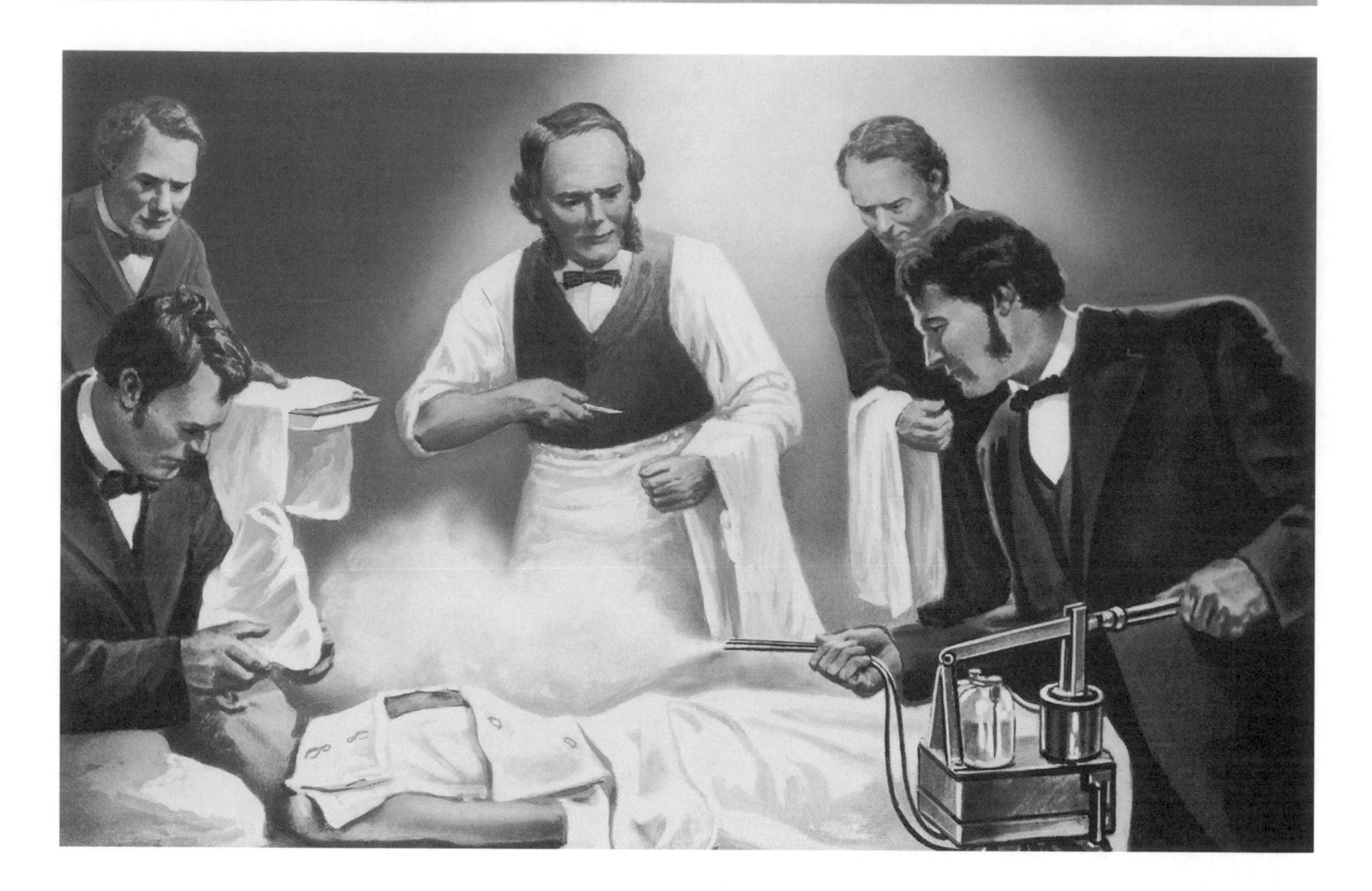

British surgeon Joseph Lister (center), inventor of antiseptic surgery.

chemist named Louis Pasteur. Pasteur had demonstrated that both fermentation of liquids and spoilage of butter were caused by "germs" carried in the air, and if these germs were prevented from entering a flask, fermentation and spoilage could be prevented. Lister saw a connection with hospital infections and concluded that it was not air itself that caused infection, but germs carried in the air. Lister reasoned that if hospital germs could be killed, infections could be prevented.

phenol: common name for hydroxybenzene (C_6H_5OH)

Pasteur had killed germs by boiling, a technique which could not be used on patients, but chemicals that killed germs could be applied to wounds, instruments, the surgeon's hands, and to bandages. Carbolic acid, now called **phenol**, was already known as a deodorizing agent and preservative, and Lister thought it might kill the germs that were causing infections. In 1865 Lister began using carbolic acid as an antiseptic during surgery and in bandaging afterward, publishing his results in 1867.

Initially his procedures were met with scorn and the idea that invisible germs were the cause of hospital infections was widely ridiculed. As his students and visitors witnessed antiseptic surgery's great success in reducing deaths from hospital infections, however, Lister's procedures became accepted. Within a few years Lister was honored in Germany and France, but not yet recognized in London. Following an appointment as professor of clinical surgery at King's College, Lister was able to directly demonstrate the success of his procedures to skeptical London surgeons. In recognition of his contribution to medicine, Lister was made a baronet in 1883 and named Baron Lister in 1897.

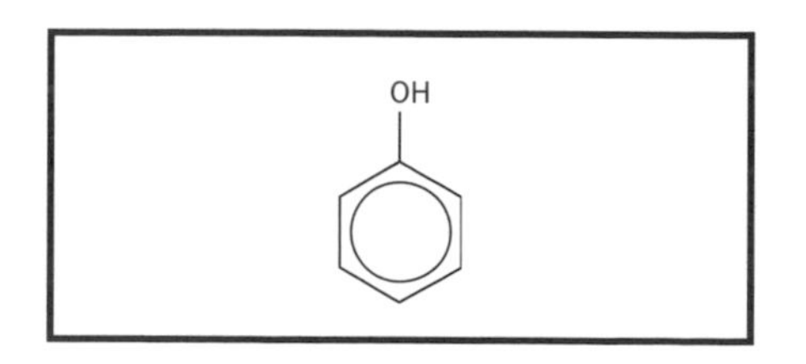

Structure of "phenol."

Although use of carbolic acid as a germ-killing antiseptic in surgery was later replaced by more effective techniques, Lister became world-renowned for demonstrating the importance of preventing microbial contamination of wounds and surgical incisions. SEE ALSO IODINE; PASTEUR, LOUIS.

Robert K. Griffith

Bibliography

Farmer, Laurence (1962). *Master Surgeon: A Biography of Joseph Lister.* New York: Harper & Brothers.

Fisher, Richard B. (1977). *Joseph Lister.* New York: Stein and Day.

Lithium

MELTING POINT: 180.5°C
BOILING POINT: 1,342°C
DENSITY: 0.534 g/cm^3
MOST COMMON IONS: +1

Lithium is a soft, silvery alkali **metal** and has the lowest density of any metal. The word "lithium" is derived from "lithos" (Greek for "stone"). Johan A. Arfvedson discovered lithium in Stockholm, Sweden, in 1817. Humphry Davy isolated it via electrolysis in 1818. Currently, lithium metal is generated by the electrolysis of a molten mixture of lithium chloride, LiCl, and potassium chloride, KCl. In nature it is never found in its elemental form. Its main sources are the minerals spodumene, petalite, lepidolite, and amblygonite. Lithium's average crustal abundance is about 18 ppm. It has the highest specific heat of any solid element and is the least reactive alkali metal toward water. Lithium burns crimson in the flame test.

metal: element or other substance the solid phase of which is characterized by high thermal and electrical conductivities

Metallic lithium has a variety of uses. It is used as an anode material in batteries and as a heat transfer agent. Magnesium-lithium **alloys** are used to produce armor plate and aerospace materials, while aluminum-lithium alloys find applications in the aircraft industry. Lithium is also used to produce chemical **reagents** such as $LiAlH_4$ (a **reducing agent**) and *n*-butyllithium (a strong base).

alloy: mixture of two or more elements, at least one of which is a metal

reagent: chemical used to cause a specific chemical reaction

reducing agent: substance that causes reduction, a process during which electrons are lost (or hydrogen atoms gained)

Compounds of lithium are also economically important. Air conditioning systems use LiCl and LiBr because they are very hygroscopic and readily absorb water from the air. Thermonuclear weapons incorporate lithium deuteride, LiD. Lithium stearate is obtained by treating tallow with lithium hydroxide, LiOH, and is used as a thickener that imparts high temperature resistance to lubricants. Carbon dioxide removal systems in submarines and spacecraft use LiOH. Lithium carbonate, Li_2CO_3, is used to increase the electrical current flow in the electrolytic production of aluminum from bauxite and to strengthen glasses by substituting for sodium ions. Although lithium carbonate has been used to treat bipolar (manic-depressive) disorder since 1949, its mechanism of operation is still not completely understood. SEE ALSO ALKALI METALS; DAVY, HUMPHRY.

Nathan J. Barrows

Bibliography

Emsley, John (2001). *Nature's Building Blocks: An A–Z Guide to the Elements.* New York: Oxford University Press.

Greenwood, Norman N., and Earnshaw, A. (1997). *Chemistry of the Elements,* 2nd edition. Boston: Butterworth-Heinemann.

English crystallographer Dame Kathleen Lonsdale, who established the molecular structure of benzene.

Lide, David R., ed. (2000). *The CRC Handbook of Chemistry & Physics,* 81st edition. New York: CRC Press.

Lonsdale, Kathleen

IRISH CRYSTALLOGRAPHER
1903–1971

Kathleen Lonsdale was born Kathleen Yardley in Newbridge, Ireland, on January 28, 1903. She was the youngest of ten children (four of whom died in infancy). In 1908 Lonsdale's mother, Jessie, separated from her husband and emigrated with her children to England. Lonsdale was a bright child, and although the older children had to leave school and go to work, she stayed in school and was allowed to enroll in Bedford College, University of London, at age sixteen, to study mathematics and, later, physics. She graduated in 1922 at the top of her class and her teacher and mentor, the eminent English physicist Sir William Henry Bragg, recruited her to join his research group in London. She began to work in x-ray crystallography (the analysis of crystal structure by means of x rays), then a new field, and became a leading scientist in this field.

In 1927 Lonsdale married Thomas Lonsdale and she and her husband lived in Leeds, England, for three years. There she was given crystals of hexamethylbenzene and hexachlorobenzene for analysis by the English chemist Sir C. K. Ingold. She subsequently confirmed that the benzene ring was flat, with all carbon-carbon angles and bond lengths identical. Hers was the first experimental proof. Ingold said of her published account of benzene structure, "[O]ne paper like this brings more certainty into organic chemistry than generations of activity by us professionals." She was the first scientist to apply Fourier analysis (a type of mathematical analysis) to the analysis of crystal structure.

In 1930 Lonsdale moved back to London. In London she brought up three children while conducting research at home. In 1931 she rejoined Bragg at the Royal Institution of Great Britain and there (using the rooms of the great Michael Faraday) continued her work in the analysis of crystals. She stayed at the Royal Institution until 1946. Her major contribution to chemistry and physics was to establish the theoretical foundation of crystallography. Lonsdale helped to create the structure factor tables that are used by present-day crystallographers, and edited the *International Tables of X-Ray Crystallography* (1935), the "crystallographer's bible."

In 1946 Lonsdale moved to University College, London (UCL), where she became professor of chemistry in 1949 and remained until her retirement in 1968. With great modesty she said she knew very little chemistry, and no organic chemistry. She gathered about herself a large research group and trained many crystallographers. She worked on many crystal structure problems, including the clarification of the structure of diamond and the nature of urinary stones.

Royal Society: The U.K. National Academy of Science, founded in 1660

Lonsdale accrued many "firsts": one of the first two women to be elected Fellow of the **Royal Society** (1945); the first woman professor at UCL (1949); the first woman to become president of the International Union of Crystallography (1966) and of the British Association for Science (1968). It is partly because of her example and influence that women have been so prominent in crystallography.

Lonsdale became a Quaker in 1936 and went to jail in England for her participation in conscientious objection during World War II. After the jail experience she became active in prison reform and later in the international peace movement. Her political activism was as significant to her as her scientific achievements. Among her books were *Is Peace Possible?* (1957) and *The Christian Life Lived Experimentally* (1976) and she saw no conflict between her faith and her science, nor between being a wife and mother and being a scientist.

Lonsdale continued to work in various ways (including scientific, educational, and charitable work) during her official retirement. She died on April 1, 1971. In 1981 the chemistry building at UCL was named the Kathleen Lonsdale Building in her honor. SEE ALSO BRAGG, WILLIAM HENRY; ORGANIC CHEMISTRY; SOLID STATE.

Peter E. Childs

Bibliography

Childs, Peter E. (2003). "Woman of Substance." *Chemistry in Britain* (January):41–43.

Julian, M. M. (1990). "Women in Crystallography." In *Women of Science*, ed. G. Kass-Simon and Patricia Farnes. Bloomington: Indiana University Press.

Laidler, Keith J. (1997). "Kathleen Lonsdale (1903–1971)." *Chem13 News* 255:1–5.

Lonsdale, Kathleen (1968). "Human Stones." *Scientific American* 219:104–111.

Internet Resources

Childs, Peter E. "Kathleen Lonsdale 1903–1971." Available from <http://www.ul.ie/~childsp/>.

Contributions of 20th Century Women to Physics. Kathleen Yardley Lonsdale. Available from <http://www.physics.ucla.edu/~cwp/>.

Low Density Lipoprotein (LDL)

Lipids are **nonpolar** molecules and are relatively insoluble in **aqueous solutions**. At low concentrations, cholesterol and cholesterol **esters**, as well as other lipids, may form microscopic droplets called chylomicrons (lipid-protein complexes) that are somewhat stable in solution. At high concentrations, the lipids would form larger droplets and clog blood vessels, so they must be transported as complexes of lipid and protein called lipoproteins. Lipoproteins are complexes of lipid and precursor protein molecules called apolipoproteins.

Some portions of the apolipoprotein molecules are nonpolar (**hydrophobic**), and these are usually oriented toward the inside (near the lipid portion) of the complex. Polar amino acid side chains in the protein portions are oriented toward the outside of the complex, where they associate with the aqueous environment, rendering the complex soluble in blood plasma. This type of structure resembles that of micelles.

Lipoprotein complexes usually have a lipid core surrounded by one or more apolipoprotein molecules. These complexes can be separated into classes according to density. They range from very low density lipoproteins (VLDL), having densities of less than 1.006 g/mL, to low density lipoproteins (LDL), having densities of between 1.019 and 1.063 g/mL, to high density lipoproteins (HDL), having densities of between 1.063 and 1.210

lipid: a nonpolar organic molecule; fatlike; one of a large variety of nonpolar hydrophobic (water-hating) molecules that are insoluble in water

nonpolar: molecule, or portion of a molecule, that does not have a permanent, electric dipole

aqueous solution: homogenous mixture in which water is the solvent (primary component)

ester: organic species containing a carbon atom attached to three moieties: an O via a double bond, an O attached to another carbon atom or chain, and a H atom or C chain; the R(C═O)OR functional group

hydrophobic: a part of a molecule that repels water

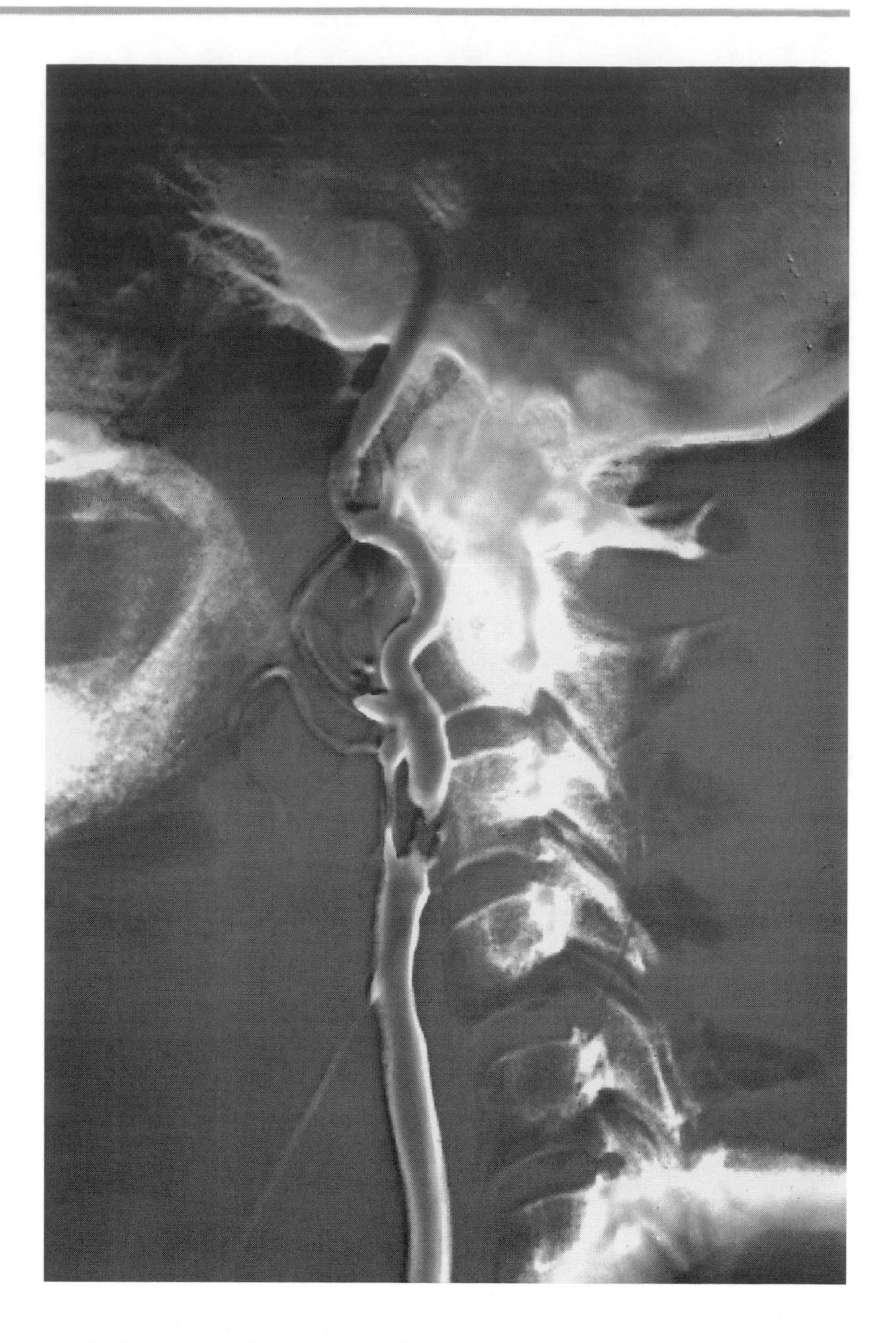
An angiogram of the carotid artery. The darker section indicates blockage. LDL, or "bad cholesterol," which at high concentrations can clog blood vessels, can be detected by the use of an angiogram.

g/mL. In general, the density of the lipoprotein increases as the proportion of apolipoprotein increases.

Small amounts of cholesterol may be transported as part of chylomicrons, but cholesterol is usually carried within lipoproteins, including low density lipoprotein (LDL), which carries cholesterol from the liver to muscle and other tissues, and high density lipoprotein (HDL), which carries cholesterol to the liver for conversion to bile acids. Physicians are especially concerned when patients have high levels of LDL (the so-called bad cholesterol) in blood; moderate exercise and low-cholesterol diets help to increase HDL (the so-called good cholesterol). Either high fat intake or problems with transport of cholesterol can lead to atherosclerosis, which in turn can contribute to heart attack (myocardial infarction) or stroke. SEE ALSO CHOLESTEROL; LIPIDS; PROTEINS.

Dan M. Sullivan

Bibliography

Boyer, Rodney F. (2002). *Concepts in Biochemistry*, 2nd edition. Pacific Grove, CA: Brooks/Cole Thomson Learning.

Devlin, Thomas M., ed. (2002). *Textbook of Biochemistry: With Clinical Correlations*, 5th edition. New York: Wiley-Liss.

McKee, Trudy, and McKee, James R. (2003). *Biochemistry: The Molecular Basis of Life*, 3rd edition. Boston: McGraw-Hill.

Lucretius

ROMAN NATURAL PHILOSOPHER
ca. 95 B.C.E.–ca. 55 B.C.E.

Little is known about Titus Lucretius Carus beyond what can be gathered from his poem *De rerum natura*. He was born in about 95 B.C.E., but the exact date is uncertain. The exact date and circumstances of his death are also uncertain, but he probably died in or before the year 55 B.C.E. We do know from his poem that he believed the teachings of the Greek atomists, ranging from those of Democritus of Abdera (ca. 460 B.C.E.–ca. 362 B.C.E.) to those of Epicurus (ca. 341 B.C.E.–270 B.C.E.). Unlike the writings of Democritus or Epicurus, Lucretius's poem was one of the few literary works not lost to European peoples after the collapse of the classical world.

De rerum natura is a poem in the Latin language that gives a summary of the teachings of the Greek atomists. His starting point is a reliance on direct human experience of the natural world. From this starting point he reasons: "Nothing can ever be created by divine power out of nothing" (Lucretius, p. 31). Accordingly, if something could be created out of nothing, things would pop in and out of existence without any pattern at all. From that deduction Lucretius develops a philosophy that does not allow for occult forces, superstition, or magic. Beliefs such as these were pervasive in the Roman world during his lifetime. That philosophy also clearly sets "atomism" against any sort of theistic religion. This religious antagonism would continue to plague atomic theories until the modern era.

According to Lucretius: "All nature as it is in itself consists of two things—bodies and the vacant space in which the bodies are situated and through which they move in different directions" (p. 39). He addresses the question of the immense variety of material things found in nature by recognizing that there must be some way for atoms to combine and at the same time maintain their individual characters: "Material objects are of two kinds, atoms and compounds of atoms. The atoms themselves cannot be swamped by any force, for they are preserved indefinitely by their absolute solidity" (p. 41). Lucretius does not suggest that we directly experience atoms. He makes no claims as to the shapes of atoms or any other of their characteristics. SEE ALSO ATOMS.

David A. Bassett

Bibliography

Lucretius (1951, reprint 1977). *On the Nature of the Universe*, tr. by Ronald Latham. New York: Penguin Books.

Luciferins *Bioluminescence; Chemiluminescence.*

Lutetium

MELTING POINT: 1,675°C
BOILING POINT: 3,315°C
DENSITY: 9.84 g/cm^3
MOST COMMON IONS: Lu^{3+}

The mixture of oxides known as ytterbia was obtained from yttria by Jean-Charles-Galissard de Marignac in 1878. From ytterbia the oxides of three elements were isolated: ytterbium (named after the town of Ytterby) by Marignac; scandium (named after Scandinavia) by L. F. Nilson in 1879; and lutetium (named after Lutetia, an ancient name of Paris) by G. Urbain, C. A. von Welsbach, and C. James in 1907. Lutetium is a rare element (comprising 7.5 (10^{-5}% of the igneous rocks of Earth's crust) and is found together with the heavy **lanthanides**. Essentially, there are two methods used to separate lutetium from monazite concentrates: (1) the extraction of **aqueous solutions** of lutetium nitrates with tri-*n*-butyl-phosphate (using kerosene as an **inert** solvent); and (2) using cationic exchange resins and solutions of EDTA, the triammonium salt of ethylenediamino-triacetate, as the eluant. The **metals** are obtained by electrolysis of the fused salts, or by metallothermic reduction of the anhydrous halides (especially the fluoride) with calcium at elevated temperatures. Lutetium is a **diamagnetic** trivalent element.

lanthanides: a family of elements (atomic number 57 through 70) from lanthanum to lutetium having from 1 to 14 4f electrons

aqueous solution: homogenous mixture in which water is the solvent (primary component)

inert: incapable of reacting with another substance

metal: element or other substance the solid phase of which is characterized by high thermal and electrical conductivities

diamagnetic: property of a substance that causes it to be repelled by a magnetic field

The lutetium halides (except the fluoride), together with the nitrates, perchlorates, and acetates, are soluble in water. The hydroxide oxide, carbonate, oxalate, and phosphate compounds are insoluble. Lutetium compounds are all colorless in the solid state and in solution. Due to its closed electronic configuration ($4f^{14}$), lutetium has no absorption bands and does not emit radiation. For these reasons it does not have any magnetic or optical importance. SEE ALSO CERIUM; DYSPROSIUM; ERBIUM; EUROPIUM; GADOLINIUM; HOLMIUM; LANTHANUM; NEODYMIUM; PRASEODYMIUM; PROMETHIUM; SAMARIUM; TERBIUM; YTTERBIUM.

Lea B. Zinner

Bibliography

Greenwood, Norman N., and Earnshaw, A. (1984). *Chemistry of the Elements.* New York: Pergamon Press.

Moeller, Therald (1975). "The Chemistry of the Lanthanides." In *Comprehensive Inorganic Chemistry,* ed. J. C. Bailar Jr. Oxford, U.K.: Pergamon.

Magnesium

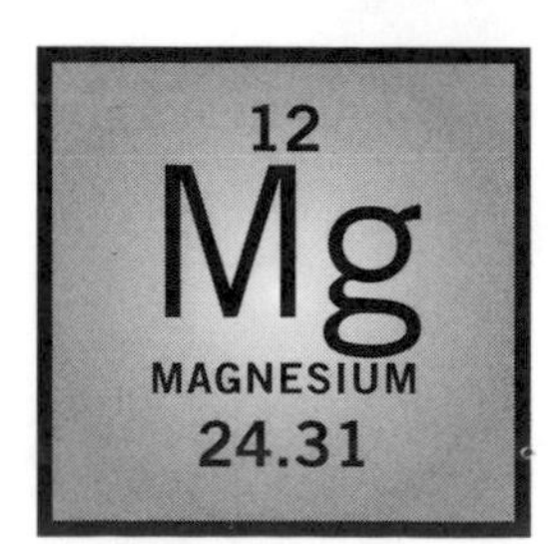

MELTING POINT: 651°C
BOILING POINT: 1,107°C
DENSITY: 1.738 g/cm^3 at 20°C
MOST COMMON IONS: Mg^{2+}

Magnesium was first recognized as an element by Joseph Black in 1755. In 1808 Sir Humphry Davy isolated the element, and in 1831 H. Bussy devised a method for producing it. Magnesium, in its combined states, is readily abundant and is the eighth most common element in Earth's crust. Magnesium **metal** is silvery white in color.

The most common method for producing elemental magnesium is in fused salt electrolytic cells, wherein magnesium chloride ($MgCl_2$) is decomposed by applying a **voltage** to elemental magnesium and chlorine gas. The magnesium chloride feed is obtained directly from seawater or from magnesium oxide deposits containing magnesite or dolomite. In these cases, the oxide is first chlorinated prior to electrolysis. Another method is to produce magnesium directly from the oxide by reducing the oxide with silicon under vacuum. The resultant Mg vapor is condensed to recover Mg metal. This process is carried out in vacuum retorts and is known as the Pidgeon process.

voltage: potential difference expressed in volts

The principal uses of Mg are for alloying with aluminum, for desulphurizing steel and pig iron, and for nodularizing the graphite in cast irons. Recently, researchers have focused on using Mg **alloys** to produce lightweight components in automobiles. As a result, Mg usage in vehicles is steadily increasing.

alloy: mixture of two or more elements, at least one of which is a metal

Compounds of magnesium, including the hydroxide, the chloride, the citrate, and the sulfate, are used in the medical field. Magnesium is an important element in both animal and plant life. On average, adults require a daily intake of about 300 milligrams (0.011 ounces) of magnesium. SEE ALSO ALKALINE EARTH METALS; BLACK, JOSEPH; DAVY, HUMPHRY; INORGANIC CHEMISTRY.

Frank Mucciardi

Bibliography

Kramer, Deborah A. (2001). "Magnesium, Its Alloys and Compounds." U.S. Geological Survey Open-File Report 01 341. Also available from <http://pubs.usgs.gov/of/of01-341/>.

Raloff, J. (1998). "Magnesium: Another Metal to Bone up On." *Science News*. 154(9):134.

Magnetic Resonance Imaging ***See Nuclear Magnetic Resonance.***

Magnetism

The magnetic properties of materials were recognized by the ancient Greeks, Romans, and Chinese, who were familiar with lodestone, an iron oxide mineral that attracts iron objects. Although the attractive or **repulsive forces** that act between magnetic materials are manifestations of magnetism familiar to everybody, the origin of magnetism lies in the atomic structure of matter. Despite the fact that magnetism can be explained only by the quantum theory developed at the beginning of the twentieth century, qualitative predictions of magnetic properties can be made within the context of classical physics. Magnetic forces originate in the motion of charged particles, such as electrons. The electrons "spin" around their axis and move in orbits around the nucleus of the atom to which they belong. Both motions generate tiny electric currents in closed loops that in turn create magnetic dipole fields, just as the current in a coil does. When placed in a magnetic field, the tiny magnetic dipole fields tend to align with the external field.

repulsive force: force that causes a repulsion between two bodies; charges of the same sign repel each other

According to their behavior in inhomogeneous magnetic fields, materials can be classified into three main categories: **diamagnetic**, paramagnetic, and ferromagnetic. *Paramagnetic materials* are attracted into a magnetic field. The main cause of this effect is the presence in the material of atoms that have a net magnetic moment composed of electron spin and orbital contributions.

diamagnetic: property of a substance that causes it to be repelled by a magnetic field

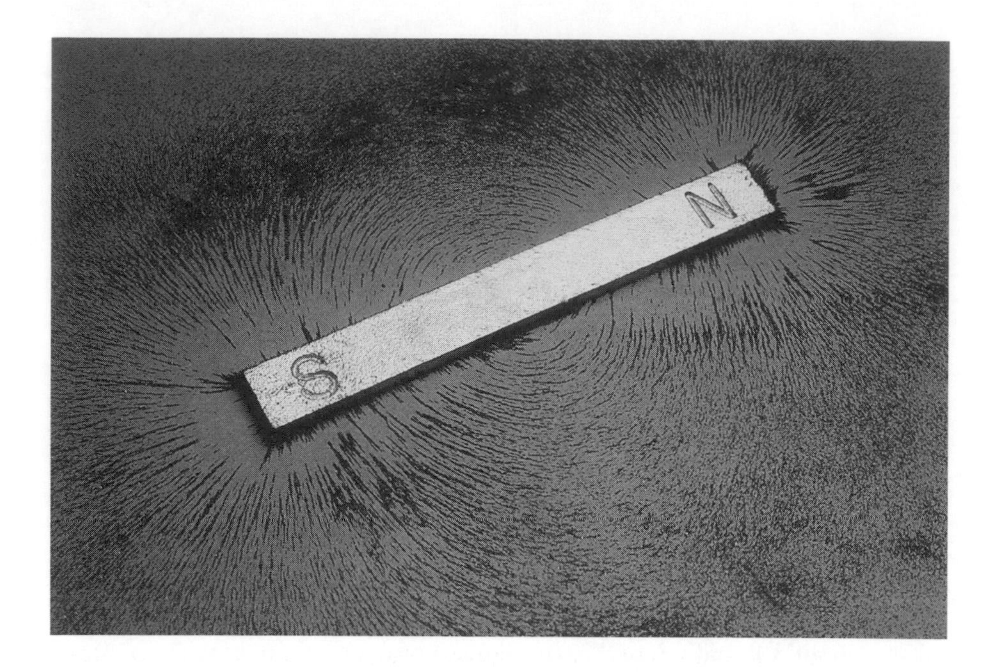

Iron filings in a circular pattern around a magnet, indicative of the field of force of the magnet.

When placed in a magnetic field, the magnetic moments of the atoms, which are otherwise randomly oriented, tend to align with the field and thus enhance the field. Paramagnetism is temperature dependent because increased thermal motion at higher temperatures impedes the alignment of the magnetic moments with the field. *Diamagnetic materials* are slightly repelled by a magnetic field. This effect occurs for materials that contain atoms in which the spin and orbital contributions to the magnetic moment cancel out. In this case, the interaction between the material and a magnetic field is caused by the occurrence of currents induced by the magnetic field in the atoms. The dipole fields corresponding to these currents are directed opposite to the applied magnetic field and cause expulsion of the material from the field. *Ferromagnetic materials* contain atoms that have magnetic moments that are aligned even in the absence of an applied magnetic field because of mutual interactions, creating a sizable net magnetic moment for domains of the material. The magnetic moments of domains can be randomly oriented unless a magnetic field is applied to the material.

alloy: mixture of two or more elements, at least one of which is a metal

transition metals: elements with valence electrons in d-sublevels; frequently characterized as metals having the ability to form more than one cation

metal: element or other substance the solid phase of which is characterized by high thermal and electrical conductivities

octahedral: relating to a geometric arrangement of six ligands equally distributed around a Lewis acid; literally, eight faces

Iron, cobalt, nickel, and their **alloys** are examples of ferromagnetic materials. These three elements are **transition metals**, and their atoms or ions have unpaired electrons in d orbitals. Rare-earth ions also have unpaired electrons situated in f orbitals. A detailed investigation of the properties of molecules that contain such **metal** ions in a magnetic field can provide significant information about how their electrons are distributed in orbitals. Typically, d orbitals of isolated atoms are degenerate (Figure 1a). This situation changes when the metal ions are part of molecules in which they experience a nonspherically symmetric environment. Figures 1b and 1c show the splitting of d orbitals for a transition metal ion that has six unpaired electrons and is situated in an environment of six atoms in an **octahedral** arrangement. Depending on the size of the splitting (the lighter shading in Figure 1) and the interelectron repulsion, the metal ion may have four unpaired electrons (Figure 1b) or no unpaired electrons (Figure 1c). This difference in electron distribution leads to significant differences in the magnetic properties of the molecules that contain such ions, with

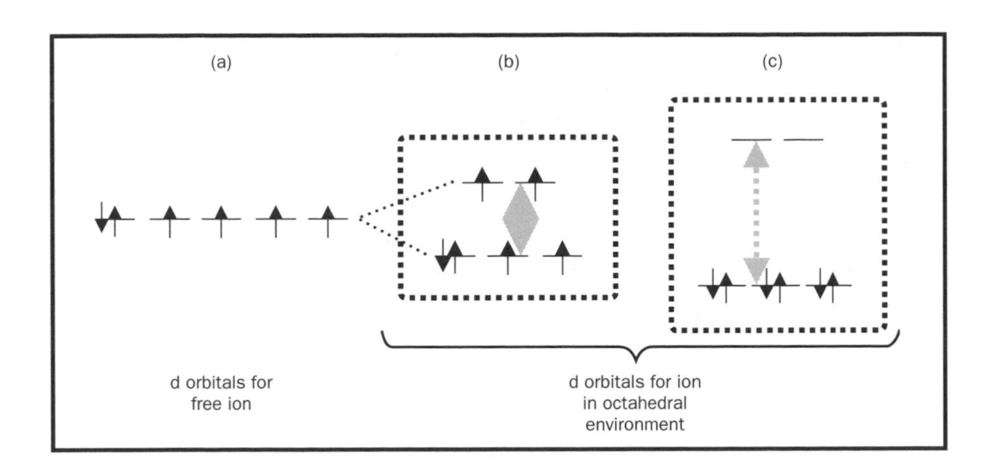

Figure 1. Splitting of d orbitals for a transition metal ion.

the former being paramagnetic and the latter being diamagnetic. When there are multiple metal sites in a molecule, the spins at different metal ions can be either ferro- (parallel) or antiferro-magnetically (antiparallel) aligned to each other. Clever use of the magnetic properties for metal ions and of the interactions between spins manifested in molecular systems enables scientists to design and synthesize molecular systems with interesting properties, such as molecular magnets.

Magnetic materials are widely used for building technological devices and scientific tools. Classical examples are electromagnets that are used in motors, clutches, and breaking systems. The electromagnet makes use of an iron core situated in a solenoid through which electric current is passed. This current creates a magnetic field at the center of the solenoid that orients the magnetic moments in the domains of the iron core, which in turn results in a significant enhancement of the magnetic field at the core of the solenoid. Electromagnets can also be used to record information on magnetic tape, which has a ferromagnetic surface.

Finally, although atomic nuclei have significantly smaller magnetic moments than electrons, the study of their interaction with magnetic fields has many important applications. They enable the scientists in the biological and medical fields to elucidate the structure of biologically relevant molecules such as proteins and to diagnose diseases using magnetic resonance imaging. SEE ALSO MAXWELL, JAMES CLERK; PHYSICAL CHEMISTRY.

Catalina Achim

Bibliography

Kittel, C. (1996). *Introduction to Solid State Physics*, 5th edition. New York: Wiley.

Miessler, G. L., and Tarr, D. A. (1999). *Inorganic Chemistry*, 2nd edition. Upper Saddle River, NJ: Prentice Hall.

Manganese

MELTING POINT: 1,246 ±3°C
BOILING POINT: 2,061°C
DENSITY: 7.21–7.44 g/cm³, depending upon allotrope
MOST COMMON IONS: Mn^{2+}, Mn^{3+}, MnO_4^{3-}, MnO_4^{2-}, MnO_4^-

Manganese is a hard, brittle, gray-white **metal** in group 7B of the Periodic Table. It was recognized as an element in 1774 by Swedish chemist Carl

metal: element or other substance the solid phase of which is characterized by high thermal and electrical conductivities

Wilhelm Scheele and isolated by his assistant Johan Gottlieb Gahn later that year. The element's name is derived from the Latin word *magnes*, meaning "magnet." This refers to the magnetic property of the common ore pyrolusite.

isotope: form of an atom that differs by the number of neutrons in the nucleus

Manganese is the twelfth most abundant transition element (1,060 parts per million of Earth's crust) with twenty-three known **isotopes**. Large nodules of manganese ore have been discovered on the ocean floor. The pure metal can be obtained by reduction of the oxide with sodium or by electrolysis.

oxidation: process that involves the loss of electrons (or the addition of an oxygen atom)

aqueous solution: homogenous mixture in which water is the solvent (primary component)

Manganese is more reactive that any of its neighbors on the Periodic Table. It reacts with water to produce hydrogen gas and dissolves in dilute acids to form Mn^{2+}. The most stable **oxidation** state of manganese is +2. The most important oxide formed is MnO_2, which decomposes to Mn_2O_3 if heated above 530°C (932°F). The deep-purple manganate (VII) salts (permanganates) are prepared in **aqueous solution** by oxidation of Mn^{2+}.

alloy: mixture of two or more elements, at least one of which is a metal

trace element: element occurring only in a minute amount

vitamins: organic molecules needed in small amounts for the normal function of the body; often used as part of an enzyme catalyzed reaction

Manganese metal is used in many **alloys**. In conjunction with aluminum and copper it forms strong ferromagnetic alloys. Ninety-five percent of all manganese ores are used in the production of steel. The element improves the strength and toughness of steel by acting as a scavenger of sulfur, preventing the formation of FeS, which induces brittleness. Biologically, manganese is an important **trace element**; it is essential to the utilization of **vitamin** B^1. Pyrolusite has been used in glassmaking since ancient Egypt, while MnO_2 is used in the manufacture of dry cells. The permanganate ion is a strong oxidizing agent and is used in quantitative analysis and medicine. SEE ALSO COORDINATION COMPOUNDS; INORGANIC CHEMISTRY; SCHEELE, CARL.

Catherine H. Banks

Bibliography

Greenwood, Norman N., and Earnshaw, A. (1997). *Chemistry of the Elements*, 2nd edition. Boston: Butterworth-Heinemann.

Lide, David R., ed. (2003). "Manganese." In *The CRC Handbook of Chemistry and Physics*, 84th edition. Boca Raton, FL: CRC Press.

Internet Resources

"Manganese." U.S. Geological Survey, Mineral Commodity Summaries. Updated January 2003. Available from <http://minerals.usgs.gov/minerals/pubs/commodity/manganese/420303.pdf>.

Manhattan Project

fission: process of splitting an atom into smaller pieces

Nuclear scientists knew in the 1930s that there was a tremendous amount of energy locked in the atomic nucleus. The problem was how to access this energy. With the discovery of nuclear **fission** in Germany by Otto Hahn, Lise Meitner, and Fritz Strassmann in late 1938 and the subsequent explanation of the process by Meitner and Otto Frisch, many scientists who had been forced to flee from Europe became concerned that Germany might somehow take advantage of this discovery and develop weapons based on nuclear energy. Germany had many competent nuclear scientists, access to heavy water in Norway, quantities of uranium oxide, and a strong engineering history. In the fall of 1939, Leo Szilard, a Hungarian-born physicist who had fled Germany for America, drafted a letter with Albert Einstein to send to President Franklin Roosevelt under Einstein's signature to warn

Roosevelt that weapons could be created using a nuclear chain reaction in uranium and that it was very likely that Germany had started working on a uranium bomb. This letter led to the formation of the Advisory Committee on Uranium. The committee did little, however, until Rudolf Peierls and Frisch, working in England, made detailed calculations about the feasibility of nuclear weapons and proposed some possible approaches to making an atomic bomb.

Although government support was relatively weak, important nuclear-science discoveries, unannounced for security reasons, were made in the United States in 1940 and 1941. Potential pathways for enriching ^{235}U, the self-fissioning **isotope** of uranium, were developed, and Glenn Seaborg and Arthur Wahl produced plutonium, an element that had great potential for use in a nuclear weapon. In December 1941, President Roosevelt authorized the formation of the Manhattan Engineer District of the Army Corps of Engineers (**"Manhattan Project"**) as the organization that would oversee the development of the atomic bomb. Groups of scientists, some of whom were already working on nuclear energy research, were organized to work on various aspects of the bomb project. One such project was called the Metallurgical Laboratory at the University of Chicago, where Italian-born physicist Enrico Fermi and other scientists worked on the construction of the first nuclear reactor, powered by uranium enriched in ^{235}U.

isotope: form of an atom that differs by the number of neutrons in the nucleus

Manhattan Project: government project dedicated to creation of an atomic weapon; directed by General Leslie Groves

The Manhattan Project officially began on September 23, 1942, when Colonel Leslie Groves was named director of the project. Groves acquired production sites at Oak Ridge, Tennessee, and Hanford, Washington, and he brought in Robert Oppenheimer, a physicist from the University of California at Berkeley, as the scientific director at Los Alamos, New Mexico. Los Alamos was to be the center of physics research, engineering, and weapons design. Oak Ridge was to be the site to enrich ^{235}U, and Hanford was to produce plutonium in nuclear reactors. Many of the country's leading chemical and engineering firms were called in to design and construct these production facilities

Oak Ridge was to produce uranium enriched in the self-fissioning isotope ^{235}U by gaseous diffusion of the **volatile** compound UF_6 at the K-25 plant, a facility more than a mile long, and by electromagnetic separation at the Y-12 plant. Nuclear reactors were built at Hanford to produce plutonium from natural uranium, ^{238}U. Processes were developed to chemically separate and purify the plutonium isotopes ^{239}Pu and ^{240}Pu. Scientists from Great Britain also played key roles in the efforts at Los Alamos, and they were instrumental in developing the functional design of the atomic bomb. Work went on at a feverish pace during 1943 and 1944, and it was estimated that an atomic weapon would be completed by early 1945.

volatile: low boiling, readily vaporized

In the spring of 1945, preparations began in the Pacific for the use of the atomic bomb. On May 8, 1945, Germany surrendered, and the project was then focused solely on Japan. On July 16, 1945, a test device code-named "Gadget" was detonated at the Alamogordo Bombing Range in New Mexico as part of Project Trinity, the first explosion of a nuclear weapon. The success of the first test of a nuclear weapon was a testament to the ability of the leadership of the Manhattan Project to carry out an unprecedented industrial project, with the world's most talented scientists

From left: J. Robert Oppenheimer, Professor. H. D. Smythe, General Nichols, and Glen Seaborg in 1946 looking at a photograph of the atomic blast at Hiroshima. The atomic bomb was developed in the Manhattan Project.

cooperating and focusing on a single goal. That the people involved in the Manhattan Project were able to achieve such a spectacular success working in a realm of utmost secrecy in isolated locations was a most singular achievement.

On August 6, 1945, after Japan refused to surrender unconditionally, the first atomic bomb, named "Little Boy," a ^{235}U-based bomb, was dropped over Hiroshima, Japan. Three days later, "Fat Man," a plutonium-based weapon, was dropped on Nagasaki.

Whether Germany really attempted to build an atomic weapon is debated even today. German scientists, led by physicist Werner Heisenberg, certainly had the talent to build a device. Germany had access to large uranium mines in Czechoslovakia and produced heavy water, a form of H_2O in which the hydrogen atoms have an extra neutron, in Norway. Most likely there never was a serious effort in Germany to build an atomic weapon, possibly as result of sabotage of the project by Heisenberg or because of a lack of interest by Adolf Hitler. SEE ALSO EINSTEIN, ALBERT; FERMI, ENRICO; HEISENBERG, WERNER; MEITNER, LISE; NUCLEAR FISSION; OPPENHEIMER, ROBERT; RADIATION; SEABORG, GLENN THEODORE.

W. Frank Kinard

Bibliography

Frayn, Michael (2000). *Copenhagen.* New York: Anchor Books.

Rhodes, Richard (1986). *The Making of the Atomic Bomb.* New York: Simon & Schuster.

Marsden, Ernest

ENGLISH PHYSICIST
Ca. 1888–1970

Ernest Marsden studied at the University of Manchester under Ernest Rutherford and Hans Geiger. Although a physicist, he would help elucidate something of value to all chemists: the internal structure of the atom. This was accomplished by observing the path of **α-particles** in Rutherford's famous "gold foil experiment," in which it was really the human eye, pressed to a short-focus telescope for hours on end in a thoroughly darkened room, that was the detector.

α-particle: subatomic particle with 2+ charge and mass of 4; a He nucleus

According to Rutherford, Marsden, a twenty-year-old undergraduate, became involved after Rutherford and Geiger decided that Marsden should begin research work. Rutherford thought that Marsden might be able to discover if α-particles could be scattered through a large angle. Geiger and Marsden spent 1909 in the "gloomy cellar" of the physics laboratories at Manchester, watching for the little sparks that announced the unlikely recoil of α-particles. About 1 in 8,000 did, and this result, published in 1909 as "On a Diffuse Reflection of the α-Particle," formed the basis for Rutherford's **nuclear** model of the atom and the discovery of the proton.

nuclear: having to do with the nucleus of an atom

Geiger and Marsden continued to study the deflection of α-particles, and in 1913 (after observing over 100,000 scintillations at a rate of 5 to 90 per minute) correlated nuclear charge with **atomic number**. In 1914 and 1915 Marsden continued to study the impact of α-particles on matter; these experiments led to Rutherford's 1919 fortuitous attainment of the alchemist's dream: the artificial transmutation of the elements.

atomic number: the number assigned to an atom of an element that indicates the number of protons in the nucleus of that atom

Marsden returned to his native New Zealand in 1915 where, on Rutherford's recommendation, he was appointed professor of physics at Victoria University in Wellington. He held various academic and governmental posts until his retirement in 1954. The national fund for the support of science in New Zealand was renamed the Sir Ernest Marsden Fund in his honor. SEE ALSO GEIGER, HANS; RUTHERFORD, ERNEST.

Mark A. Pichaj

Bibliography

Chown, Marcus (2001). *The Magic Furnace: The Search for the Origins of Atoms.* Oxford, U.K.: Oxford University Press.

da Costa Andrade, Edward Neville (1964). *Rutherford and the Nature of the Atom.* Garden City, NY: Doubleday.

Gamow, George (1958). *Matter, Earth and Sky.* Englewood Cliffs, NJ: Prentice-Hall.

Shamos, Morris H., ed. (1959). *Great Experiments in Physics.* New York: Holt, Rinehart & Winston.

Martin, Archer John Porter

ENGLISH BIOCHEMIST
1910–2002

Very few chemical reactions produce clean, pure products with no trace of starting materials or impurities. Most generate a mixture whose individual components must be purified before the results can be identified. In the nineteenth and early twentieth centuries, purification of a chemical reaction product often required repetitive crystallizations, distillation, or solvent extraction.

British chemist Archer John Porter Martin, co-recipient, with Richard L. M. Synge, of the 1952 Nobel Prize in chemistry, "for their invention of partition chromatography."

fractional distillation: separation of liquid mixtures by collecting separately the distillates at certain temperatures

phase: homogeneous state of matter

chromatography: the separation of the components of a mixture in one phase (the mobile phase) by passing through another phase (the stationary phase) making use of the extent to which the components are absorbed by the stationary phase

volatile: low boiling, readily vaporized

Archer John Porter Martin grew up in London, England, and from an early age demonstrated an aptitude for chemistry. As a child, he designed and built an apparatus for distillation from old coffee tins packed with charcoal, some as tall as five feet. He entered Cambridge University with the intention of pursuing a degree in chemical engineering. However, he was influenced by J. B. S. Haldane to specialize in biochemistry. At Cambridge his childhood experience with **fractional distillation** became valuable.

Martin continued with his explorations of multi**phase** separation technology and went to work as a research chemist for the Wool Industries Research Association in Leeds. It was there that he met Richard Lawrence Millington Synge and began to collaborate with Synge on the problem of separating acetylamino acids. Eventually, Martin and Synge came up with the idea that, instead of using a counterflow extraction process with solvents moving against one another, they could partition one phase (hold one phase stationary using an appropriate support). The result was the invention of liquid-liquid partition **chromatography**, first reported in the *Biochemistry Journal* in 1941.

In their landmark paper, Martin and Synge also indicated that partition chromatography that used a carrier gas as the mobile phase was possible. In his Nobel lecture of 1952, Martin casually revealed that he, in collaboration with A. T. James, had devised a mechanism for gas-liquid chromatography. The use of a gas as the mobile phase did place limits on the types of material that could be analyzed, as the compounds had to be **volatile** and better detectors were needed, but these difficulties proved to be surmountable. Today, gas-liquid chromatography is probably the single most widely used analytical tool in chemistry. SEE ALSO ANALYTICAL CHEMISTRY; SYNGE, RICHARD LAURENCE MILLINGTON.

Todd W. Whitcombe

Bibliography

Martin, Archer J. P., and Synge, Richard L. M. (1941). *Biochemistry Journal* 35: 1358–1366. American Chemical Society and the Chemical Heritage Foundation.

Shetty, Prabhakara H. (1993). "Archer John Porter Martin." In *Nobel Laureates in Chemistry 1901–1992*, ed. Laylin K. James. Washington, DC: American Chemical Society and the Chemical Heritage Foundation.

Internet Resources

Nobel e-Museum. "Archer John Porter Martin—Biography." Available from <http://www.nobel.se>.

Materials Science

metallurgy: the science and technology of metals

After World War II, the application of materials became less empirical and more founded in scientific principles. The term "materials science" emerged in the 1960s to reflect this trend and the realization that solutions to many of the world's most challenging technological problems were increasingly materials-limited. Within the field of engineering, the term "materials science and engineering" has come to describe the subfield concerned with materials applications. This longer term represents a blend of scientific fundamentals and practical engineering. The foundations of materials science are physical chemistry, polymer chemistry, and condensed matter physics. The foundations of materials engineering include the fields of **metallurgy**

A contemporary automobile is comprised of a wide range of materials, from the traditional to the advanced. Science has introduced the manufacturing of parts such as moldable polymers.

and ceramic engineering. Many common themes in the fields of chemical engineering and materials science have led to the creation of academic departments that encompass both areas.

Material possessions have traditionally represented human wealth and defined social relationships. The eras of early human civilization (the Stone Age, the Bronze Age, and the Iron Age) have been named in terms of the materials from which tools and weapons were made. The Bronze Age (approximately 2000 B.C.E. to 1000 B.C.E.), in fact, represents the foundation of metallurgy. Although we do not use the term "pottery age," domestic vessels made from baked clay have been valuable in providing clues to daily life in ancient cultures, and glass articles from ancient Mesopotamia have been traced back to 4000 B.C.E.

Contemporary culture is sometimes described as "plastic," a somewhat critical reference to the pervasive use of polymeric materials in modern life. Others suggest that the current era is rightfully called the "Silicon Age," in honor of the far-ranging impacts of modern electronics based on silicon technology. In any case, modern products, such as automobiles, contain a full spectrum of materials, from the traditional to the advanced.

An underlying principle of materials science is that the properties (or characteristics) of materials are generally understood in terms of the microscopic or atomic structures of the materials. Another underlying principle is that the selection of optimal materials for specific modern technological applications requires consideration of the ways in which those materials are processed.

Types of Materials

Engineers generally build things from a limited "menu" of materials—namely, **metals**, polymers, and ceramics. This menu follows directly from the three types of primary chemical bonding: metallic, covalent, and ionic.

metal: element or other substance the solid phase of which is characterized by high thermal and electrical conductivities

Most of the elements in the Periodic Table (in the pure state) are metallic in nature. Aluminum, copper, and iron are examples. The metallic bond involves a mobile "gas" of electrons. This gas of negatively charged electrons binds together the positively charged atomic cores. The electron gas is also responsible for the electrical conductivities and optical absorption that are characteristic of metals.

Polymers are high molecular weight solids that are an important part of everyday life. An example is polyethylene $(C_2H_4)_n$, where *n* is the "degree of polymerization," a number of around 1,000 (representing the fact that polyethylene is composed of a large number of ethylene molecules bound together by covalent bonding). All polymers are composed of a relatively small number of elements in the Periodic Table (primarily carbon and hydrogen and a few other "nonmetallic" elements such as nitrogen and fluorine). Each **covalent bond** involves electron sharing between adjacent atoms, with the result that polymers do not have "free" electrons for electrical conduction and are electrical insulators. The use of polymeric insulation for electrical wiring is a practical example of this. The lack of free electrons endows some polymers with optical transparency ("clear plastic" wrap is an excellent example). The alternative name for a polymer substance—"plastic"—comes from the extensive formability of many polymers.

covalent bond: bond formed between two atoms that mutually share a pair of electrons

We can define ceramics by what they are not: They are nonmetallic and inorganic. Ceramics are chemical combinations of at least one metallic element and at least one nonmetallic one. A simple example is aluminum oxide (Al_2O_3). Such chemical combinations represent, in fact, a fundamental tendency in nature. For example, metals tend to combine chemically with nonmetallic elements in their environments. The rusting of iron is a familiar and costly example. It is also interesting to note that the **melting point** of aluminum is 660°C (1,220°F), whereas the melting point of aluminum oxide is 2,020°C (3,668°F). The chemical stability associated with the ionic bonds between aluminum and oxygen (involving electron transfer from aluminum to oxygen to produce Al^{3+} and O^{2-} ions) makes ceramics temperature-resistant and chemically **inert**.

melting point: temperature at which a substance in the solid state undergoes a phase change to the liquid state

inert: incapable of reacting with another substance

The category of ceramics is often broadened to "ceramics and glasses" because of the wide use of silicate glasses, distinctive materials that are chemically similar to ceramics. Silicon dioxide, SiO_2, is a ceramic compound and the basis of a large family of silicate ceramics. Clay minerals and the many clayware ceramics are the most traditional examples. SiO_2 is readily obtained in relatively pure form in common sand deposits. (These deposits, and the presence of SiO_2 in many geological minerals, are the reason that silicon and oxygen together account for roughly 75 percent of the elements in Earth's crust.) Upon heating, many of these silicate materials can be melted and, after cooling, retain the liquidlike structure of the melt. Common window and container glass is made in this way, with a typical composition, by weight, of (roughly): 75 percent SiO_2, 15 percent Na_2O, and 10 percent CaO. Thus, ceramics and glasses are of one category (combinations of ionically bonded positive and negative ions). Their differences are at the atomic scale. Ceramics are crystalline substances, in which the ions are arranged in a regular and repeating order. Glasses are noncrystalline substances, in which the ions are situated in irregular, liquidlike fashion.

In defining the previous three materials (metals, polymers, and ceramics/glasses), we found that each category conveniently related to one of the primary types of chemical bonding: metallic, covalent, and ionic, respectively. To be precise, atomic bonding is seldom "pure." There is generally some covalent nature (electron sharing) to the ionic bonding in ceramics and glasses. The bonding between the adjacent atoms in large polymeric molecules is highly covalent, but the bonding between molecules is often "secondary." For example, there are weak attractions between adjacent polyethylene molecules that involve polarization, not electron transfer or sharing. This weak secondary bonding is the primary reason that commercial "plastics" are characteristically weak and deformable in comparison to metals and ceramics/glasses.

Among the materials available for modern structural applications, a fourth category is generally included—namely, "composites." Composite materials are defined as microscopic-scale combinations of individual materials belonging to the previous three categories (metals, polymers, ceramics/glasses). A good example is fiberglass, a composite of glass fibers (a few micrometers in diameter) embedded in a polymer matrix. Over the past several decades, fiberglass products have become commonplace. The advantage of composites is that they display the best properties of each component, producing products superior to products made of a single component. In the case of fiberglass, the high strength of the small diameter glass fibers is combined with the flexibility of the polymer matrix.

Although most engineered materials can be put into one of the four categories described above, a sorting of the same materials based on electrical conductivity rather than atomic bonding demands an additional, fifth category. We noted above that metals are typically good electrical conductors and that polymers and ceramics/glasses are typically electrical insulators. Composites tend to have properties that are averages of those of their individual components. As an example, fiberglass is an electrical insulator because both glass fibers and the polymer matrix tend to be insulators. Since the middle of the twentieth century, "semiconductors," with intermediate levels of electrical conductivity, have played an increasingly critical role in modern technology. The primary example is elemental silicon, which, as noted above, is a central component of modern, solid-state electronics. Silicon is in column IVA of the Periodic Table. Its neighbor in column IVA, germanium, is also a semiconductor and also widely used in electronic devices. Chemical compounds of the elements near column IVA often display semiconduction—for example, gallium arsenide (GaAs), which is used as a high temperature rectifier and a laser material. The chemical bonding in the various elemental and compound semiconductors is generally strongly covalent. In summary, a full list of the types of engineered materials contains five categories. (See Table 1.)

From Structure to Properties

An underlying principle of materials science is that structure (on the atomic or microscopic scale) leads to properties (on the macroscopic scale of real world, engineering applications). We have already seen that the natures of ceramics and glasses are very different because ceramics have a crystalline atomic arrangement and glasses are noncrystalline. Similarly, transparent glass

TYPES OF MATERIALS

Material Type	Bonding Character	Examples
Metal	Metallic	Iron (Fe); Brass (Cu and Zn)
Polymer	Covalent and secondary	Polyethylene $[(C_2H_4)_n]$
Ceramic/glass	Ionic/covalent	Silica (SiO_2): crystalline and noncrystalline
Composites	(determined by components)	Fiberglass (glass fibers in polymer matrix)
Semiconductors	Covalent or covalent/ionic	Silicon (Si); Gallium Arsenide (GaAs)

Table 1.

becomes opaque when it has many microscopic air bubbles that scatter light and prevent a clear image from being transmitted through the material. Examples of the structure–property relationship arise throughout the field of materials science.

Processing and Selecting Materials

The use of materials in modern technology depends on our ability to make those materials. Processing is dependent on the nature of the material, and the specific processing technique can, in turn, have an effect on the properties of the material. Given the wide range of materials described in Table 1, and the fact that an individual material's properties are dependent on the way in which it is manufactured, the selection of materials for a given application needs to be done in a systematic way. The selection process and the final decision are dependent on a range of factors, including desired properties, ability to be manufactured, and cost.

Conclusion

Chemical bonding and electrical conductivity provide five major categories of engineered materials: metals, polymers, ceramics/glasses, composites, and semiconductors. The properties of these materials are dependent on atomic- and microscopic-scale structure, as well as on the way in which a given material is processed. Materials science enables the selection of the optimal material for a given application. SEE ALSO CERAMICS; GLASS; PHYSICAL CHEMISTRY; POLYMERS, SYNTHETIC; SEMICONDUCTORS.

James F. Shackelford

Bibliography

Ashby, Michael F. (1999). *Materials Selection in Mechanical Design,* 2nd edition Oxford, U.K.: Butterworth-Heinemann.

Callister, W. D., Jr. (2000). *Materials Science and Engineering—An Introduction,* 5th edition. New York: John Wiley and Sons.

Shackelford, James F. (2000). *Introduction to Materials Science for Engineers,* 5th edition. Upper Saddle River, NJ: Prentice-Hall.

Maxwell, James Clerk

SCOTTISH PHYSICIST
1831–1879

James Clerk Maxwell is generally regarded as one of the outstanding physicists of the nineteenth century. He made important advances in the theory of electricity and magnetism, as well as in thermodynamics and the **kinetic theory** of gases. Many modern ideas about these topics are still based on his work from the mid-1800s.

Maxwell was born in Edinburgh, Scotland, and his father greatly encouraged him in his intellectual pursuits. At the age of fourteen, while a student at the Edinburgh Academy, he wrote a paper on ovals and geometric figures with more than two foci. His paper was read to the **Royal Society** of Edinburgh by an adult member because it was considered inappropriate for a young boy to present it to the society himself. Although some of the ideas in this paper had been discussed earlier by the renowned French mathematician René Descartes, it was still an amazing achievement for a teenage boy.

At sixteen, Maxwell entered Edinburgh University, where he studied physics, mathematics, and logic. Three years later he went to Cambridge University, from which he graduated in 1854 with a degree in mathematics.

In 1856 Maxwell became professor of **natural philosophy** at Marischal College in Aberdeen. There he became interested in the theory of gases and in the study of electricity and magnetism. His position as professor, however, was eliminated in 1860 when Marischal and another college merged.

Maxwell spent the next five years at King's College in London. He successfully applied statistical methods to describe the movements of the tiny invisible particles of a gas, an approach adopted a century earlier by the Swiss mathematician Daniel Bernoulli, but with less sophisticated mathematics. The Austrian physicist Ludwig Boltzmann also studied the problem of gas behavior at the same time as Maxwell, and the names of both men are usually associated with the kinetic theory of gases.

Because of his overwhelming interest in the science of electricity, Maxwell was drawn to the writings of the English physicist Michael Faraday, who had begun publishing his three-volume *Experimental Researches in Electricity* in 1839. Faraday's approach was almost entirely experimental, and Maxwell saw this as an opportunity to treat the subject in mathematical terms. Beginning in the 1850s, Maxwell published several papers on electricity, including the analogy between electricity and heat from a mathematical point of view. These research efforts culminated in his important writings in the 1860s and 1870s on electromagnetic theory and his identification of light as an electromagnetic wave. Maxwell's theoretical conclusions about electromagnetism are summarized in a set of four equations known as Maxwell's equations, which first appeared in his *Treatise on Electricity and Magnetism* in 1873 and were later cast in their modern form by other physicists.

In 1865 Maxwell resigned his position in London and returned to his family estate Glenair in Scotland, where he continued his scientific work for five years. In 1870, however, a new chair and laboratory of physics were established at Cambridge University, and Maxwell eventually accepted an offer after two other physicists had refused. Maxwell continued his work in electricity and magnetism, organized the new laboratory, and edited the papers of Henry Cavendish for whom the laboratory was named. Early in 1879 Maxwell's health began to decline, and he died several months later during his forty-ninth year. SEE ALSO BOLTZMANN, LUDWIG; CAVENDISH, HENRY; FARADAY, MICHAEL; MAGNETISM; PHYSICAL CHEMISTRY.

Richard E. Rice

Scottish physicist James Clerk Maxwell, who identified light as an electromagnetic wave.

kinetic theory: theory of molecular motion

Royal Society: The U.K. National Academy of Science, founded in 1660

natural philosophy: study of nature and the physical universe

Bibliography

Cropper, William H. (2001). "The Scientist as Magician: James Clerk Maxwell." In *Great Physicists: The Life and Times of Leading Physicists from Galileo to Hawking.* New York: Oxford University Press.

Internet Resources

Haley, Christopher. "James Clerk Maxwell (1831–1879) Mathematical Physicist." Available from <http://65.107.211.206/science/maxwell1.html>.

O'Connor, J. J., and Robertson, E. F. "James Clerk Maxwell." Available from <http://www-gap.dcs.st-and.ac.uk/~history/Mathematicians/Maxwell.html>.

O'Connor, J. J., and Robertson, E. F. "A Visit to James Clerk Maxwell's House." Available from <http://www-gap.dcs.st-and.ac.uk/~history/HistTopics/Maxwell_House.html>.

Measurement

British mathematician and physicist William Thomson (1824–1907), otherwise known as Lord Kelvin, indicated the importance of measurement to science:

> When you can measure what you are speaking about and express it in numbers, you know something about it; but when you cannot measure it, when you cannot express it in numbers, your knowledge is of a meager and unsatisfactory kind; it may be the beginning of knowledge, but you have scarcely in your thoughts advanced to the state of science, whatever the matter may be.

combustion: burning, the reaction with oxygen

Possibly the most striking application of Kelvin's words is to the explanation of **combustion** by the French chemist Antoine Lavoisier (1743–1794). Combustion was confusing to scientists of the time because some materials, such as wood, seemed to decrease in mass on burning: Ashes weigh less than wood. In contrast, others, including iron, increased in mass: Rust weighs more than iron. Lavoisier was able to explain that combustion results when oxygen in the air unites with the material being burned, after careful measurement of the masses of the reactants—air and the material to be burned—and those of the products. Because Lavoisier was careful to capture all products of combustion, it was clear that the reason wood seemed to lose mass on burning was because one of its combustion products is a gas, carbon dioxide, which had been allowed to escape.

Lavoisier's experiments and his explanations of them and of the experiments of others are often regarded as the beginning of modern chemistry. It is not an exaggeration to say that modern chemistry is the result of careful measurement.

Most people think of measurement as a simple process. One simply finds a measuring device, uses it on the object to be measured, and records the result. Careful scientific measurement is more involved than this and must be thought of as consisting of four steps, each one of which is discussed here: choosing a measuring device, selecting a sample to be measured, making a measurement, and interpreting the results.

Choosing a Measuring Device

The measuring device one chooses may be determined by the devices available and by the object to be measured. For example, if it were necessary to

A sundial indicates time based on the position of the Sun.

determine the mass of a coin, obviously inappropriate measuring devices would include a truck scale (reading in units of 20 pounds, with a 10-ton capacity), bathroom scale (in units of 1 pound, with a 300-pound capacity), and baby scale (in units of 0.1 ounce, with a 30-pound capacity). None of these is capable of determining the mass of so small an object. Possibly useful devices include a centigram balance (reading in units of 0.01 gram, with a 500-gram capacity), milligram balance (in units of 0.001 gram, with a 300-gram capacity), and analytical balance (in units of 0.00001 gram, with a 100-gram capacity). Even within this limited group of six instruments, those that are suitable differ if the object to be measured is an approximately one-kilogram book instead of a coin. Then only the bathroom scale and baby scale will suffice.

In addition, it is essential that that the measuring device provide reproducible results. A milligram balance that yields successive measurements of 3.012, 1.246, 8.937, and 6.008 grams for the mass of the same coin is clearly faulty. One can check the reliability of a measuring device by measuring a standard object, in part to make sure that measurements are reproducible. A common measuring practice is to intersperse samples of known value within a group of many samples to be measured. When the final results are tallied, incorrect values for the known samples indicate some fault, which may be that of the measuring device, or that of the experimenter. In the example of measuring the masses of different coins, one would include several "standard" coins, the mass of each being very well known.

Selecting a Sample

There may be no choice of sample because the task at hand may be simply that of measuring one object, such as determining the mass of a specific coin. If the goal is to determine the mass of a specific kind of coin, such as a U.S. penny, there are several questions to be addressed, including the following. Are uncirculated or worn coins to be measured? Worn coins may have less mass because copper has worn off, or more mass because copper oxide weighs more than copper and dirt also adds mass. Are the coins of just one year to be measured? Coin mass may differ from year to year. How many coins should be measured to obtain a representative sample? It is likely that there is a slight variation in mass among coins and a large enough number of coins should be measured to encompass that variation. How many sources (banks or stores) should be visited to obtain samples? Different batches of new coins may be sent to different banks; circulated coins may be used mostly in vending machines and show more wear as a result.

The questions asked depend on the type of sample to be measured. If the calorie content of breakfast cereal is to be determined, the sampling questions include how many factories to visit for samples, whether to sample unopened or opened boxes of cereal, and the date when the breakfast sample was manufactured, asked for much the same reason that similar questions were advanced about coins. In addition, other questions come to mind. How many samples should be taken from each box? From where in the box should samples be taken? May samples of small flakes have a different calorie content than samples of large flakes?

These sampling questions are often the most difficult to formulate but they are also the most important to consider in making a measurement. The purpose of asking them is to obtain a sample that is as representative as possible of the object being measured, without repeating the measurement unnecessarily. Obviously, a very exact average mass of the U.S. penny can be obtained by measuring every penny in circulation. This procedure would be so time-consuming that it is impractical, in addition to being expensive.

Making a Measurement

As mentioned above, making a measurement includes verifying that the measuring device yields reproducible results, typically by measuring standard samples. Another reason for measuring standard samples is to calibrate the measuring instrument. For example, a common method to determine the viscosity of a liquid—its resistance to flow—requires knowing the density of that liquid and the time that it takes for a definite volume of liquid to flow through a thin tube, within a device called a viscometer. It is very difficult to construct duplicate viscometers that have exactly the same length and diameter of that tube. To overcome this natural variation, a viscometer is calibrated by timing the flow of a pure liquid whose viscosity is known—such as water—through it. Careful calibration involves timing the flow of a standard volume of more than one pure liquid.

Calibration not only accounts for variations in the dimensions of the viscometer. It also compensates for small variations in the composition of the glass of which the viscometer is made, small differences in temperatures, and even differences in the gravitational acceleration due to different positions on Earth. Finally, calibration can compensate for small variations in technique from one experimenter to another.

These variations between experimenters are of special concern. Different experimenters can obtain very different values when measuring the same sample. The careful experimenter takes care to prevent bias or difference in technique from being reflected in the final result. Methods of prevention include attempting to measure different samples without knowing the identity of each sample. For instance, if the viscosities of two colorless liquids are to be measured, several different aliquots of each liquid will be prepared, the aliquots will be shuffled, and each **aliquot** will be measured in order. As much of the measurement as possible will be made mechanically. Rather than timing flow with a stopwatch, it is timed with an electronic device that starts and stops as liquid passes definite points.

aliquot: specific volume of a liquid used in analysis

Finally, the experimenter makes certain to observe the measurement the same way for each trial. When a length is measured with a meter stick or a volume is measured with a graduated cylinder, the eye of the experimenter is in line with or at the same level as the object being measured to avoid parallax. When using a graduated device, such as a thermometer, meter stick, or graduated cylinder, the measurement is estimated one digit more finely than the finest graduation. For instance, if a thermometer is graduated in degrees, 25.4°C (77.7°F) would be a reasonable measurement made with it, with the ".4" estimated by the experimenter.

Each measurement is recorded as it is made. It is important to not trust one's memory. In addition, it is important to write down the measurements made, not the results from them. For instance, if the mass of a sample of sodium chloride is determined on a balance, one will first obtain the mass of a container, such as 24.789 grams, and then the mass of the container with the sodium chloride present, such as 32.012 grams. It is important to record both of these masses and not just their difference, the mass of sodium chloride, 7.223 grams.

Interpreting Results

Typically, the results of a measurement involve many values, the observations of many trials. It is tempting to discard values that seem quite different from the others. This is an acceptable course of action if there is good reason to believe that the errant value was improperly measured. If the experimenter kept good records while measuring, notations made during one or more trials may indicate that an individual value was poorly obtained—for instance, by not zeroing or leveling a balance, neglecting to read the starting volume in a buret before titration, or failing to cool a dried sample before obtaining its mass.

Simply discarding a value based on its deviation from other values, without sound experimental reasons for doing so, may lead to misleading results besides being unjustified. Consider the masses of several pennies determined with a milligram balance to be: 3.107, 3.078, 3.112, 2.911, 3.012, 3.091, 3.055, and 2.508 grams. Discarding the last mass because of its deviation would obscure the facts that post-1982 pennies have a zinc core with copper **cladding** (representing a total of about 2.4% copper), whereas pre-1982 pennies are composed of an **alloy** that is 95 percent copper. There are statistical tests that help in deciding whether to reject a specific value or not.

cladding: protective material surrounding a second material

alloy: mixture of two or more elements, at least one of which is a metal

It is cumbersome, however, to report all the values that have been measured. Reporting solely the average or mean value gives no indication of how carefully the measurement has been made or how reproducible the repeated measurements are. Care in measurement is implied by the number of significant figures reported; this corresponds to the number of digits to

A scale is a method for determing mass.

which one can read the measuring devices, with one digit beyond the finest graduation, as indicated earlier.

The reproducibility of measurements is a manifestation of their precision. Precision is easily expressed by citing the range of the results; a narrow range indicates high precision. Other methods of expressing precision include relative average deviation and standard deviation. Again, a small value of either deviation indicates high precision; repeated measurements are apt to replicate the values of previous ones.

When several different quantities are combined to obtain a final value—such as combining flow time and liquid density to determine viscosity—standard propagation-of-error techniques are employed to calculate the deviation in the final value from the deviations in the different quantities.

Both errors and deviations combine in the same way when several quantities are combined, even though error and deviation are quite different concepts. As mentioned above, deviation indicates how reproducible successive measurements are. Error is a measure of how close an individual value—or an average—is to an accepted value of a quantity. A measurement with small error is said to be accurate. Often, an experimenter will believe that high precision indicates low error. This frequently is true, but very precise measurements may have a uniform error, known as a systematic error. An example would be a balance that is not zeroed, resulting in masses that are uniformly high or low.

The goal of careful measurement ultimately is to determine an accepted value. Careful measurement technique—including choosing the correct measuring device, selecting a sample to be measured, making a measurement, and interpreting the results—helps to realize that goal. SEE ALSO International System of Units; Lavoisier, Antoine.

Robert K. Wismer

Bibliography

Youden, W. J. (1991). *Experimentation and Measurement.* NIST Special Publication 672. Washington, DC: National Institute of Standards and Technology.

Meitner, Lise

AUSTRIAN PHYSICIST
1878–1968

On any list of scientists who should have won a Nobel Prize but did not, Lise Meitner's name would be near the top. She was the physicist who first realized that the atomic nucleus could be split to form pairs of other atomic nuclei—the process of **nuclear fission**. Although she received many honors for her work, the greatest of all was to elude her because of the unprofessional conduct of her colleague Otto Hahn.

nuclear: having to do with the nucleus of an atom

fission: process of splitting an atom into smaller pieces

Born in Vienna, Meitner decided early on that she had a passion for physics. At that time, education for female children in the Austro-Hungarian Empire terminated at fourteen, as it was argued that girls did not need any more education than that to become a proper wife and mother. Willing to support his daughter's aspirations, her father paid for private tutoring so she could cover in two years the eight years of education normally needed for university entrance. In 1901 Meitner was one of only four women admitted to the University of Vienna, and in 1905 she graduated with a Ph.D. in physics.

As a student, Meitner had become fascinated with the new science of radioactivity, but she realized that she would have to travel to a foreign country to pursue her dream of working in this field. She applied for work with Marie Curie, but was rejected. However, she did eventually receive an offer from the University of Berlin, which had just hired a young scientist by

Austrian physicist Lise Meitner standing with Otto Hahn (l.). Meitner discovered nuclear fission, but was never honored as such.

theoretical physics: branch of physics dealing with the theories and concepts of matter, especially at the atomic and subatomic levels

the name of Otto Hahn. Having a chemical background, Hahn was looking for a collaborator with a **theoretical physics** background. Unfortunately, the chemistry institute at the university was run by Emil Fischer who had banned women from the institute's premises. Reluctantly, Fischer agreed to let Meitner work in a small basement room. During this time, she received no salary and relied on her family for enough money to cover her living expenses. Meitner and Hahn's research during this time period resulted in the discovery of the element protactinium.

The post–World War I government in Germany was much more favorable to women, and Meitner became the first woman to serve as a physics professor in that country. By the 1930s scientists were bombarding heavy elements with neutrons and it was claimed that new superheavy elements formed as a result of this process. Using such a procedure, Meitner and Hahn thought they had discovered nine new elements. Meitner was puzzled by all the new elements for which claims were made.

Unfortunately, the Nazi Party's rise to power changed everything for Meitner. Because she was a Jew by birth, although a later convert to Christianity, Meitner's situation became increasingly precarious. With help from a Dutch scientist, Dirk Coster, she escaped across the German border into Holland and then made her way to Stockholm, where the director of the Nobel Institute for Experimental Physics reluctantly offered her a position. Stockholm had one advantage for Meitner, an overnight mail service to Germany so she could keep in regular contact with Hahn.

On December 19, 1938, Hahn sent Meitner a letter describing how one of the new elements had chemical properties strongly resembling those of barium and asking if she could provide an explanation. The physicist Otto Frisch visited Meitner, his aunt, for Christmas to help dispel her loneliness. While there, the two went for the now famous "walk in the snow." During an extended conversation in the woods, they came to realize that if the nucleus was considered a liquid drop, the impact of a subatomic particle could cause the atom to fission. If so, it was possible that the barium-like element was actually barium itself.

Meitner immediately contacted Hahn and his colleague Fritz Strassmann. Through experiment they confirmed that the so-called new element was indeed barium. They reported their discovery of nuclear fission to the world's scientific press, barely mentioning the names of Meitner and Frisch. In fact, Hahn never admitted that it was Meitner who had made the critical conceptual breakthrough. In 1944 Hahn was awarded the Nobel Prize in chemistry for his contribution to the discovery of nuclear fission.

Although nominated several times, Meitner never did receive the Nobel Prize for physics that many scientists considered her due. Only now, with element 109 having been named Meitnerium (symbol Mt) has she finally received some recognition for her crucial work. Meitner retired to England where she died at the age of eighty-nine. SEE ALSO BARIUM; CURIE, MARIE SKLODOWSKA; FISCHER, EMIL HERMANN; NUCLEAR FISSION; PROTACTINIUM; RADIATION.

Marelene Rayner-Canham
Geoffrey W. Rayner-Canham

Bibliography

McGrayne, Sharon Bertsch (1993). *Nobel Prize Women in Science: Their Lives, Struggles and Momentous Discoveries.* New York: Birch Lane Press.

Rayner-Canham, Marelene, and Rayner-Canham, Geoffrey (1998). *Women in Chemistry: Their Changing Roles from Alchemical Times to the Mid-Twentieth Century.* Washington, DC: American Chemical Society and the Chemical Heritage Foundation.

Sime, Ruth Lewin (1996). *Lise Meitner: A Life in Physics.* Berkeley: University of California Press.

Internet Resources

"Lise Meitner Online." Available from <http://www.users.bigpond.com/Sinclair/fission/LiseMeitner.html>.

Membrane

All living creatures are made of cells. One cellular component, the membrane, plays a crucial role in almost all cellular activities. The primary function of all cell membranes is to act as barriers between the intracellular and extracellular environments, and as sites for diverse biochemical activities. The cell itself is encapsulated by its own membrane, the plasma membrane. Although the composition of membranes varies, in general, **lipid** molecules make up approximately 40 percent of their dry weight; proteins, approximately 60 percent. The lipids and proteins are held together by **noncovalent** interactions.

Among several possible stable arrangements of protein and lipid molecules in membranes, the bilayer model, first described over seventy years ago, characterizes most biological membranes. An important feature of this model is that the **hydrophilic** groups of the lipid molecules are oriented toward the surfaces of the bilayer, and the **hydrophobic** groups toward the interior. In 1972 Jonathan Singer and Garth Nicolson postulated a unified theory of membrane structure called the fluid-mosaic model. They proposed that the matrix, or continuous part, of membrane structure is a fluid bilayer, and that globular amphiphilic proteins are embedded in a single monolayer,

lipid: a nonpolar organic molecule; fatlike; one of a large variety of nonpolar hydrophobic (water-hating) molecules that are insoluble in water

noncovalent: having a structure in which atoms are not held together by sharing pairs of electrons

hydrophilic: having an affinity for water

hydrophobic: repelling water

> ***RECOGNITION SITES***
>
> Glycolipids and glycoproteins can act as recognition sites in a variety of processes involving recognition between cell types or recognition of cellular structures by other molecules. Recognition events are important in normal cell growth, fertilization, transformation of cells, and other processes.

ester: organic species containing a carbon atom attached to three moieties: an O via a double bond, an O attached to another carbon atom or chain, and a H atom or C chain; the R(C-O)OR functional group

nonpolar: molecule, or portion of a molecule, that does not have a permanent, electric dipole

hydrogen bond: interaction between H atoms on one molecule and lone pair electrons on another molecule that constitutes hydrogen bonding

aqueous solution: homogenous mixture in which water is the solvent (primary component)

receptor: area on or near a cell wall that accepts another molecule to allow a change in the cell

with some proteins spanning the thickness of both monolayers. Both proteins and lipids are mobile and, thus, the membrane can be viewed as a two-dimensional solution of proteins in lipids.

The major class of lipids in plasma membranes is phospholipids. Phospholipids consist of a glycerol backbone and two fatty acids joined by **ester** linkage to the first two carbons of glycerol, and a phosphate group joined to the third. Different groups can be esterified to the phosphate, and these groups define the different classes of phospholipids. In addition, the fatty acids have varying chain lengths and degrees of unsaturation. The presence of the **nonpolar** acyl chain regions and the polar head groups gives the phospholipid molecules their amphipathic character, which allows them to assume the bilayer arrangements of membranes. In addition to phospholipids, two other kinds of lipids are found in the membranes of animal cells: glycolipids and cholesterol. Glycolipids usually make up only a small fraction of the lipids in the membrane but have been shown to possess many biological functions, one of which is their capacity to function as recognition sites. Cholesterol is an important component of plasma membranes and has been shown to play a key role in the control of membrane fluidity.

Several membrane functions are believed to be largely mediated by proteins. Membrane proteins have been put into two general categories: peripheral and integral. Peripheral proteins (or extrinsic proteins) are those that do not penetrate the bilayer to any significant degree and are associated with it by virtue of noncovalent interactions (ionic interactions and **hydrogen bonds**) between membrane surfaces and protein surfaces. Integral proteins (or intrinsic proteins), in contrast, possess hydrophobic surfaces that readily penetrate the lipid bilayer, as well as other surfaces that prefer contact with aqueous medium. These proteins can either insert into the membrane or extend all the way across it and expose themselves to the **aqueous solutions** on both sides.

One of the main functions of the plasma membrane is to separate cytoplasm from extracellular surroundings. In fact, membranes are highly selective permeability barriers, as they contain specific channels and pumps that enable the transport of substances across membranes. These transport systems to a large degree regulate the molecular and ionic composition of intracellular media. Membranes also control the flow of information between cells, and between cells and their extracellular environments, and they contain specific **receptors** that make membranes sensible to external stimuli. In addition, some membranes conduct and pass on signals that can be chemical or electrical, as in the transmission of nerve impulses. Thus, membranes play a central role in signal transduction processes and in biological communication. SEE ALSO CHOLESTEROL; LIPIDS; PHOSPHOLIPIDS; TRANSMEMBRANE PROTEIN.

Michèle Auger

Bibliography

Garrett, Reginald H., and Grisham, Charles M. (2002). *Principles of Biochemistry with a Human Focus.* Fort Worth, TX: Harcourt College Publishers.

Harrison, Roger, and Lunt, George G. (1980). *Biological Membranes: Their Structure and Function.* New York: Wiley.

Singer, S. Jonathan, and Nicolson, Garth L. (1972). "The Fluid Mosaic Model of the Structure of Cell Membranes." *Nature* 175:720–731.

Mendeleev, Dimitri

RUSSIAN CHEMIST
1834–1907

Russian chemist Dimitri Mendeleev, who devised the atomic mass-based Periodic Table.

Dimitri Ivanovich Mendeleev (or Mendeleyev or Mendelejeff) was born in Tobolsk, Siberia, on January 27, 1834. He was the fourteenth and youngest child of the family. His father was the director of the Tobolsk Gymnasium (high school). Tragedy plagued the family in Mendeleev's early years. His father became blind and was forced to retire from his job, and then unexpectedly died. His mother supported the family by managing a glass factory, but in 1848 it burned to the ground. His mother moved the family first to Moscow and then to St. Petersburg. In 1850 Mendeleev began his training as a teacher, following in his father's footsteps at the Pedagogical Institute in St. Petersburg. A few months after this, his mother and older sister died of tuberculosis.

When Mendeleev graduated, he moved to Simferopol on the Crimean Peninsula to assume a post as a science teacher, but the school was soon closed because of the Crimean War. He returned to St. Petersburg and received a master's degree in 1856 after presenting his thesis "Research and Theories on Expansion of Substances Due to Heat."

The years 1859 to 1861, when the Ministry of Public Instruction sent him abroad to study, shaped Mendeleev's career as a scientist. He studied **gas density** with the chemist Henri Victor Regnault in Paris and **spectroscopy** with the physicist Gustav Kirchhoff in Heidelberg. It was while working in Heidelberg that Mendeleev discovered the principle of critical temperature for gases. Once a gas is heated to a temperature above its critical point, no amount of pressure will turn it into a liquid. His work went unnoticed, and the discovery of critical temperatures is usually attributed to the Irish physicist and chemist Thomas Andrews.

gas density: weight in grams of a liter of gas

spectroscopy: use of electromagnetic radiation to analyze the chemical composition of materials

Mendeleev also attended the 1860 Karlsruhe Congress, the first international chemistry conference. Many of the leading chemists of the day were in attendance, and one of the central questions addressed was the appropriate method for calculating **atomic weight**. Different chemists used different systems, leading to widespread confusion over everything from nomenclature to chemical formulas. Mendeleev heard the Italian chemist Stanislao Cannizzaro present Amedeo Avogadro's hypothesis that equal volumes of gas under equal temperature and pressure contained equal numbers of molecules.

atomic weight: weight of a single atom of an element in atomic mass units (AMU)

Mendeleev returned to St. Petersburg determined to make a name for himself and build on the innovations to which he had been exposed. He became a professor of chemistry at the Technological Institute in 1863. His attention to science also extended to practical application, and he often worked as a consultant to the government on farming, mining, and oil production.

In 1866 Mendeleev became professor of general chemistry at the University of St. Petersburg. Finding that no modern organic chemistry textbook existed in Russian, Mendeleev decided to write one (it became a classic work, going through many editions). It was in the course of this project that he made his most important contribution to chemistry. *Principles of*

Table 1. Eka aluminum was predicted by Mendeleev and discovered by the French chemist Paul-Émile Lecoq de Boisbaudran in 1875 and named gallium.

Property	Eka Aluminum	Gallium
Atomic weight	±68	69.9
Density	5.9	5.93
Melting point	Low	30.1°C

Chemistry was not a mere compilation of facts; it presented chemistry as a unified study. At its heart was the relationship of the elements.

On February 14, 1869, Mendeleev began work on the chapter that would discuss the elements. He already believed that there was some underlying principle connecting the elements. He transcribed his notes onto a set of cards, one for each element containing everything he knew about that element. He arranged and rearranged the cards until he was struck by a similarity between his arrangements and those of the card game patience (solitaire), in which cards are sorted by suit and then in descending numerical order. Exhausted, Mendeleev fell asleep. When he awoke, he devised a grouping of the elements by common property in ascending order of atomic weight. He called his innovation the Periodic Table of the Elements.

Within weeks, Mendeleev's Periodic Table was presented to the Russian Chemical Society and was published in the *Journal of Russian Physical Chemistry*; it was published later the same year in the prestigious German journal *Zeitschrift für Chemie.* Revised and expanded tables appeared in the *Annalen der Chemie* in 1872. Since the German journals were known to every research chemist, Mendeleev's Periodic Table became widely known almost at once. Although details of the tables were subject to argument, and many newly discovered elements were later added, the basic principle of organization behind the table was quickly accepted.

The true insight that informed Mendeleev's work was shown not just in what he had included in the Periodic Table, but also in what he had left out. He did not assume that all elements were known. Where there was a significant gap in atomic weights between the elements in the table, he left a gap in the table. He posited that there were undiscovered elements that existed in the gaps and even predicted the characteristics of three of them. He called these *eka boron*, *eka aluminum*, and *eka silicon* (*eka* being Sanskrit for "first"). See Tables 1 through 3 for the properties of these elements.

When these elements were eventually discovered, and because his system agreed with one developed independently by the German chemist Lothar Meyer in 1864, Mendeleev achieved widespread fame. The Periodic Table of the Elements provided a unifying system for classifying and understanding the elements and their function in the composition of matter.

Mendeleev received the Davy Medal (with Meyer in 1882) and the Copley Medal (in 1905), but Russia's Imperial Academy of Sciences refused to acknowledge his work. He resigned his university position in 1890 and was

Table 2. Eka boron was predicted by Mendeleev and discovered by the Swedish physicist Lars Fredrik Nilson in 1879 and named scandium.

Property	Eka Boron	Scandium
Atomic weight	44	44.1
Density of oxide	3.5	3.8

Property	Eka Silicon	Germanium
Atomic weight	72	72.32
Specific gravity	5.5	5.47
Valence	4	4

Table 3. Eka silicon was predicted by Mendeleev and discovered by the German chemist Clemens Winkler in 1886 and named germanium.

appointed director of the Bureau of Weights and Measures, holding this job until his death on January 20, 1907. SEE ALSO AVOGADRO, AMEDEO; CANNIZZARO, STANISLAO; MEYER, LOTHAR; PERIODIC TABLE.

Andrew Ede

Bibliography

Kelman, Peter, and Stone, Harris (1970). *Mendeleyev: Prophet of Chemical Elements.* Englewood Cliffs, NJ: Prentice-Hall.

Mendeleyev, Dmitry Ivanovich (1969). *Principles of Chemistry.* New York: Kraus Reprint.

Strathern, Paul (2000). *Mendeleyev's Dream.* New York: Berkley Publishing Group.

Internet Resources

ChemNet (Chemistry Department, Moscow State University). "Dmitriy Mendeleev Online." Available from <http://www.chem.msu.su/eng/misc/mendeleev>.

Mendelevium

MELTING POINT: **827°C**
BOILING POINT: **Unknown**
DENSITY: **Unknown**
MOST COMMON IONS: Md^{2+}, Md^{3+}

Mendelevium was discovered in 1955 by Albert Ghiorso, Bernard G. Harvey, Gregory R. Choppin, Stanley G. Thompson, and Glenn T. Seaborg via the bombardments of a minute quantity of a rare, radioactive **isotope** of einsteinium (^{253}Es) with **α-particles** in the 60-inch cyclotron of the University of California, Berkeley, which produced ^{256}Md. Only 17 atoms were detected. Md is the first element to be produced and chemically identified on a one-atom-at-a-time basis. Mendelevium-256 decayed by electron capture (with a 1.3-hour half-life) to the known daughter nuclide fermium-256 (^{256}Fm), which decayed primarily by spontaneous **fission** (with a half-life of 2.6 hours). The atoms recoiling from the target were caught in a thin gold catcher foil that was quickly dissolved, and the resulting solution was passed through a cation exchange resin column which sorbed the atoms of Md and its known daughter Fm. Es and Fm were then identified by the order of their elution from the column with alpha-hydroxyisobutyrate solution relative to the known elution positions of Es and Cf tracers.

isotope: form of an atom that differs by the number of neutrons in the nucleus

α-particle: subatomic particle with 2+ charge and mass of 4; a He nucleus

fission: process of splitting a heavy atom into smaller pieces

Mendelevium is the heaviest element whose initial **atomic number** assignment was based on chemical separation. It was named after Dimitri Mendeleev, the great Russian chemist. Mendelevium isotopes of masses 245 through 260 have been reported. All are radioactive, decaying by α-particle emission, electron capture, and/or spontaneous fission, with half-lives ranging from 0.35 second for mass 245 to 31.8 days for mass 260, the heaviest isotope. The ground state electronic configuration of Md is believed to be $[Rn]5f^{13}7s^2$, by analogy to its **lanthanide** homologue thulium (element 69). Its most stable ion in **aqueous solution** is Md^{3+}, although Md^{2+} can be

atomic number: the number assigned to an atom of an element that indicates the number of protons in the nucleus of that atom

lanthanides: a family of elements (atomic number 57 through 70) from lanthanum to lutetium having from 1 to 14 4f electrons

aqueous solution: homogenous mixture in which water is the solvent (primary component)

reducing agent: substance that causes reduction, a process during which electrons are lost (or hydrogen atoms gained)

metal: element or other substance the solid phase of which is characterized by high thermal and electrical conductivities

prepared with strong **reducing agents**. The **metal** is believed to be divalent because of its high volatility relative to that of other actinide metals, but this has not been experimentally verified. SEE ALSO ACTINIUM; BERKELIUM; EINSTEINIUM; FERMIUM; LAWRENCIUM; MENDELEEV, DIMITRI; NEPTUNIUM; NOBELIUM; PLUTONIUM; PROTACTINIUM; RADIOACTIVITY; RUTHERFORDIUM; SEABORG, GLENN THEODORE; THORIUM; TRANSMUTATION; URANIUM.

Darleane C. Hoffman

Bibliography

Ghiorso, A.; Harvey, B. G.; Choppin, G. R.; Thompson, S. G.; Seaborg, G.T. (1955). "New Element Mendelevium, Atomic Number 101." *Physical Review* 98:1518–1519.

Hoffman, Darleane C.; Ghiorso, Albert; and Seaborg, Glenn T. (2000). *The Transuranium People: The Inside Story.* Singapore: World Scientific Publishing.

Hoffman, Darleane C., and Lee, Diana M. (1999). "Chemistry of the Heaviest Elements—One Atom at a Time." *Journal of Chemical Education* 76(3):331–347.

Seaborg, Glenn T., and Loveland, Walter D. (1990). *The Elements beyond Uranium.* New York: Wiley.

Menten, Maud

CANADIAN BIOCHEMIST
1879–1960

When biochemists are asked to name a mathematical relationship, it is almost certain that they will choose the Michaelis–Menten equation. This equation enables biochemists to study quantitatively the way in which an enzyme speeds up a biochemical reaction. It was discovered by the German-born American biochemist Leonor Michaelis (1875–1949) and his assistant Maud Leonora Menten.

Though both discoverers deserve recognition, Menten faced the additional challenge of being a woman scientist at a time when professional advancement for women was very difficult. Born in Port Lambton, Ontario, Canada, Menten graduated from the University of Toronto with a B.A. in 1904 and an M.B. in medicine in 1907. For the 1907 to 1908 year, she was appointed a fellow at the Rockefeller Institute for Medical Research, New York, where she studied the effect of radium on tumors. Returning to Canada, Menten continued her medical studies, and in 1911 she became one of the first women in Canada to receive a medical doctorate.

catalysis: the action or effect of a substance in increasing the rate of a reaction without itself being converted

The pivotal year in Menten's life was 1912, when she crossed the Atlantic Ocean to spend a year working with Michaelis at the University of Berlin. While there, they developed the Michaelis–Menten hypothesis that provided a general explanation of the enzyme **catalysis** of biochemical reactions. From the hypothesis, they deduced the mathematical relationship that also bears their name. Their discovery changed scientists' approach to the study of biochemical reactions and helped shape the future of the subject.

Returning to North America, Menten performed doctoral research in biochemistry at the University of Chicago, receiving a Ph.D. in 1916. Despite her strong qualifications and the renown she received for the equation coformulated with Michaelis, she was unable to find any suitable employment in Canada. As a result, in 1918 she joined the medical school at the

University of Pittsburgh as a pathologist. She was appointed assistant professor of pathology in 1923 and promoted to associate professor in 1925. At the same time, she served as a clinical pathologist at the Children's Hospital at Pittsburgh, where she insisted on knowing about every interesting or puzzling case admitted to the hospital. Besides this, Menten maintained an active research program, authoring or coauthoring over seventy research papers. Among her other important discoveries were the use of electric fields to determine differences in human hemoglobin (a process called **electrophoresis**) and the development of a dye reaction to study enzymes in the kidney.

electrophoresis: migration of charged particles under the influence of an electric field, usually in solution; cations, positively charged species, will move toward the negative pole and anions, the negatively charged species, will move toward the positive pole

Menten accomplished much by working long 18-hour days. Medical science, however, was not her whole life. She was fluent in several languages, had her oil paintings exhibited in major exhibitions, and was an avid mountain climber. Although Menten did make tremendous contributions to medical science while in Pittsburgh, it was not until a year before her retirement at the age of seventy that the university promoted her to the highest rank of full professor. Formal retirement nonetheless did not slow down Menten. Returning to Canada in 1950, she conducted cancer research at the British Columbia Research Institute until ill health caused her to resign in 1954. SEE ALSO ENZYMES; RADIUM.

Marelene Rayner-Canham
Geoffrey W. Rayner-Canham

Bibliography

Rayner-Canham, Marelene, and Rayner-Canham, Geoffrey (1998). *Women in Chemistry: Their Changing Roles from Alchemical Times to the Mid-Twentieth Century.* Washington, DC: American Chemical Society and the Chemical Heritage Foundation.

Stock, Aaron H., and Carpenter, Anna-Mary (1961). "Prof. Maud Menten (Obituary)." *Nature* 189 (4769):965.

Internet Resources

Skloot, Rebecca. "Some Called Her Miss Menten." Available from <http://www.health.pitt.edu/pittmed/oct_2000>.

Mercury

MELTING POINT: −38.87°C
BOILING POINT: 359.6°C
DENSITY: 13.54 g/cm^3
MOST COMMON IONS: Hg_2^{2-}, Hg^{2+}

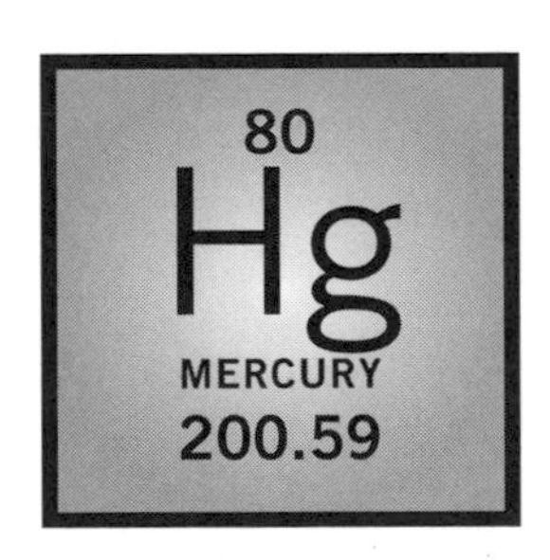

Mercury is at room temperature a silver-white, **volatile** liquid **metal**. It is reputed to have been known in ancient Egypt. Dioscorides, a Greek physician who flourished ca. 60 C.E., recounted the condensation of mercury vapor after the heating of cinnabar, the major ore of mercury. In the modern era mercury is produced via a variation on the procedure used by the ancients: The bright red ore (cinnabar) is now heated in oxygen, with lime, or with iron.

volatile: low boiling, readily vaporized

metal: element or other substance the solid phase of which is characterized by high thermal and electrical conductivities

$$HgS(s) + O_2(g) \rightarrow SO_2(g) + Hg$$

$$HgS(s) + Fe \rightarrow Hg + FeS$$

$$4HgS(s) + 4CaO(s) \rightarrow 4Hg + 3CaS(s) + CaSO_4(s)$$

A landfill with aluminum drums containing mercury waste. Mercury pollution is a major threat to the public health.

oxidation: process that involves the loss of electrons (or the addition of an oxygen atom)

organometallic compound: compound containing a metal (transition) attached to one or more organic moieties

Mercury has three **oxidation** states: 0, 1+ (mercurous), and 2+ (mercuric). It forms few simple compounds. It does form several simple, water-soluble mercuric compounds: mercuric chloride, $HgCl_2$; mercuric nitrate, $Hg(NO_3)_2$; and mercuric acetate, $Hg(CH_3COO)_2$. The mercurous chloride, Hg_2Cl_2, is insoluble in water. Relatively stable **organometallic compounds** are formed with aliphatic and organic compounds. Methylmercury (CH_3–Hg^+) is the major polluting form of mercury. Methylmercury reacts with thiol groups in enzymes.

The mining of mercury has declined in recent decades, as major international concern over the health threat of mercury's extensive pollution of the environment has mounted. Much American freshwater fish is contaminated. The U.S. Environmental Protection Agency estimates 3,000 uses of mercury. Mercury usage is down in the chloroalkali industry, in which mercury is the cathode material used in the electrolysis of sodium chloride solutions, which produce sodium hydroxide and chlorine. An abundance of

500 ppb (0.5 μg/g) in Earth's crust gives rise to a discharge into the atmosphere of mercury on **combustion** of fossil fuels and the manufacture of metals and cement. SEE ALSO HEAVY METAL TOXINS; INORGANIC CHEMISTRY.

Robert A. Bulman

combustion: burning, the reaction with oxygen

Bibliography

Magos, L. (1987). "Mercury." In *Handbook of the Toxicity of Inorganic Compounds*, ed. Hans G. Seiler, Helmut Sigel, and Astrid Sigel. New York: Marcel Dekker.

Metal Alloy *See Steel.*

Methylphenidate

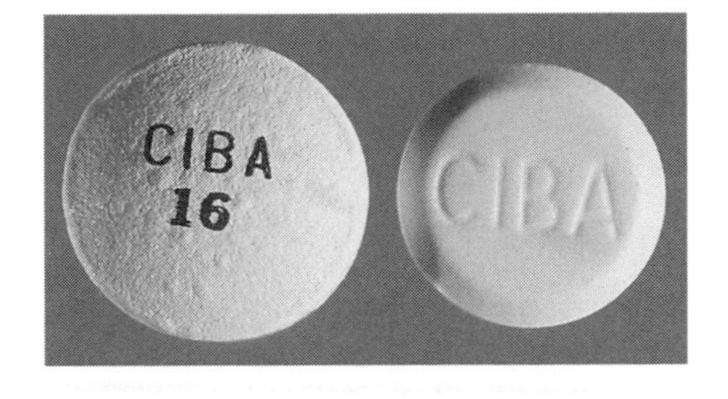

Ritalin, the trademarked name of methylphenidate, is commonly prescribed to children ages 7–18 with attention-deficit hyperactivity disorder.

Methylphenidate is a mild stimulant prescribed to individuals (usually for children, and sometimes controversially) who have behavioral problems characterized by hyperactivity and short attention span. The National Institute of Mental Health estimates that approximately 3–5 percent of the general population has attention-deficit disorder (ADD) or attention-deficit hyperactivity disorder (ADHD). The administration of methylphenidate to children diagnosed with hyperactivity and/or attention-deficit disorder can have a calming effect on the children and can enable them to focus on schoolwork. Methylphenidate is also used to treat narcolepsy, a sleep disorder characterized by a permanent and overwhelming feeling of sleepiness and fatigue.

Methylphenidate is similar to **amphetamine** and, like amphetamine, stimulates the central nervous system (CNS), which consists of the brain and spinal cord. Stimulant drugs affect mood and alertness, and depress food appetite by increasing levels of several neurotransmitters in the brain. Although the exact therapeutic mode of action of methylphenidate is not known, the drug has been shown to elevate levels of some of these neurotransmitters, primarily dopamine and norepinephrine (noradrenaline).

amphetamine: class of compounds used to stimulate the central nervous system

Dopamine and norepinephrine are **excitatory** neurotransmitters. When nerve cells in the brain are stimulated, neurotransmitters stored in **vesicles** in nerve cell endings are released to extracellular spaces (synapses). The liberated chemical messengers can then interact with **receptors** on an adjacent nerve cell and can generate a new nerve signal (a nerve impulse). When levels of dopamine or norepinephrine are depressed, regions of the brain that rely on the two substances to regulate nerve impulse conduction are unable to function properly. Research has shown that children with ADD or ADHD have lower levels of dopamine in the CNS. It is believed that drugs acting as CNS stimulants, such as amphetamine, cocaine, and methylphenidate, compensate for lowered levels of excitatory neurotransmitters (i.e., dopamine and norepinephrine) in the brain. Indeed, administration of methylphenidate to healthy adult men has been found to increase dopamine levels.

excitatory: causing cells to become active

vesicle: small compartment in a cell that contains a minimum of one lipid bilayer

receptor: area on or near a cell wall that accepts another molecule to allow a change in the cell

Different theories have been proposed to explain how methylphenidate increases levels of dopamine in the brain. One such theory propounds that methylphenidate causes dopamine to remain longer in extracellular spaces. Once a neurotransmitter has transmitted its message, it dissociates from the receptor to which it was bound and is taken up by the nerve cell from which it was originally released. Researchers at Brookhaven National Laboratory have demonstrated that methylphenidate inhibits this "reuptake" of dopamine by nerve cells. As a result, extracellular dopamine levels are increased, and the neurotransmitter continues to be available to initiate nerve impulses.

Ritalin is the brand name of a formulation of methylphenidate that is available in tablet form. Ritalin tablets are most often prescribed to children, aged 7 to 18, who have been diagnosed with ADHD. However, individuals who suffer from anxiety or panic disorders are warned not to take the drug, as Ritalin may aggravate symptoms of agitation and/or anxiety. Nervousness and insomnia are the most common side effects associated with Ritalin. There have also been reports of the onset of Tourette's syndrome, a **neurological** disorder characterized by repeated and involuntary body movements (tics), or at least of symptoms resembling those of Tourette's, in patients taking Ritalin, and therefore patients with this disorder are advised not to take Ritalin. In April 2002 the National Institutes of Health reported that a combination of Ritalin and Clonidine, a drug often used to treat **hypertension**, is more effective in the treatment of ADHD than either drug alone. Furthermore, Clonidine is reported to have a tic-suppressing effect, and it is believed that the drug may counteract the apparent tic-accentuating effect of methylphenidate.

neurologic: of or pertaining to the nervous system

hypertension: condition in which blood pressure is abnormally high

During the early 1990s, reports of abuse of methylphenidate began to appear. Nonmedical use of the drug for its stimulant effects impelled the U.S. Drug Enforcement Administration (DEA) to regulate the manufacture, distribution, and prescription of methylphenidate. Because methylphenidate is related to amphetamine, it can be addictive and result in physical and psychological dependence. SEE ALSO NEUROCHEMISTRY; NEUROTRANSMITTERS; PHARMACEUTICAL CHEMISTRY.

Nanette M. Wachter

Bibliography

Goldman, Larry S., et al. (1998) "Diagnosis and Treatment of Attention-Deficit/Hyperactivity Disorder in Children and Adolescents." *Journal of the American Medical Association* 279(14):1100-1107.

Internet Resources

National Institute of Drug Abuse. National Institutes of Health. Available from <http://www.nida.nih.gov/>.

National Institute of Neurological Disorders and Stroke. Available from <http://www.ninds.nih.gov/>.

Meyer, Lothar

GERMAN CHEMIST
1830–1895

Lothar Meyer was the son and grandson of physicians, so it was only natural that initially he decided on a career as a physician. At the age of twenty-one, he began his studies in medicine at the University of Zurich and received his M.D. in 1854. By then Meyer had become interested in the chemistry of the body and went on to study under Robert Bunsen at Heidelberg, where he learned how to analyze gases. He used these techniques to study the absorption of oxygen and carbon monoxide by the blood, and was able to establish that they both reacted in a similar fashion with the same constituent present in the blood. Meyer also determined that carbon monoxide was able to displace oxygen from the blood. However, he was unable to identify the particular component in the blood responsible for binding. This substance was identified as hemoglobin eight years later by Felix Hoppe-Seyler, a professor of physiological chemistry at the University of Strasbourg in France. For this work, Meyer received his Ph.D. in 1858 at the University of Breslau, and he became the director of the chemical laboratory in the physiology institute there until 1866.

German chemist Lothar Meyer, known for the discovery of the periodic properties of the elements.

In 1864 Meyer published *Modern Theories of Chemistry*, which went through five editions and was translated into English, French, and Russian. This book contained a prototype of his 1870 Periodic Table, which consisted of only twenty-eight elements arranged in six families that had similar chemical and physical characteristics. Above all he used a number referred to as the combining power of each element, later termed the **valence**, to link together a particular family. For example, carbon, silicon, tin, and lead were assigned to the same family because each exhibited a combining power of four. He also recognized the following from the observation that atomic weights usually increase by a certain amount between family members: A missing element existed between silicon and tin (later this gap was filled by germanium, discovered in 1886 by the German chemist Clemens Winkler). By 1868 he had expanded his table to include fifty-three elements, but this version was not made public until 1895. This was unfortunate because in 1869 the Russian chemist Dimitri Mendeleev published his version of the Periodic Table in a paper entitled, "The Relation of the Properties to the Atomic Weights of the Elements." As well as postulating his table, Mendeleev described how it could be used to predict not only the **atomic weight** of missing elements, but also their actual properties.

valence: combining capacity

atomic weight: weight of a single atom of an element in atomic mass units (AMU)

The most famous of Mendeleev's predictions involved eka-boron (scandium), eka-aluminium (gallium), and eka-silicon (germanium). For example, for eka-silicon he predicted its atomic weight, its density, the compounds it would form, and details about their physical properties. When thirteen years later germanium was discovered and it was determined that Mendeleev's predictions had been correct, scientists began to recognize the importance of the Periodic Table, and its discovery was quite naturally associated with Mendeleev, who encouraged this association.

Even in the twenty-first century, although historians recognize that others, especially Meyer, should be given considerable credit for the discovery of the periodic properties of the elements, most textbooks credit only Mendeleev. SEE ALSO Bunsen, Robert; Mendeleev, Dimitri; Periodic Table.

John E. Bloor

Bibliography

Van Spronsen, Johannes W. (1969). "The Priority Conflict between Mendeleev and Meyer." *Journal of Chemical Education* 46:136–139.

Internet Resources

Beavon, Rod. Translation of part of Meyer's 1870 paper on the Periodic Table. Available from <http://www.rod.beavon.clara.net/lotharme.htm>.

Millikan, Robert

AMERICAN PHYSICIST
1868–1953

American physicist Robert Millikan, recipient of the 1923 Nobel Prize in physics, "for his work on the elementary charge of electricity and on the photoelectric effect."

Born in Morrison, Illinois, Robert Andrew Millikan was the second son of the Reverend Silas Franklin Millikan and Mary Jane Andrews. When Millikan was seven, his family moved to Maquoketa, Iowa, where he attended high school. In 1886 he entered Oberlin College in Ohio. In 1887 he enrolled in several classics classes there, and because he did quite well in Greek, at the end of his sophomore year, he was asked to teach an introductory-level physics class. He enjoyed teaching physics and accepted a two-year teaching post at Oberlin upon graduation in 1891. It was during this period that he developed an even keener interest in physics.

In 1893 Millikan began his doctoral work at Columbia University, receiving a Ph.D. in 1895. After traveling to Germany, he eventually accepted a faculty position at the University of Chicago. It was as a teacher and textbook author that Millikan first made his mark. He wrote or cowrote a number of elementary physics texts that became the classics in this field. However, while valued activities, they did not lead to his promotion to full professor. Determined to ascend in academic rank, Millikan began his research into the charge on the electron.

At the time, the debate over whether or not atoms were real had almost played out, but the questions surrounding the true nature of the electron were still unanswered. Although the work of the English physicist J. J. Thomson had elucidated the charge-to-mass ratio, determining that the electron had a discrete, fixed charge and mass remained.

Being an experimentalist, Millikan used a tiny, submillimeter drop of oil suspended between capacitor plates to measure the incremental charge on an electron. His reasoned that the oil drop would pick up a charge due to friction as it entered the region between the plates. By ionizing the atmosphere and monitoring the motion of multiple drops, he was able to compare the time that the drop took to fall under the influence of gravity and with the electrical plates off, against the time that it took for the drop to climb under the influence of applied **voltage**. The interaction of the drop with the electric field always occurred in discrete units, indicating that the electron charge was a single value, and that it was the same value for all different forms of electricity.

voltage: potential difference expressed in volts

Millikan's oil-drop experiment settled the argument and determined accurately (within one part in a thousand) both the charge and, by virtue of the charge-to-mass ratio, the mass of the electron. Both numbers allowed the Danish physicist Niels Bohr to finally calculate Rydberg's constant and provided the first and most important proof of the new **atomic theory**.

atomic theory: physical concept that asserts that matter is composed of microscopically tiny particles called atoms and that various elements differ from each other by having different atoms

Millikan went on to demonstrate the photoelectron effect, providing a valuable proof of Albert Einstein's equations. His experiments also aided both Einstein and Bohr in their later research efforts. In 1923 he was awarded a Nobel Prize in physics for both his work in determining the charge on the electron and exploring the photoelectric effect. SEE ALSO BOHR, NIELS; EINSTEIN, ALBERT; THOMSON, JOSEPH JOHN.

Todd W. Whitcombe

Bibliography

Kargon, R. (1982). *The Rise of Robert Millikan. Portrait of a Life in American Science.* Ithaca, NY: Cornell University Press.

Millikan, R. A. (1950). *The Autobiography of Robert A. Millikan.* New York: Ayer Company Publishers.

Internet Resources

Millikan, R. A. (1913). "On the Elementary Electrical Charge and the Avogadro Constant." *Physical Review* 2:109–143. Available from <http://www.aip.org/history/gap/PDF/millikan.pdf>.

Millikan, R. A. (1924). "The Electron and the Light-Quanta from the Experimental Point of View." Nobel Lecture, May 23, 1924. Available from <http://www.nobel.se/physics/laureates/1923/millikan-lecture.pdf>.

"Robert A. Millikan—Biography." Nobel e-Museum. Available from <http://www.nobel.se>.

Minerals

Minerals are the building blocks of rocks. A mineral may be defined as any naturally occurring inorganic solid that has a definite chemical composition (that can vary only within specified limits) and possesses a crystalline structure. The study of minerals is known as mineralogy, which dates back to prehistory. The use of minerals in the construction of primitive weapons and as suppliers of color for ancient artists makes mineralogy one of the oldest of the human arts.

Minerals may be characterized by the fundamental patterns of their crystal structures. A crystal structure is commonly identified by its fundamental repeating unit, which upon protraction into three dimensions generates a macroscopic crystal. Crystal structures can be divided into crystal systems, which can be further subdivided into crystal classes—a total of thirty-two crystal classes, which are sometimes referred to as point classes.

More commonly, minerals are described or classified on the basis of their chemical composition. Although some minerals, such as graphite or diamond, consist primarily of a single element (in this instance, carbon), most minerals occur as ionic compounds that consist of orderly arrangements of cations and **anions** and have a specific crystalline structure determined by the sizes and charges of the individual ions. Cations (positively charged ions) are formed by the loss of negatively charged electrons from atoms. Anions consist of a single element, the atoms of which have become negatively charged via the acquisition of electrons, or they consist of several elements, the atoms bound together by **covalent bonds** and bearing an overall negative charge. Pyrite (FeS_2) is a mineral that contains a sulfide ion as its anion. Gypsum [$CaSO_4-2(H_2O)$] contains the polyatomic anion known as sulfate (SO_4^{2-}) as well as two waters of hydration (water molecules that are part of the crystalline structure).

anion: negatively charged chemical unit, like Cl^-, CO_3^{2-}, or NO_3^-

covalent bond: bond formed between two atoms that mutually share a pair of electrons

It has been noted that the chemical composition of minerals could vary within specified limits. This phenomenon is known as solid solution. For example, the chemical composition of the mineral dolomite is commonly designated as $CaMg(CO_3)_2$, or as $(Ca, Mg)CO_3$. This does not mean that dolomite has calcium and magnesium existing in a one-to-one ratio. It signifies that dolomite is a carbonate mineral that has significant amounts of

This seacliff in Wales shows strata of banded liassic limestone and shale.

both cations (calcium and magnesium ions) in an infinite variety of proportions. When minerals form, ions of similar size and charge, such as calcium and magnesium ions, can substitute for each other and will be found in the mineral in amounts that depend on the proportions that were present in solution, or in the melt (liquid magma) from which the mineral formed. Thus, many minerals can exist in solid solution. When solid solutions exist, names are often given to the end-members. In the case of the calcium and magnesium carbonates, one end-member, $CaCO_3$ is named calcite or aragonite, depending on the crystalline symmetry, whereas the other end-member, $MgCO_3$, is referred to as magnesite.

Because minerals are naturally occurring substances, the abundance of minerals tends to reflect the abundance of elements as they are found in Earth's crust. Although about 4,000 minerals have been named, there are forty minerals that are commonly found and these are referred to as the rock-forming minerals.

The most abundant element in Earth's crust is oxygen, which makes up about 45 percent of the crust by mass. The second most abundant element is silicon, which accounts for another 27 percent by mass. The next six most abundant elements, in order of abundance, are aluminum, iron, calcium, magnesium, sodium, and potassium, which collectively comprise about 26 percent, leaving only about 2 percent for all other elements. If one classifies minerals according to the commonly accepted system that is based on their anions, it is not surprising that silicates (having anions that are polyatomic combinations of oxygen and silicon) are the most common mineral group.

Silicates

In order to understand the chemical structures and formulas of the silicate minerals, one must begin with the basic building block of all silicates: the silica tetrahedron. A silica tetrahedron is an anionic species, which consists of a silicon atom covalently bound to four oxygen atoms. The silicon atom is in the geometric center of the tetrahedron and at each of the four points of the

The natural matrix of the Kimberlite diamond.

tetrahedron is an oxygen atom. The structure has an overall charge of negative four and is represented as SiO_4^{4-}. The mineral olivine, a green-colored mineral as the name suggests, has the formula $(Mg, Fe)_2SiO_4$. When olivine is a gem-quality crystal it is referred to as peridot. As the formula suggests, olivine is really a group of minerals that vary in composition, from almost pure end-member forsterite (Mg_2SiO_4) to almost pure fayalite (Fe_2SiO_4).

All of the silicate minerals arise from various combinations of silica tetrahedra and a sense of their variety may be gleaned from the understanding that the oxygen atoms at the tetrahedral vertices may be shared by adjacent tetrahedra in such a way as to generate larger structures, such as single chains, double chains, sheets, or three-dimensional networks of tetrahedra. Various **cations** occurring within solid solutions neutralize the negative charges on the silicate backbone. The variation in geometric arrangements generates a dazzling array of silicate minerals, which includes many common gemstones.

cation: positively charged ion

The pyroxene group and the amphibole group, respectively, are representatives of silicate minerals having single-chain and double-chain tetrahedral networks. Pyroxenes are believed to be significant components of Earth's mantle, whereas amphiboles are dark-colored minerals commonly found in continental rocks.

Clays have sheet structures, generated by the repetitious sharing of three of the four oxygen atoms of each silica tetrahedron. The fourth oxygen atom of the silica tetrahedron is important as it has a capacity for cation exchange. Clays are thus commonly used as natural ion-exchange resins in water purification and desalination. Clays can be used to remove sodium ions from seawater, as well as to remove calcium and magnesium ions in the process of water softening. Because the bonds between adjacent sheets of silicon tetrahedra are weak, the layers tend to slip past one another rather easily, which contributes to the slippery texture of clays.

Clays also tend to absorb (or release) water. This absorption or release of water significantly changes clay volume. Consequently, soils that contain significant amounts of water-absorbing clays are not suitable as building construction sites.

Clays are actually secondary minerals—meaning that they are formed chiefly by the weathering of primary minerals. Primary minerals are those that form directly by precipitation from solution or magma, or by deposition from the vapor **phase**. In the case of clays their primary or parent minerals are feldspars, the mineral group with the greatest abundance in Earth's crust. Feldspars and clays are actually aluminosilicates. The formation of an aluminosilicate involves the replacement of a significant portion of the silicon in the tetrahedral backbone by aluminum.

phase: homogeneous state of matter

The feldspar minerals have internal arrangements that correspond to a three-dimensional array of silica tetrahedra that arises from the sharing of all four oxygen atoms at the tetrahedral vertices, and are sometimes referred to as framework silicates. Feldspars, rich in potassium, typically have a pink color and are responsible for the pinkish color of many of the feldspar-rich granites that are used in building construction. The feldspathoid minerals are similar in structure to feldspars but contain a lesser abundance of silica. Lapis lazuli, now used primarily in jewelry, is a mixture of the feldspathoid lazurite and other silicates, and was formerly used in granulate form as the paint pigment ultramarine.

Zeolites are another group of framework silicates similar in structure to the feldspars. Like clays they have the ability to absorb or release water. Zeolites have long been used as molecular sieves, due to their ability to absorb molecules selectively according to molecular size.

One of the most well-known silicate minerals is quartz (SiO_2), which consists of a continuous three-dimensional network of silica and oxygen without any atomic substitutions. It is the second most abundant continental mineral, feldspars being most abundant. The network of covalent bonds (between silicon and oxygen) is responsible for the well-known hardness of quartz and its resistance to weathering. Although pure quartz is clear and without color, the presence of small amounts of impurities may result in the formation of gemstones such as amethyst.

Nonsilicate Minerals

Although minerals of other classes are relatively scarce in comparison to the silicate minerals, many have interesting uses and are important economically. Because of the great abundance of oxygen in Earth's crust, the oxides are the most common minerals after the silicates. Litharge, for example, is a yellow-colored oxide of lead (PbO) and is used by artists as a pigment. Hematite (Fe_2O_3), a reddish-brown ore, is an iron oxide and is also used as a pigment. Other important classes of nonsilicate minerals include sulfides, sulfates, carbonates, halides, phosphates, and hydroxides. Some minerals in these groups are listed in Table 1.

Although minerals are often identified by the use of sophisticated optical instruments such as the polarizing microscope or the x-ray diffractometer, most can be identified using much simpler and less expensive methods. Color can be very helpful in identifying minerals (although it can also be misleading). A very pure sample of the mineral carborundum (Al_2O_3) is colorless but the presence of small amounts of impurities in carborundum may yield the deep red gemstone ruby or the blue gemstone sapphire. The streak

EXAMPLES OF COMMON NONSILICATE MINERALS AND THEIR USES

Mineral	Formula	Economic Use
Pyrite	FeS_2	sulfuric acid production
Anhydrite	$CaSO_4$	plaster
Calcite	$CaCO_3$	lime
Halite	$NaCl$	table salt
Turquoise	$CuAl_6(PO_4)_4(OH)_8$	gemstone
Bauxite	$Al(OH)_3.nH_2O$	aluminum ore
Rutile	TiO_2	jewelry, semiconductor

SOURCE: Tarbuck, Edward J., and Lutgens, Frederick K. (1999). *Earth: An Introduction to Physical Geology*, 6th edition. Upper Saddle River, NJ: Prentice Hall.

Table 1. Examples of common nonsilicate minerals and their uses.

of a mineral (the color of the powdered form) is actually much more useful in identifying a mineral than is the color of the entire specimen, as it is less affected by impurities. The streak of a mineral is obtained by simply rubbing the sample across a streak plate (a piece of unglazed porcelain), and the color of the powder is then observed. Virtually all mineral indexes used to identify minerals, such as those found in *Dana's Manual of Mineralogy*, list streaks of individual minerals.

Streak is used along with other rather easily determined mineral properties, such as hardness, specific gravity, cleavage, double refraction, the ability to react with common chemicals, and the overall appearance, to pinpoint the identity of an unknown mineral. Mineral hardness is determined by the ability of the sample to scratch or be scratched by readily available objects (a knife blade, a fingernail, a glass plate) or minerals of known hardness. Hardness is graded on the Moh's scale of hardness, which ranges from a value of one (softest) to ten (hardest). The mineral talc (used in talcum powder) has a hardness of one, whereas diamond has a hardness of ten. A fingernail has a hardness of 2.5; therefore quartz, which has a hardness of seven, would be able to scratch talc or a fingernail, but quartz could not scratch diamond or topaz, which has a hardness of eight. Conversely, topaz or diamond would be able to scratch quartz. Specific gravity is the ratio of the weight of a mineral to the weight of an equal volume of water and is thus in concept similar to density. The cleavage of a mineral is its tendency to break along smooth parallel planes of weakness and is dependent on the internal structure of the mineral. A mineral may exhibit double refraction. That is, the double image of an object will be seen if one attempts to view that object through a transparent block of the mineral in question. Calcite is a mineral that exhibits double refraction. Some minerals react spontaneously with common chemicals. If a few drops of hydrochloric acid are placed on a freshly broken surface of calcite, the calcite will react vigorously. **Effervescence**, caused by reaction of the calcite with hydrochloric acid to form the gas carbon dioxide, is observed. In contrast, dolomite will effervesce in hydrochloric acid only upon the first scratching the surface of the dolomite.

effervescence: bubbling or foaming

Minerals are a part of our daily lives. They comprise the major part of most soils and provide essential nutrients for plant growth. They are the basic building blocks of the rocks that compose the surface layer of our planet. They are used in many types of commercial operations, and the mining of minerals is a huge worldwide commercial operation. They are also used in water purification and for water softening. Finally, minerals are perhaps most valued for their great beauty. SEE ALSO GEMSTONES; INORGANIC CHEMISTRY; MATERIALS SCIENCE; ZEOLITES.

Mary L. Sohn

Bibliography

Dana, James D.; revised by Cornelius S. Hulburt Jr. (1959). *Dana's Manual of Mineralogy*, 17th edition. New York: Wiley.

Dietrich, Richard V., and Skinner, Brian J. (1979). *Rocks and Rock Minerals.* New York: Wiley.

Tarbuck, Edward J., and Lutgens, Frederick K. (1999). *Earth: An Introduction to Physical Geology*, 6th edition, Upper Saddle River, NJ: Prentice Hall.

Mole Concept

In chemistry the mole is a fundamental unit in the Système International d'Unités, the SI system, and it is used to measure the amount of substance. This quantity is sometimes referred to as the *chemical amount.* In Latin *mole* means a "massive heap" of material. It is convenient to think of a chemical mole as such.

Visualizing a mole as a pile of particles, however, is just one way to understand this concept. A sample of a substance has a mass, volume (generally used with gases), and number of particles that is proportional to the chemical amount (measured in moles) of the sample. For example, one mole of oxygen gas (O_2) occupies a volume of 22.4 L at standard temperature and pressure (STP; 0°C and 1 atm), has a mass of 31.998 grams, and contains about 6.022×10^{23} molecules of oxygen. Measuring one of these quantities allows the calculation of the others and this is frequently done in stoichiometry.

The *mole* is to the *amount of substance* (or chemical amount) as the *gram* is to *mass.* Like other units of the SI system, prefixes can be used with the mole, so it is permissible to refer to 0.001 mol as 1 mmol just as 0.001 g is equivalent to 1 mg.

Formal Definition

According to the National Institute of Standards and Technology (NIST), the Fourteenth Conférence Générale des Poids et Mesures established the definition of the mole in 1971.

> The mole is the amount of a substance of a system which contains as many elementary entities as there are atoms in 0.012 kilogram of carbon-12; its symbol is "mol." When the mole is used, the elementary entities must be specified and may be atoms, molecules, ions, electrons, other particles, or specified groups of such particles.

One Interpretation: A Specific Number of Particles

When a quantity of particles is to be described, mole is a grouping unit analogous to groupings such as pair, dozen, or gross, in that all of these words represent specific numbers of objects. The main differences between the mole and the other grouping units are the magnitude of the number represented and how that number is obtained. One mole is an amount of substance containing Avogadro's number of particles. Avogadro's number is equal to 602,214,199,000,000,000,000,000 or more simply, $6.02214199 \times 10^{23}$.

Unlike pair, dozen, and gross, the exact number of particles in a mole cannot be counted. There are several reasons for this. First, the particles are too small and cannot be seen even with a microscope. Second, as naturally occurring carbon contains approximately 98.90% carbon-12, the sample would need to be purified to remove every atom of carbon-13 and carbon-14. Third, as the number of particles in a mole is tied to the mass of exactly 12 grams of carbon-12, a balance would need to be constructed that could determine if the sample was one atom over or under exactly 12 grams. If the first two requirements were met, it would take one million machines counting one million atoms each second more than 19,000 years to complete the task.

Obviously, if the number of particles in a mole cannot be counted, the value must be measured indirectly and with every measurement there is some degree of uncertainty. Therefore, the number of particles in a mole, Avogadro's constant (N_A), can only be approximated through experimentation, and thus its reported values will vary slightly (at the tenth decimal place) based on the measurement method used. Most methods agree to four significant figures, so N_A is generally said to equal 6.022×10^{23} particles per mole, and this value is usually sufficient for solving textbook problems. Another key point is that the formal definition of a mole does not include a value for Avogadro's constant and this is probably due to the inherent uncertainty in its measurement. As for the difference between Avogadro's constant and Avogadro's number, they are numerically equivalent, but the former has the unit of mol^{-1} whereas the latter is a pure number with no unit.

A Second Interpretation: A Specific Mass

Atoms and molecules are incredibly small and even a tiny chemical sample contains an unimaginable number of them. Therefore, counting the number of atoms or molecules in a sample is impossible. The multiple interpretations of the mole allow us to bridge the gap between the submicroscopic world of atoms and molecules and the macroscopic world that we can observe.

To determine the chemical amount of a sample, we use the substance's *molar mass*, the mass per mole of particles. We will use carbon-12 as an example because it is the standard for the formal definition of the mole. According to the definition, one mole of carbon-12 has a mass of exactly 12 grams. Consequently, the molar mass of carbon-12 is 12 g/mol. However, the molar mass for the element carbon is 12.011 g/mol. Why are they different? To answer that question, a few terms need to be clarified.

On the Periodic Table, you will notice that most of the atomic weights listed are not round numbers. The atomic weight is a weighted average of the atomic masses of an element's natural isotopes. For example, bromine has two natural isotopes with atomic masses of 79 u and 81 u. The unit *u* represents the atomic mass unit and is used in place of grams because the value would be inconveniently small. These two isotopes of bromine are present in nature in almost equal amounts, so the atomic weight of the element bromine is 79.904. (i.e., nearly 80, the arithmetic mean of 79 and 81). A similar situation exists for chlorine, but chlorine-35 is almost three times as abundant as chlorine-37, so the atomic weight of chlorine is 35.4527. Technically, atomic weights are ratios of the average atomic mass to the unit *u* and that is why they do not have units. Sometimes atomic weights are given the unit *u*, but this is not quite correct according to the International Union of Pure and Applied Chemistry (IUPAC).

To find the molar mass of an element or compound, determine the atomic, molecular, or formula weight and express that value as g/mol. For bromine and chlorine, the molar masses are 79.904 g/mol and 35.4527 g/mol, respectively. Sodium chloride (NaCl) has a formula weight of 58.443 (atomic weight of Na + atomic weight of Cl) and a molar mass of 58.443 g/mol. Formaldehyde (CH_2O) has a molecular weight of 30.03 (atomic weight of C + 2 [atomic weight of H]) + atomic weight of O] and a molar mass of 30.03 g/mol.

The concept of molar mass enables chemists to measure the number of submicroscopic particles in a sample without counting them directly simply by determining the chemical amount of a sample. To find the chemical amount of a sample, chemists measure its mass and divide by its molar mass. Multiplying the chemical amount (in moles) by Avogadro's constant (N_A) yields the number of particles present in the sample.

Occasionally, one encounters gram-atomic mass *(GAM)*, gram-formula mass *(GFM)*, and gram-molecular mass *(GMM)*. These terms are functionally the same as molar mass. For example, the GAM of an element is the mass in grams of a sample containing N_A atoms and is equal to the element's atomic weight expressed in grams. GFM and GMM are defined similarly. Other terms you may encounter are formula mass and molecular mass. Interpret these as formula weight and molecular weight, respectively, but with the units of *u*.

Avogadro's Hypothesis

Some people think that Amedeo Avogadro (1776–1856) determined the number of particles in a mole and that is why the quantity is known as Avogadro's number. In reality Avogadro built a theoretical foundation for determining accurate atomic and molecular masses. The concept of a mole did not even exist in Avogadro's time.

Much of Avogadro's work was based on that of Joseph-Louis Gay-Lussac (1778–1850). Gay-Lussac developed the law of combining volumes that states: "In any chemical reaction involving gaseous substances the volumes of the various gases reacting or produced are in the ratios of small whole numbers." (Masterton and Slowinski, 1977, p. 105) Avogadro reinterpreted Gay-Lussac's findings and proposed in 1811 that (1) some molecules were diatomic and (2) "equal volumes of all gases at the same temperature and pressure contain the same number of molecules" (p. 40). The second proposal is what we refer to as Avogadro's hypothesis.

The hypothesis provided a simple method of determining relative molecular weights because equal volumes of two different gases at the same temperature and pressure contained the same number of particles, so the ratio of the masses of the gas samples must also be that of their particle masses. Unfortunately, Avogadro's hypothesis was largely ignored until Stanislao Cannizzaro (1826–1910) advocated using it to calculate relative atomic masses or atomic weights. Soon after the 1st International Chemical Congress at Karlsrule in 1860, Cannizzaro's proposal was accepted and a scale of atomic weights was established.

To understand how Avogadro's hypothesis can be used to determine relative atomic and molecular masses, visualize two identical boxes with oranges in one and grapes in the other. The exact number of fruit in each box is not known, but you believe that there are equal numbers of fruit in each box (Avogadro's hypothesis). After subtracting the masses of the boxes, you have the masses of each fruit sample and can determine the mass ratio between the oranges and the grapes. By assuming that there are equal numbers of fruit in each box, you then know the average mass ratio between a grape and an orange, so in effect you have calculated their relative masses (atomic masses). If you chose either the grape or the orange as a standard, you could eventually determine a scale of relative masses for all fruit.

A Third Interpretation: A Specific Volume

By extending Avogadro's hypothesis, there is a specific volume of gas that contains N_A gas particles for a given temperature and pressure and that volume should be the same for all gases. For an ideal gas, the volume of one mole at STP (0°C and 1.000 atm) is 22.41 L, and several real gases (hydrogen, oxygen, and nitrogen) come very close to this value.

The Size of Avogadro's Number

To provide some idea of the enormity of Avogadro's number, consider some examples. Avogadro's number of water drops (twenty drops per mL) would fill a rectangular column of water 9.2 km (5.7 miles) by 9.2 km (5.7 miles) at the base and reaching to the moon at perigee (closest distance to Earth). Avogadro's number of water drops would cover the all of the land in the United States to a depth of roughly 3.3 km (about 2 miles). Avogadro's number of pennies placed in a rectangular stack roughly 6 meters by 6 meters at the base would stretch for about 9.4×10^{12} km and extend outside our solar system. It would take light nearly a year to travel from one end of the stack to the other.

History

Long before the mole concept was developed, there existed the idea of chemical equivalency in that specific amounts of various substances could react in a similar manner and to the same extent with another substance. Note that the historical equivalent is not the same as its modern counterpart, which involves electric charge. Also, the historical equivalent is not the same as a mole, but the two concepts are related in that they both indicate that different masses of two substances can react with the same amount of another substance.

The idea of chemical equivalents was stated by Henry Cavendish in 1767, clarified by Jeremias Richter in 1795, and popularized by William Wollaston in 1814. Wollaston applied the concept to elements and defined it in such a way that one equivalent of an element corresponded to its atomic mass. Thus, when Wollaston's equivalent is expressed in grams, it is identical to a mole. It is not surprising then that the word "mole" is derived from "molekulargewicht" (German, meaning "molecular weight") and was coined in 1901 or 1902. SEE ALSO AVOGADRO, AMEDEO; CANNIZZARO, STANISLAO; CAVENDISH, HENRY; GAY-LUSSAC, JOSEPH-LOUIS.

Nathan J. Barrows

Bibliography

Atkins, Peter, and Jones, Loretta (2002). *Chemical Principles*, 2nd edition. New York: W. H. Freeman and Company.

Lide, David R., ed. (2000). *The CRC Handbook of Chemistry & Physics*, 81st edition. New York: CRC Press.

Masterton, William L., and Slowinski, Emil J. (1977). *Chemical Principles*, 4th edition. Philadelphia: W. B. Saunders Company.

Internet Resources

National Institute of Standards and Technology. "Unit of Amount of Substance (Mole)." Available from <http://www.nist.gov>.

Molecular Geometry

Molecules, from simple diatomic ones to macromolecules consisting of hundreds of atoms or more, come in many shapes and sizes. The term "molecular geometry" is used to describe the shape of a molecule or polyatomic ion as it would appear to the eye (if we could actually see one). For this discussion, the terms "molecule" and "molecular geometry" pertain to polyatomic ions as well as molecules.

Molecular Orbitals

When two or more atoms approach each other closely enough, pairs of valence shell electrons frequently fall under the influence of two, and sometimes more, nuclei. Electrons move to occupy new regions of space (new orbitals—molecular orbitals) that allow them to "see" the nuclear charge of multiple nuclei. When this activity results in a lower overall energy for all involved atoms, the atoms remain attached and a molecule has been formed. In such cases, we refer to the interatomic attractions holding the atoms together as covalent bonds. These molecular orbitals may be classified according to strict mathematical (probabilistic) determinations of atomic behaviors. For this discussion, the two most important classifications of this kind are *sigma* (σ) and *pi* (π). Though we may be oversimplifying a highly complex mathematics, it may help one to visualize sigma molecular orbitals as those that build up electron density along the (internuclear) axis connecting bonded nuclei, and pi molecular orbitals as those that build up electron density above and below the internuclear axis.

sigma molecular orbital

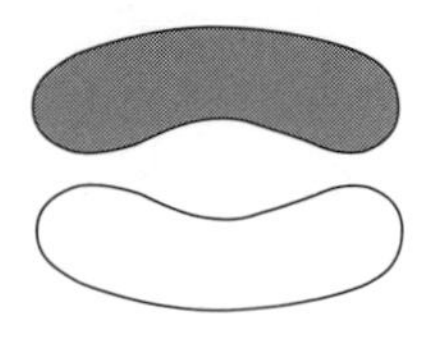

pi molecular orbital

Bonding Theories

This discussion will examine two approaches chemists have used to explain bonding and the formation of molecules, the molecular orbital (MO) theory and the valence bond (VB) theory. At their simplest levels, both approaches ignore nonvalence shell electrons, treating them as occupants of molecular orbitals so similar to the original (premolecular formation) atomic orbitals that they are localized around the original nuclei and do not participate in bonding. The two approaches diverge mainly with respect to how they treat the electrons that are extensively influenced by two or more nuclei. Though the approaches differ, they must ultimately converge because they describe the same physical reality: the same nuclei, the same electrons.

Molecular orbital theory. In MO theory, there are three types of molecular orbitals that electrons may occupy.

1. Nonbonding molecular orbitals. Nonbonding molecular orbitals closely resemble atomic orbitals localized around a single nucleus. They are called nonbonding because their occupation by electrons confers no net advantage toward keeping the atoms together.

2. Bonding molecular orbitals. Bonding molecular orbitals correspond to regions where electron density builds up between two, sometimes more, nuclei. When these orbitals are occupied by electrons, the electrons "see" more positive nuclear charge than they would if the atoms had not come together. In addition, with increased electron density in the spaces between the nuclei, nucleus-nucleus repulsions are minimized. Bonding orbitals allow for increased electron-nucleus attraction and decreased nucleus-nucleus repulsion, therefore electrons in such orbitals tend to draw atoms together and bond them to each other.

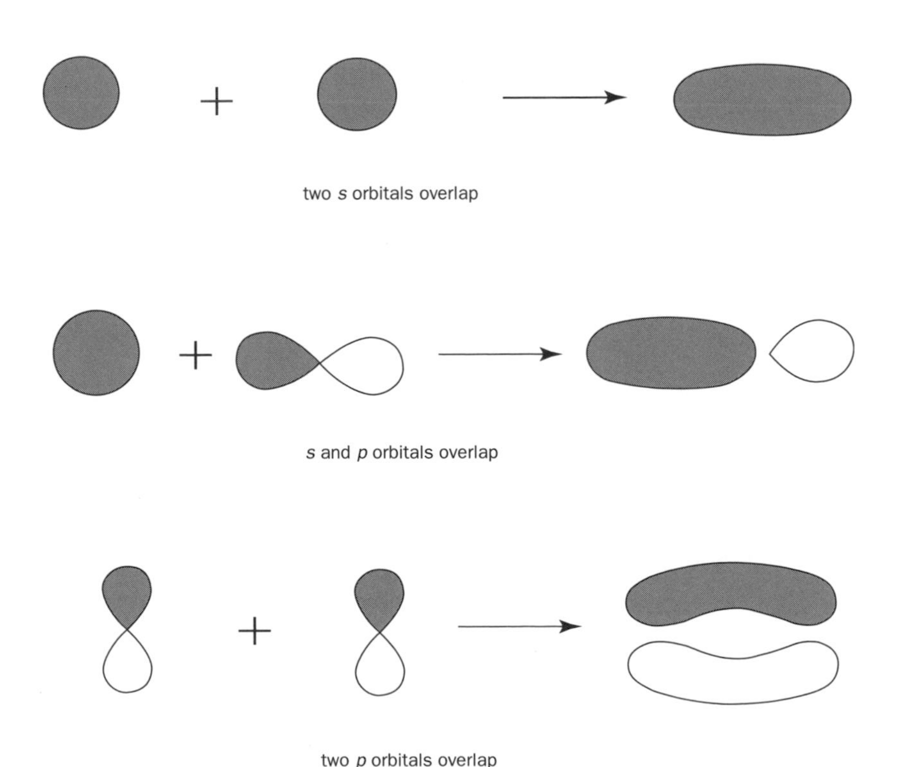

3. Antibonding molecular orbitals. One antibonding molecular orbital is formed for each bonding molecular orbital that is formed. Antibonding orbitals tend to localize electrons outside the regions *between* nuclei, resulting in significant nucleus-nucleus repulsion—with little, if any, improvement in electron-nucleus attraction. Electrons in antibonding orbitals work against the formation of bonds, which is why they are called antibonding.

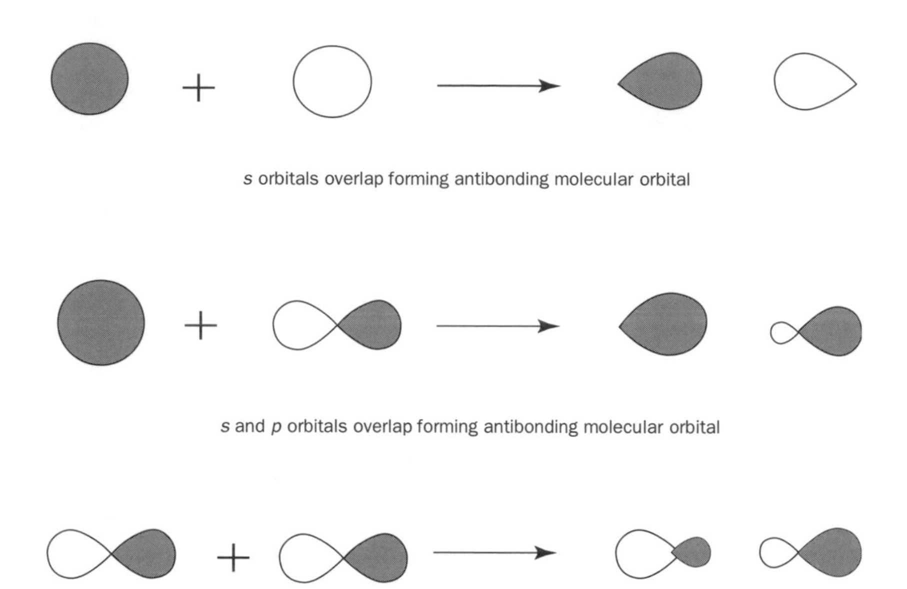

s orbitals overlap forming antibonding molecular orbital

s and *p* orbitals overlap forming antibonding molecular orbital

p orbitals overlap forming antibonding molecular orbital

According to MO theory, atoms remain close to one another (forming molecules) when there are more electrons occupying lower energy sigma and/or pi *bonding* orbitals than occupying higher energy antibonding orbitals; such atoms have a lower overall energy than if they had not come together. However, when the number of bonding electrons is matched by the number of antibonding electrons, there is actually a *dis*advantage to having the atoms stay together, therefore no molecule forms.

Valence bond theory. Valence bond (VB) theory assumes that atoms form covalent bonds as they share pairs of electrons via overlapping valence shell orbitals. A single covalent bond forms when two atoms share a pair of electrons via the sigma overlap of two atomic orbitals—a valence orbital from each atom. A double bond forms when two atoms share two pairs of electrons, one pair via a sigma overlap of two atomic orbitals and one via a pi overlap. A triple bond forms by three sets of orbital overlap, one of the sigma type and two of the pi type, accompanied by the sharing of three pairs of electrons via those overlaps. (When a pair of valence shell electrons is localized at only one atom, that is, when the pair is not shared between atoms, it is called a lone or nonbonding pair.)

Let us apply this greatly simplified picture of VB theory to three diatomic molecules: H_2, F_2, and HF. VB theory says that an H_2 molecule forms when a 1*s* orbital containing an electron that belongs to one atom overlaps a 1*s* orbital with an electron of opposite spin belonging to the other, creating a sigma molecular orbital containing two electrons. The two nuclei share the pair of electrons and draw together, giving both electrons access to the positive charge of both nuclei. Diatomic fluorine, F_2, forms similarly, via the sigma overlap of singly occupied $2p$ orbitals. The HF molecule results from the sharing of a pair of electrons whereby an electron in a hydrogen $1s$ orbital experiences sigma overlap with an electron in a fluorine $2p$ orbital.

Molecular Geometries

This VB approach allows us to return to the focus of our discussion. The geometry of a molecule or polyatomic ion is determined by the positions of individual atoms and their positions relative to one another. It can get very complicated. However, let us start with some simple examples and your imagination will help you to extend this discussion to more complicated ones. What happens when two atoms are bonded together in a diatomic molecule? The only possible geometry is a straight line. Hence, such a molecular geometry (or shape) is called "linear." When we have three bonded atoms (in a triatomic molecule), the three atoms may form either a straight line, creating a linear molecule, or a bent line (similar to the letter V), creating a "bent," "angular," "nonlinear," or "V-shaped" molecule. When four atoms bond together, they may form a straight or a zigzag line, a square or other two-dimensional shape in which all four atoms occupy the same flat plane, or they may take on one of several three-dimensional geometries (such as a pyramid, with one atom sitting atop a base formed by the other three atoms). With so many possibilities, it may come as a surprise that we can "predict" the shape of a molecule (or polyatomic ion) using some basic assumptions about electron-electron repulsions.

We start by recognizing that, ultimately, the shape of a molecule is the equilibrium geometry that gives us the lowest possible energy for the system. Such a geometry comes about as the electrons and nuclei settle into positions that minimize nucleus-nucleus and electron-electron repulsions, and maximize electron-nucleus attractions.

Modern computer programs allow us to perform complex mathematical calculations for multiatomic systems with high predictive accuracy. However, without doing all the mathematics, we may "predict" molecular geometries quite well using VB theory.

Valence shell electron pair repulsion approach. In the valence shell electron pair repulsion (VSEPR) approach to molecular geometry, we begin by seeing the valence shell of a bonded atom as a spherical surface. Repulsions among pairs of valence electrons force the pairs to locate on this surface *as far from each other as possible.* Based on such considerations, somewhat simplified herein, we determine where all the electron pairs on the spherical surface of the atom "settle down," and identify which of those pairs correspond to bonds. Once we know which pairs of electrons bond (or glue) atoms together, we can more easily picture the shape of the corresponding (simple) molecule.

However, in using VSEPR, we must realize that in a double or triple bond, the sigma and pi orbital overlaps, and the electrons contained therein, are located in the same basic region between the two atoms. Thus, the four electrons of a double bond or the six electrons of a triple bond are not independent of one another, but form coordinated "sets" of four or six electrons that try to get as far away from other sets of electrons as possible. In an atom's valence shell, a lone pair of electrons or, collectively, the two, four, or six electrons of a single, double, or triple bond each form a set of electrons. It is repulsions among *sets* of valence shell electrons that determine the geometry around an atom.

Consider the two molecules carbon dioxide (CO_2) and formaldehyde (H_2CO). Their Lewis structures are

:Ö═C═Ö:

and

H—C═Ö:
    |
    H

In CO_2, the double bonds group the carbon atom's eight valence electrons into two sets. The two sets get as far as possible from each other by residing on opposite sides of the carbon atom, creating a straight line extending from one set of electrons through the carbon nucleus to the other. With oxygen atoms bonded to these sets of electrons, the oxygen–carbon–oxygen axis is a straight line, making the molecular geometry a straight line. Carbon dioxide is a linear molecule.

In H_2CO, the carbon atom's eight valence electrons are grouped into three sets, corresponding to the two single bonds and the one double bond. These sets minimize the repulsions among themselves by becoming as distant from one another as possible—each set pointing at a vertex of a triangle surrounding the carbon atom in the center. Attaching the oxygen and hydrogen atoms to their bonding electrons has them forming the triangle with the carbon remaining in the center; all four atoms are in the same plane. Formaldehyde has the geometry of a trigonal (or triangular) planar molecule, "planar" emphasizing that the carbon occupies the same plane as the three peripheral atoms.

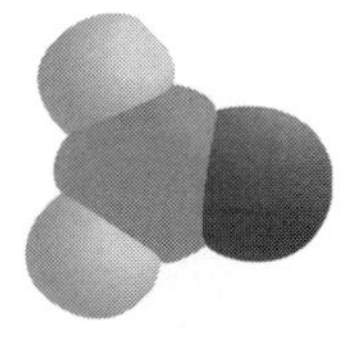

COMMONLY ENCOUNTERED ELECTRON GEOMETRIES

Number of Sets	Most Common "Set" Geometry	Appearance
2	Linear	
3	Trigonal (Triangular) Planar	
4	Tetrahedral	
5	Trigonal Pyramidal	
6	Octahedral	

We may extend this approach to central atoms with four, five, six, or even more sets of valence shell electrons. The most common geometries found in small molecules appear in Table 1.

Table 1.

Until now, this article has focused on all the electrons in a central atom's valence shell, including sets not engaged in bonding. Though all such sets must be included in the conceptualization of the electron-electron repulsions, a molecule's geometry is determined solely by where its atoms are: A molecule's geometry is identified by what people would see if they could see atoms. In the carbon dioxide and formaldehyde examples, the molecules have the same overall geometries as the electron sets, because in both cases all sets are attached to peripheral atoms: Carbon dioxide is a linear molecule and formaldehyde is a trigonal (or triangular) planar one.

On the other hand, a water molecule (H_2O)

H—Ö:
|
H

has four sets of electrons around the O atom (two lone pairs and those making up two sigma bonds) that assume a tetrahedral arrangement, but the molecular geometry as determined by the positions of the three atoms is a bent, or V-shaped, molecule, with a H–O–H angle approaching the tetrahedral angle of 109.5°.

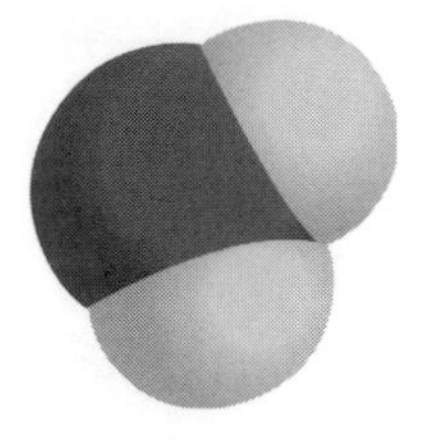

Similarly, a hydronium ion (H_3O^+)

```
H—Ö—H
   |
   H
```

has four sets of electrons around the central O atom (one lone pair and those making up three sigma bonds) in a tetrahedral arrangement, but the molecular geometry as determined by the four atoms is a trigonal (three-pointed base) pyramidal ion with the O atom "sitting" atop the three H atoms. The hydronium ion also has a H–O–H angle approaching the tetrahedral angle of 109.5°.

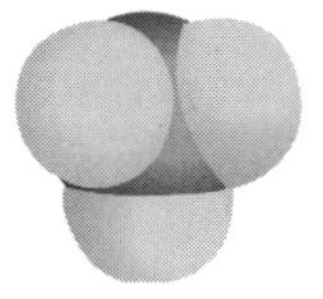

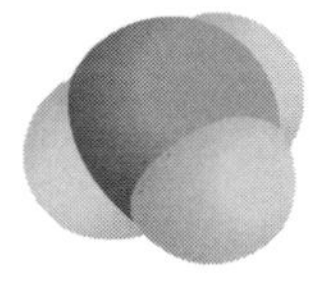

Table 2 outlines the most common molecular geometries for different combinations of lone pairs and up to four total sets of electrons that have assumed positions around a central atom, and the hybridizations (see below) required on the central atom.

Hybridization. Finally, what does valence bond theory say about the atomic orbitals demanded by VSEPR? For example, though the regions occupied by sets of electrons having a tetrahedral arrangement around a central atom make angles of 109.5° to one another, valence *p*-orbitals are at 90° angles.

To reduce the complex task of finding orbitals that "fit" VSEPR, we base their descriptions on mathematical combinations of "standard" atomic orbitals, a process called *hybridization;* the orbitals thus "formed" are *hybrid orbitals.* The number of hybrid orbitals is equal to the number of "standard" valence atomic orbitals used in the mathematics. For example, combining two *p*-orbitals with one *s*-orbital creates three unique and equivalent sp^2 (*s*-*p*-two) hybrid orbitals pointing toward the vertices of a triangle surrounding the atom.

ELECTRON SETS, HYBRIDIZATION AND MOLECULAR GEOMETRIES

Number of Electrons Sets	Electron "Set" Geometry	Number of Lone Pairs	Molecular Geometry	Hybridization	Appearance
2	Linear	–	Linear	sp	
		0	Trigonal Planar		
3	Trigonal (Triangular) Planner	1	Bent or V-shaped	sp^2	
		2	Linear		
		0	Tetrahedral		
4	Tetrahedral	1	Trigonal Pyramid	sp^3	
		2	Bent or V-shaped		
		3	Linear		

Table 2.

Valence electron sets (lone pairs and electrons in sigma bonds) are "housed," at least in part, in hybrid orbitals. This means that an atom surrounded by three electron sets uses three hybrid orbitals, as in formaldehyde. There, the central carbon atom uses hybrid orbitals in forming the C–H single bonds and the sigma portion of the C=O double bond. The carbon's remaining unhybridized *p*-orbital overlaps a *p*-orbital on the oxygen, creating the pi bond that completes the carbon-oxygen double bond. The H–C–O and H–C–H angles are 120°, as is found among sp^2 hybridized orbitals in general. The hybridizations required for two, three, and four electron sets are given in Table 2, along with their corresponding electron geometries. SEE ALSO ISOMERISM; LEWIS STRUCTURES; MOLECULES; NUCLEAR MAGNETIC RESONANCE.

Mark Freilich

Molecular Modeling

A model is a semblance or a representation of reality. Early chemical models were often mechanical, allowing scientists to visualize structural features of molecules and to deduce the stereochemical outcomes of reactions. The disadvantage of these simple models is that they only partly represent (model) most molecules. More sophisticated physics-based models are needed; these other models are almost exclusively computer models.

Two major categories of physics-based, computational molecular models exist: macroscopic and microscopic. Macroscopic models describe the coarse-grained features of a system or a process but do not describe the atomic or molecular features. Microscopic or atomistic models take full account of all atoms in the system.

Atomistic modeling can be done in two ways: by applying theory or by using fitting procedures. The fitting procedures are attempts to rationalize connections between molecular structure and physicochemical properties (quantitative structure property relationships, QSPR), or between molecular structure and biological response (quantitative structure activity relationships, QSAR). Usually, a molecule's biological response is regressed onto a set of molecular descriptors.

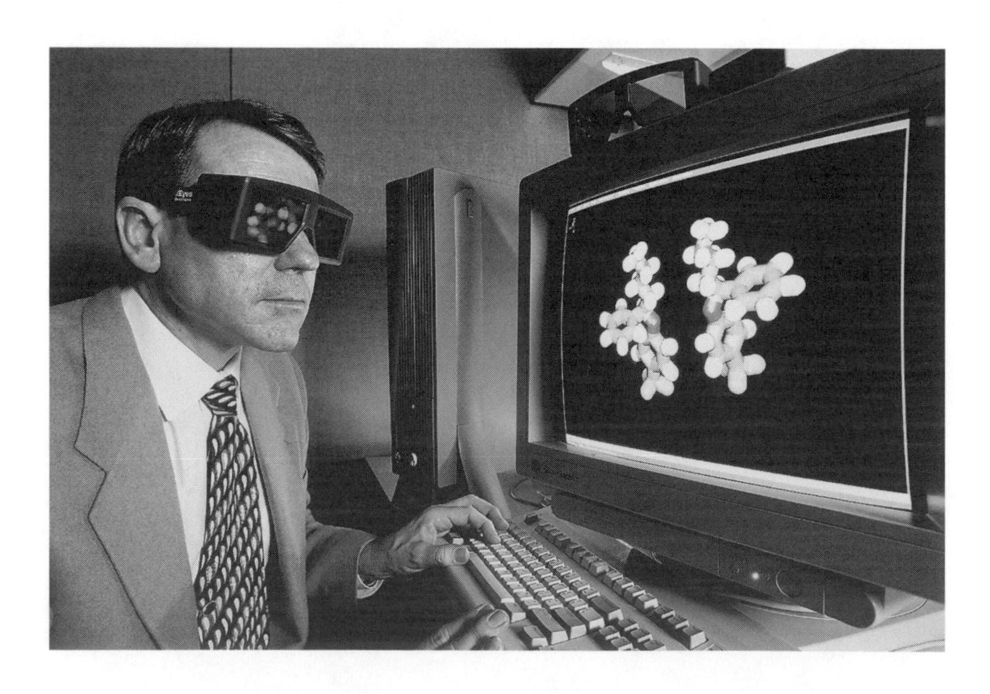

Michael Chaney of Lilly Research Labs wearing 3-D glasses to view a computer model of Fluoxetine, or Prozac. The glasses dim and brighten in response to the flashing of the computer monitor.

$$\log(1/C) = b_0 + \sum_i b_i D_i$$

Here C is the minimum concentration of a compound that elicits a response to an assay of some sort (e.g., an LD_{50}, or something else), b_0 is a constant, b_i is the least-squares multiple regression coefficient, and D_i is the molecular descriptor. For example, the best model for determining the retention index (RI) of drug molecules on a gas-liquid **chromatography** column was found to be RI = 9.92 MW − 3.11 (number of ring atoms) + 139 (number of ring nitrogens) + 296 (total σ charge) + 921 Σ atomic IDs on nitrogen − 335 ^6XCH − 211 ^3XC − $49^2\kappa_\alpha$ − 1958 (X and κ are topological and topographical descriptors). One can then predict an unknown drug's RI very accurately by substituting the values of the descriptors for that drug into the above equation. There are no rules about what kind of descriptors may or may not be used, but descriptors often include information about molecular size, shape, electronic effects, and lipophilicity. Microscopic modeling based on fitting methodologies requires the use of existing data to create a model. The model is then used to predict the properties or activities of as yet unknown molecules.

chromatography: the separation of the components of a mixture in one phase (the mobile phase) by passing it through another phase (the stationary phase) making use of the extent to which the components are absorbed by the stationary phase

The other approach to microscopic molecular modeling implements theory, and uses various sampling strategies to explore a molecule's potential energy surface (PES). Knowing a molecule's PES is convenient, because one can interpret directly from it the molecule's shape and reactivity. Popular modeling tools used for determining PES are shown in Figure 1.

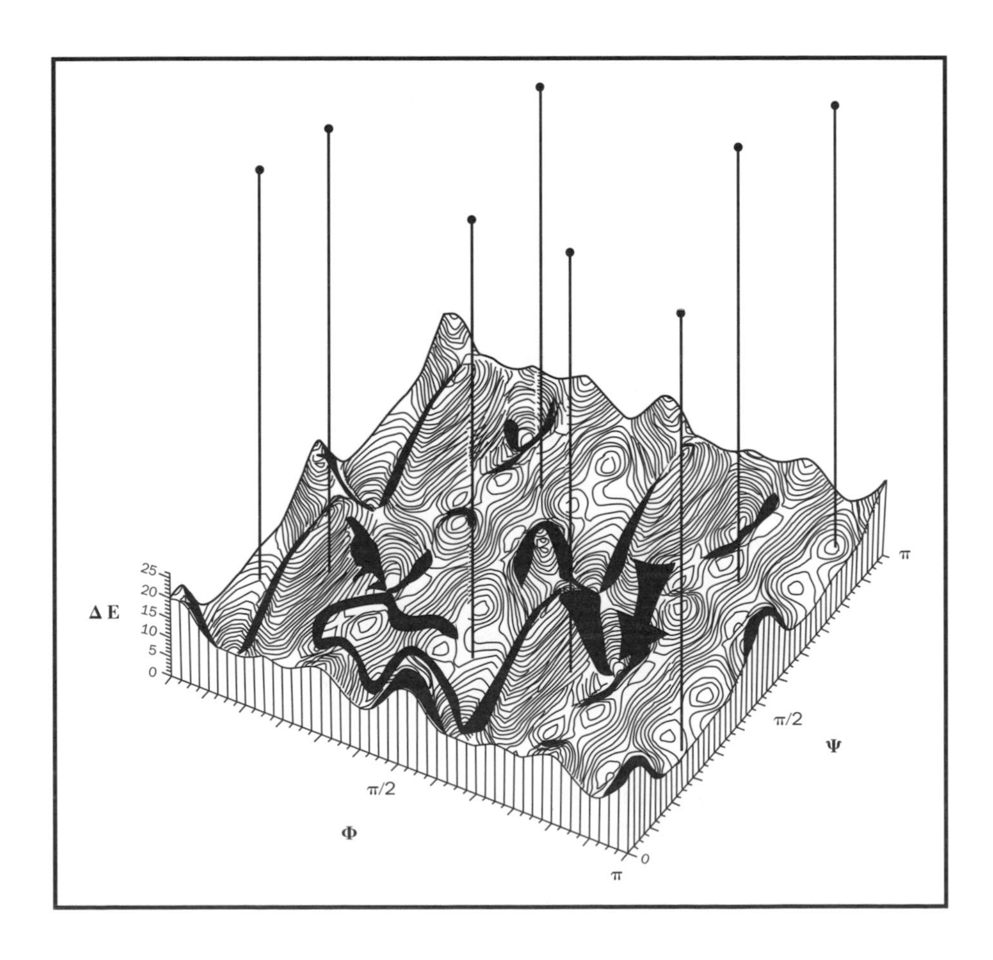

Figure 1. A PES illustrating the differences between modeling methods. Energy minimization is portrayed by the large black arrow. Only minima near the initially guessed structure are located. Monte Carlo methods randomly select many points on the PES (dropped lines), whereas Molecular Dynamics moves over the surface (winding line)

Quantum Mechanics (QM). The objective of QM is to describe the spatial positions of electrons and nuclei. The most commonly implemented QM method is the molecular orbital (MO) theory, in which electrons are allowed to flow around fixed nuclei (the Born-Oppenheimer approximation) until the electrons reach a self-consistent field (SCF). The nuclei are then moved, iteratively, until the energy of the system can go no lower. This energy minimization process is called geometry optimization.

Molecular Mechanics (MM). Molecular mechanics is a non-QM way of computing molecular structures, energies, and some other properties of molecules. MM relies on an empirical force field (EFF), which is a numerical recipe for reproducing a molecule's PES. Because MM treats electrons in an implicit way, it is a much faster method than QM, which treats electrons explicitly. A limitation of MM is that bond-making and bond-breaking processes cannot be modeled (as they can with QM).

Newtonian: based on the physics of Isaac Newton

Molecular Dynamics (MD). Energy-minimized structures are motionless and, accordingly, incomplete models of reality. In molecular dynamics, atomic motion is described with **Newtonian** laws: $F_i(t)=m_i a_i$, where the force F_i exerted on each atom a_i is obtained from an EFF. Dynamical properties of molecules can be thus modeled. Because simulation periods are typically in the nanosecond range, only inordinately fast processes can be explored.

Monte Carlo (MC). The same EFFs used in the MM and MD methods are used in the Monte Carlo method. Beginning with a collection of particles, the system's initial energy configuration is computed. One or more particles are randomly moved to generate a second configuration, whose energy is "accepted" for further consideration, or "rejected," based on energy criteria. Millions of structures on the PES are sampled randomly. Averaged energies and averaged properties are thus obtained.

Most atomistic modeling involves the exploration of a complex and otherwise unknown PES. Simple energy minimization with QM or MM locates a single, stable structure (the local minimum) on the PES, which may or may not be the most stable structure possible (the global minimum). MD and MC sampling methods involve a more complete searching of the PES for low energy states and, accordingly, are more time-consuming. SEE ALSO Quantum Chemistry; Theoretical Chemistry.

Kenneth B. Lipkowitz
Jonathan N. Stack

Bibliography

Leach, A. R. (2001). *Molecular Modeling: Principles and Applications*, 2nd edition. Englewood Cliffs, NJ: Prentice-Hall.

Lipkowitz, Kenneth B., and Boyd, D. B., eds. (1990–2002). *Reviews in Computational Chemistry*, Vols. 1–21. New York: Wiley-VCH.

Von Rague Schleyer, P., ed. (1998). *Encyclopedia of Computational Chemistry*, Vols. 1–5. Chichester, U.K.: John Wiley.

Molecular Orbital Theory

The molecular orbital (MO) theory is a way of looking at the structure of a molecule by using *molecular* orbitals that belong to the molecule as a whole

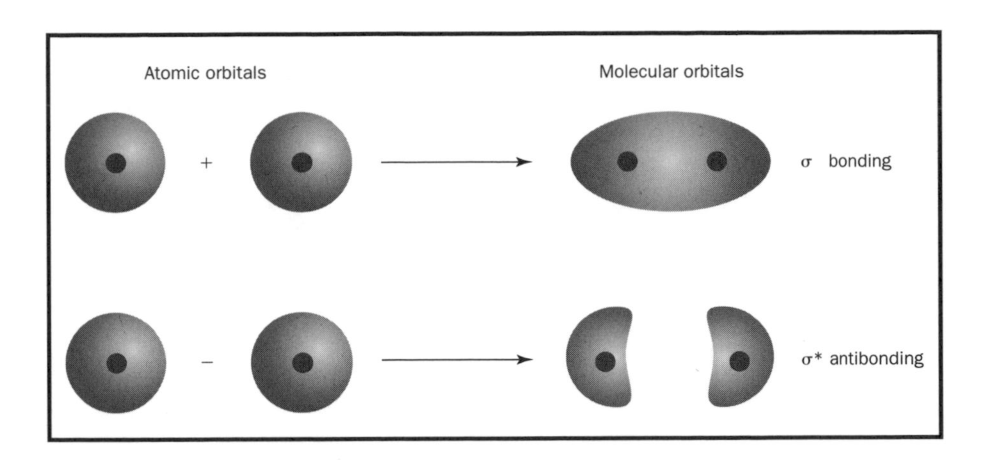

Figure 1. Combination of two 1*s* atomic orbitals to form a sigma bonding orbital or a sigma-starred antibonding orbital.

rather than to the individual atoms. When simple bonding occurs between two atoms, the pair of electrons forming the bond occupies an MO that is a mathematical combination of the wave functions of the atomic orbitals of the two atoms involved. The MO method originated in the work of Friedrich Hund and Robert S. Mulliken.

When atoms combine to form a molecule, the number of orbitals in the molecule equals the number of orbitals in the combining atoms. When two very simple atoms, each with one atomic orbital, are combined, two molecular orbitals are formed. One is a *bonding* orbital, lower in energy than the atomic orbitals, and derived from their sum. It is called *sigma.* The other is an *antibonding* orbital, higher in energy than the atomic orbitals, and resulting from their difference. It is called *sigma-starred* (σ^*). (See the diagram in Figure 1.)

The basic idea might best be illustrated by considering diatomic molecules of hydrogen and helium. The energy diagrams are shown in Figure 2. Each hydrogen atom has one 1*s* electron. In the H_2 molecule the two hydrogen electrons go into the lowest energy MO available, the sigma orbital. In the case of helium, each helium atom has two electrons, so the He_2 molecule would have four. Two would go into the lower energy bonding orbital, but the other two would have to go into the higher energy sigma-starred orbital.

Figure 2. Molecular orbital energy diagrams for (a) H_2 showing both electrons in the bonding sigma MO; and (b) He_2 in which two of the electrons are in the antibonding sigma-starred MO

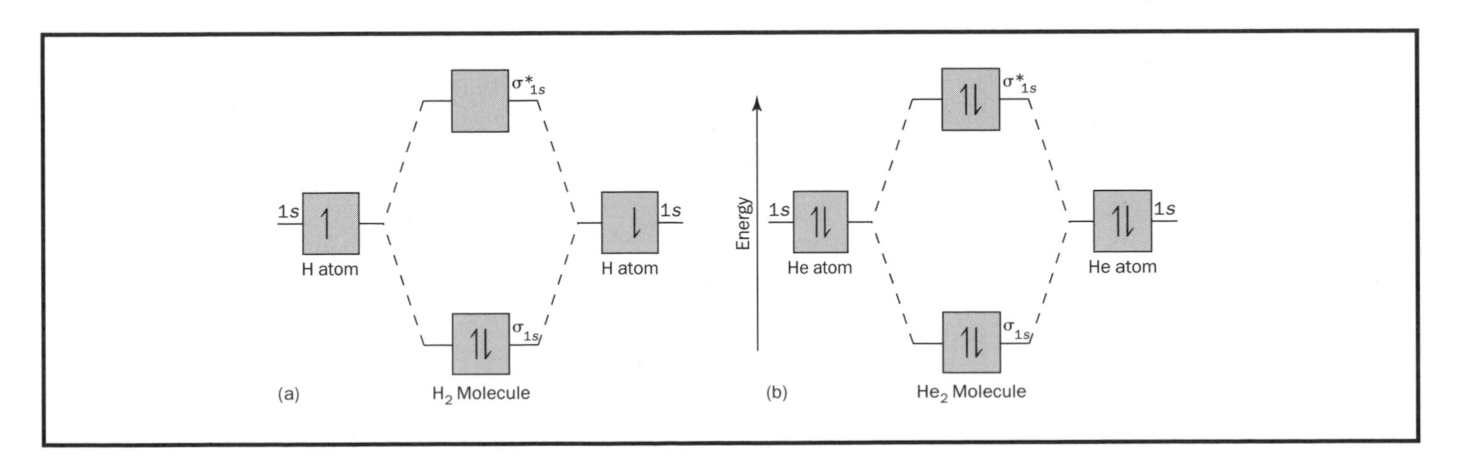

Figure 3. Combination of *p* atomic orbitals to form (a) sigma MOs by end-to-end interactions or (b) pi MOs by sideways interaction.

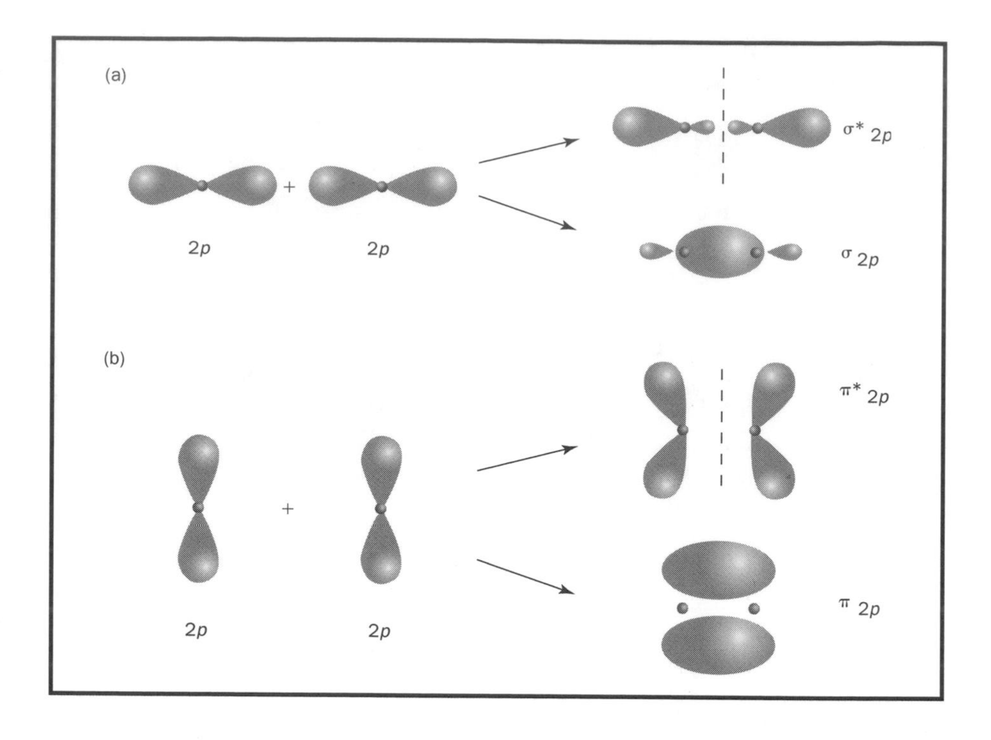

Bond Order

The bond order for a molecule can be determined as follows: bond order = $\frac{1}{2}$ (bonding electrons − antibonding electrons).Therefore, the H_2 molecule has a bond order of $\frac{1}{2}(2 - 0) = 1$. In other words, there is a single bond connecting the two H atoms in the H_2 molecule. In the case of He_2, on the other hand, the bond order is $\frac{1}{2}(2 - 2) = 0$. This means that He_2 is not a stable molecule.

Multiple Bonds

Double or triple bonds involve two or three pairs of bonding electrons. Single bonds are always sigma bonds, but in multiple bonds the first bond is sigma, while any second or third bonds are pi bonds. The overlap of *p* orbitals can yield either pi or sigma MOs, as shown in Figure 3. When they overlap end to end, they form sigma orbitals, but when they overlap side to side, they form pi orbitals.

Consider now the oxygen molecule. The Lewis structure for oxygen is :Ö::Ö: The double bond is necessary in order to satisfy the octet rule for both oxygen atoms. The measured bond length for oxygen supports the presence of a double bond. Yet we know that this Lewis formula cannot be the correct structure for oxygen because oxygen is paramagnetic, which means that the oxygen molecule must have unpaired electrons.

Look now at the MO diagram for oxygen, which is shown in Figure 4. It still indicates a bond order of 2 [$\frac{1}{2}(10 - 6) = 2$], but it also shows two unpaired electrons.

The MO theory also works well for larger molecules such as N_2O, CO_2, and BF_3 and for ions such as NO_3^- and CO_3^{2-}, in which the bonding MOs are delocalized, involving three or more atoms. It is especially useful for aro-

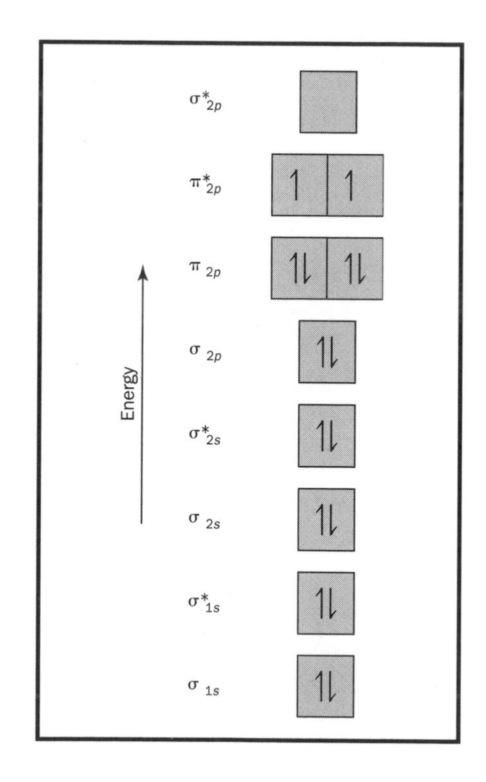

Figure 4. MO energy diagram for O_2. Eight electrons from each oxygen atom add up to 16 electrons in the O_2 molecule. They combine to form the molecular orbitals indicated above.

matic molecules such as benzene. In this case all six C atoms in the ring are equally involved in a delocalized pi electron cloud that envelops the entire molecule. The MO theory can even be extended to complex ions and to solids, including materials such as superconductors and semiconductors. SEE ALSO BONDING; LEWIS STRUCTURES; VALENCE BOND THEORY.

Doris K. Kolb

Bibliography

Chang, Raymond (2002). *Chemistry*, 7th edition. New York: McGraw Hill.

deKock, Roger L., and Gray, Harry B. (1980). *Chemical Structure and Bonding*. Menlo Park, CA: Benjamin/Cummings.

Levine, Ira N. (1991). *Quantum Chemistry*, 4th edition. Englewood Cliffs, NJ: Prentice Hall.

Solomons, T. W. Graham (1997). *Fundamentals of Organic Chemistry*, 5th edition. New York: John Wiley.

Molecular Structure

The Rise and Reemergence of Atomism

Throughout history, humans have created models to help them explain the observed character of substances and phenomena in the material world. The ancient philosophers Democritus and **Lucretius** were among the first to speculate that matter was discontinuous, and that small, indivisible particles not only made up substances but also gave them their observed properties. The Greeks called these particles "atoms" (the English equivalent), a word that meant indivisible. Lucretius imagined that the particles that made up vapor had smooth surfaces and could not interconnect, giving vapors (gases) their extreme mobility. Liquids, on the other hand, were thought to be made up of particles, each particle having a few hooks. These few hooks would get entwined but would not immobilize the particles, thereby causing the particles to cling, yet still be fluid. The particles that made up solids, by contrast, were thought to have many hooks, resulting in the extremely sturdy nature of solid materials. The hypothesis of finite particles implied empty space between them. Yet, the majority of Greek philosophers did not believe that nothingness (the vacuums between particles) could exist, so the idea of atoms did not last long in the ancient times. Ironically, the objection was not to the existence of particles, but to the vacancies that must exist between them.

Lucretius: Roman poet of first century B.C.E., also known as Titus Carus; author of *De rerum natura*

Most cultures have linked properties of matter with religious and/or superstitious ideas. The term "gold" derives from an Old English word meaning "something shiny and yellow like the Sun"; it served not only as the name of the **metal** but also identified its properties. Polished gold nearly captures the sunlight it reflects, and the astronomical, astrological, medical, and religious attributes of the Sun were thought to be present in gold metal. For thousands of years, substances were said to contain essences or essential parts that gave them their characters. In a sense modern ideas about molecular structure do something similar. Chemists construct explanations for observed, **macroscopic phenomena** (e.g., reactivity) by describing the assemblages, shapes, and motions of submicroscopic particles.

metal: element or other substance the solid phase of which is characterized by high thermal and electrical conductivities

macroscopic phenomena: events observed with human vision unassisted by instrumentation

The theory of atoms did not reemerge until the seventeenth century. The discovery of elements rapidly led to the idea that nonelementary substances were made up of molecules that were, in turn, collections of elemental atoms. During the first years of chemical analyses, different substances were observed to have different compositions; the deduction was made that substances were different because their compositions were different. One type of mineral might be 34 percent iron and 66 percent oxygen. Each sample of that mineral would give the same results (34% iron and 66% oxygen). A different mineral, that is, one with different properties, might be 56 percent iron and 44 percent oxygen. Although there was still no concept of bonding between atoms or of molecular geometry at the beginning of the nineteenth century, chemists had developed the idea that different molecules were different collections of atoms.

Isomerism and the Development of Molecular Structural Models

elemental analysis: determination of the percent of each atom in a specific molecule

Scientific theories are sometimes discarded. When information that contradicts a theory is reliable, the theory must be changed to fit the new data. As the **elemental analysis** of compounds expanded greatly during the early 1800s, observations that different substances were of the same elemental composition were inevitable. In his *History of Chemistry* (1830), Thomas Thomson drew illustrations of varying hypothetical particle arrangements, using symbols that were used at that time (those of John Dalton), as a way to explain why two acids of the same elemental composition could have different physical and chemical properties (see Figure 1). These are believed to be the earliest recorded representations of molecular structure that showed varying arrangements of the same atoms; the phenomenon would soon be called isomerism (from the Greek *iso*, meaning same, and *meros*, meaning part). In 1828 Friedrich Wöhler (1800–1882) synthesized urea, $(NH_2)_2C{=}O$ or CH_4N_2O, that was indistinguishable from that that had been isolated from urine. He prepared this organic substance from the clearly inorganic (mineralogical) starting material ammonium cyanate, $NH_4(+)\ NCO(-)$, also CH_4N_2O, the result of the combination of ammonium chloride and silver cyanate. Urea and ammonium cyanate are **constitutional isomers**, and together illustrate the fact that fixed arrangements of atoms, molecular structures, must be invoked to explain observed phenomena.

constitutional isomer: form of a substance that differs by the arrangement of atoms along a molecular backbone

The constitution of a molecule (number of, kind of, and connectivities of atoms) may be represented by a two-dimensional "map" in which the interatomic linkages (bonds) are drawn as lines. There are two constitutional isomers that are represented by the molecular formula C_2H_6O: ethanol and dimethyl ether. The differences in connectivities, which are not evident in the common constitutional inventory C_2H_6O, can be conveyed by typographical line formulas (CH_3CH_2OH for ethanol and CH_3OCH_3 for dimethyl ether), or by structural representations (see Figure 2). As the number and kinds of atoms in substances increase, the number of constitutional isomers increases.

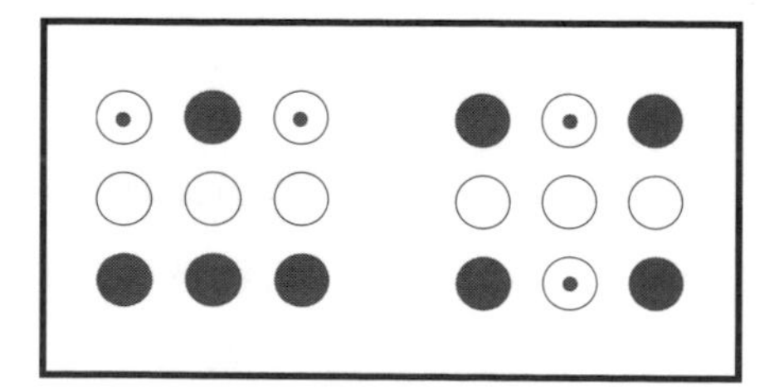

Figure 1. Early representations for isomers using Daltonian symbols.

By the mid-1850s, a new theory of molecular structure had emerged. Given a unique collection of atoms, it was not the identities of the atoms that distinguished one molecule from another, but rather the connectivity,

```
    H  H           H       H
    |  |           |       |
H—C—C—O—H      H—C—O—C—H
    |  |           |       |
    H  H           H       H
```

Figure 2. Structural representations for ethanol (CH_3CH_2OH) and dimethyl ether (CH_3OCH_3)

or bonding, of those atoms. The nature of the chemical bond was unknown, and the phenomenon of chemical bonding was described as "chemical affinity." Because it was observed that the passing of electricity through some substances, such as water, could "break" the molecules apart into their elements (electrolysis), the electrostatic attractions of charged particles (ions) were used to contribute to an explanation of chemical affinity. Just as the hypothesis of the varying connectivities of atoms emerged as a response to observations that could not be explained, variation in the three-dimensional arrangements of atoms in space was proposed to reconcile other observed phenomena. Jacobus van't Hoff (1852–1911) and Joseph-Achille Le Bel (1847–1930) proposed (independently of one another, in 1874) that molecules of the same connectivity yet different physical properties (e.g., optical activity) might be explained if, in the case of four different particles, the arrangement (configuration) of the particles was tetrahedral. Macroscopically or microscopically, a tetrahedral array of four different things gives rise to two and only two different arrangements that are nonsuperimposable mirror images (enantiomers; see Figure 3). Distinct molecular structural units that have the same connectivities but varying three-dimensional arrangements are also isomers. The term "stereoisomer" was introduced by Viktor Meyer in 1888 to describe molecules that differ only in their three-dimensional arrangements.

Connectivity and stereoisomerism give chemists a way to uniquely differentiate one molecular structure from another. The molecular formula C_4H_9Br, for instance, represents five different substances (see Figure 4). Predictably, although there is only one compound for each of the connectivities designated 1-bromobutane, 2-bromo-2-methylpropane, and 1-bromo-2-methylpropane, there are two compounds represented by the connectivity designated 2-bromobutane (carbon 2 has four different groups attached, and thus two three-dimensional arrangements of the molecule, whose geometries are labeled *R* and *S*, exist). There are no other isomers of C_4H_9Br that are predicted, and none that are observed.

Although the arrangement of molecular atoms around a given point is fixed, molecules are not static objects. The sequence of links in a chain, for instance, is constant, but the chain can be twisted and knotted into countless shapes. In the case of a molecule, twists do not affect the identity of a substance, but the overall molecular shape is part of molecular structure and can have an impact on the observed properties. According to Ernest Eliel and Samuel Wilen (1994, p. 102), *configurational stereoisomers* result from "arrangements of atoms in space of a molecule with a defined constitution, without regard to arrangements that differ only by rotation about one or more single bonds, providing that such a rotation is so fast as not to allow isolation of the species so differing." Conformational stereoisomers are

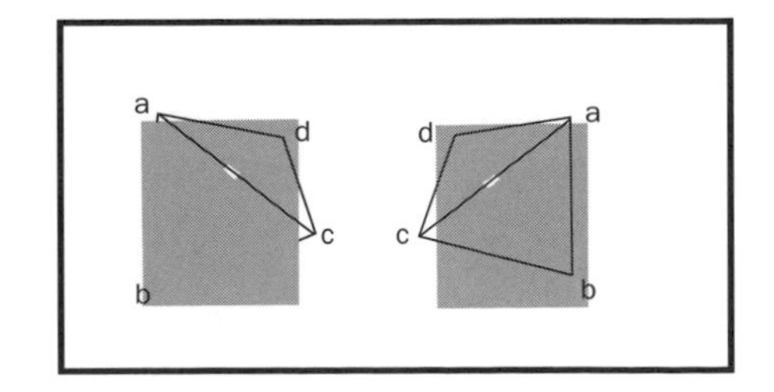

Figure 3. Tetrahedral arrays of four different objects create enantiomorphic shapes.

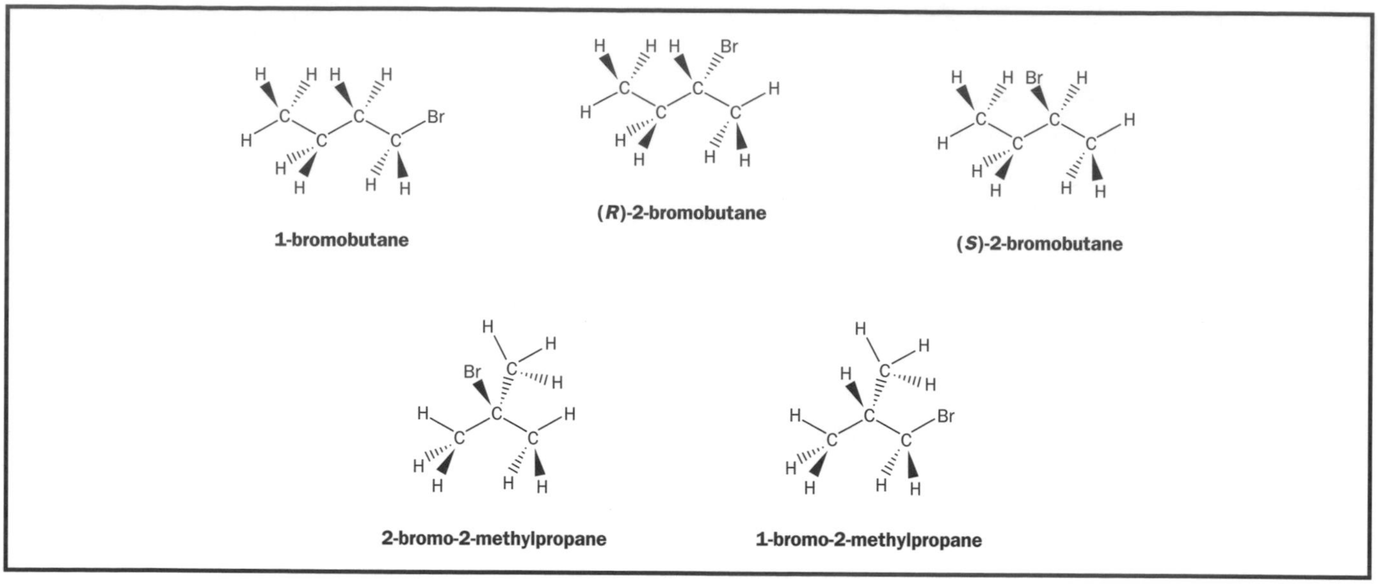

1-bromobutane

(*R*)-2-bromobutane

(*S*)-2-bromobutane

2-bromo-2-methylpropane

1-bromo-2-methylpropane

Figure 4. Structural representations for the five different C_4H_9Br molecules.

molecular identity: "fingerprint" of a molecule describing the structure

molecular shapes resulting from bond rotations that do not affect **molecular identity**. The drawings shown in Figure 5 represent some of the different conformational shapes that the single molecule (*S*)-2-bromobutane can assume.

The overall geometry of a molecule was recognized as contributing to its chemical reactivity in the 1950s, and methods used to determine molecular structure have grown dramatically since that time. Throughout the early 1900s, direct experimental evidence of the three-dimensional arrangements of atoms was becoming available as a result of x-ray diffraction crystallography. **Nuclear** magnetic resonance **spectroscopy** (first used in the 1960s) and atomic force microscopy (in the 1980s) are two techniques of many that are now used to gather experiment-based information about molecular structure. What might have taken years to determine in 1950, and what was impossible to know about extremely large biopolymers (e.g., **DNA**, enzymes, and polysaccharides at a cell surface) as late as 1990 can now sometimes be determined in a matter of seconds.

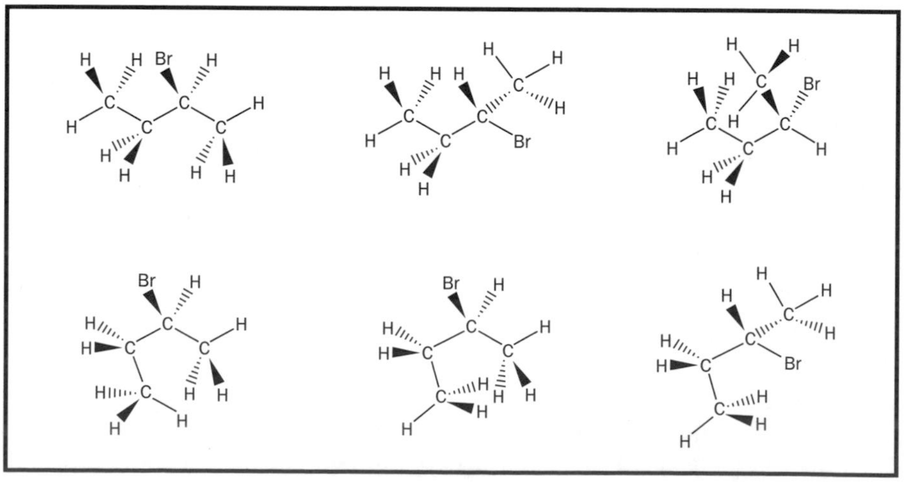

nuclear: having to do with the nucleus of an atom

spectroscopy: use of electromagnetic radiation to analyze the chemical composition of materials

DNA: deoxyribonucleic acid—the natural polymer that stores genetic information in the nucleus of a cell

Figure 5. Conformational isomers of (*S*)-2-bromobutane.

Molecular environment influences molecular structure. The shape that a molecule assumes within a **crystal lattice** is necessarily different from its shape in water and will vary according to solvent and other environmental factors (e.g., temperature and pH). Beginning in the late 1980s the significance of the **noncovalent aggregation** of large numbers of molecular entities began to be understood. A protein, for instance, folds into its three-dimensional shape because water is present; without water, the shape is quite different. Thus, molecular structure is determined by a combination of extrinsic as well as intrinsic factors. The field of molecular structure and reactivity that deals with large aggregations of molecules and how they influence each other is called supramolecular chemistry.

crystal lattice: three-dimensional structure of a regular solid

noncovalent aggregation: nonspecific interaction leading to the association of molecules

Molecular Structural Theory

The electron was discovered in 1900, and it took about twenty years for the electronic nature of the chemical bond to come into wide acceptance. Particle-based models for atomic and molecular structure soon gave way to the **quantum mechanical** view, in which electrons are not treated as localized, discrete particles (electrons orbiting around a nucleus), but as **delocalized** areas of wavelike charge, each possessing a given probability of being found in a given location near an atomic nucleus (an orbital). The chemical bonding in molecules, which began the twentieth century as shared electron pairs between atoms, evolved to become a matter of molecular orbitals. Molecular orbitals describe three-dimensional arrangements of the atomic nuclei in a molecule and the probability that any given electron of a given energy will occupy a given location with respect to those nuclei. Single bonds are explained by the overlap of **atomic orbitals** along the **internuclear** axis of two atoms. Multiple bonds are the combination of sigma plus pi bonding, the latter corresponding to the overlap of atomic orbitals that is not along the internuclear axis. A rough guide to the bonding molecular orbitals in methane is depicted in Figure 6. The eight **valence** shell electrons (four from carbon, four from the four hydrogens) are

quantum mechanical: theoretical model to describe atoms and molecules by wave functions

delocalized: of a type of electron that can be shared by more than one orbital or atom

atomic orbital: mathematical description of the probability of finding an electron around an atom

internuclear: distance between two nuclei

valence: combining capacity

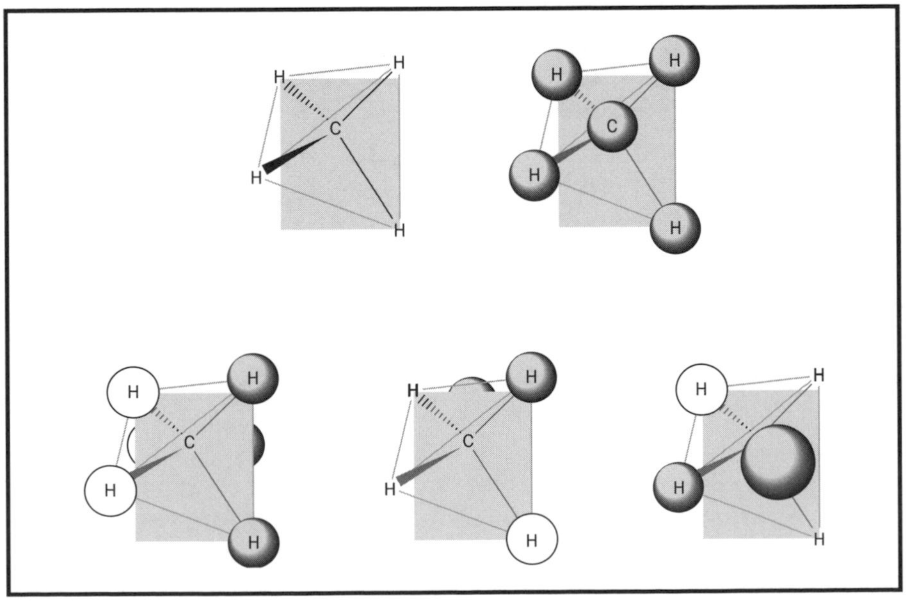

Figure 6. Representations for the bonding molecular orbitals in methane.

distributed among four molecular orbitals. One of the four orbitals is composed of favorable bonding interactions between the 2*s*-orbital of carbon and the four 1*s*-orbitals of the hydrogen atoms, whereas the other three are the equally likely combinations of one of the three 2*p*-orbitals of carbon and the 1*s* orbitals of hydrogen atoms. Computer-based models for chemical bonding are as important to modern molecular structural theory as experimental measurements. SEE ALSO ISOMERISM; LE BEL, JOSEPH-ACHILLE; MOLECULES; NUCLEAR MAGNETIC RESONANCE; VAN'T HOFF, JACOBUS; WÖHLER, FRIEDRICH.

Brian P. Coppola

Bibliography

Eliel, Ernest L., and Wilen, Samuel H. (1994). *Stereochemistry of Organic Compounds.* New York: Wiley.

Thomson, Thomas (1830). *The History of Chemistry.* London: Colburn and Bentley.

Molecules

A molecule is the smallest entity of a pure compound that retains its characteristic chemical properties, and consequently has constant mass and atomic composition. It is an assembly of nonmetallic atoms held together into specific shapes by **covalent bonds**. As much as a car is a single unit made up of many parts, a molecule is a unit made up of atoms bonded around each other in certain fixed geometries. Shapes influence the physical and chemical properties and consequently much of the chemistry of a molecule.

covalent bond: bond formed between two atoms that mutually share a pair of electrons

While molecules may be monoatomic (such as the **inert** gases helium, neon, or krypton), most molecules are diatomic, triatomic, or polyatomic, consisting of two or more atoms (some molecules may be a collection of thousands of atoms). A diatomic molecule may be **homonuclear** (e.g., O_2 or N_2) or **heteronuclear** (e.g., CO or NO). Similarly, a triatomic molecule may be homonuclear (e.g., O_3) or heteronuclear (e.g., HCN).

inert: incapable of reacting with another substance

homonuclear: having identical nuclei

heteronuclear: having different nuclei

The modern concept of the covalent bond has resulted in the ability to predict the geometry and hence the properties of matter such as reactivity, toxicity, and solubility. A fundamental challenge in chemistry is to determine the arrangement of atoms in a molecule in order to elucidate its bonding, geometry, and properties.

Historical Development

Since Roman times matter had been viewed by some as discrete particles somehow linked together. Early in the eighteenth century the behavior of gases was viewed as a function of **kinetic theory**. Kinetic theory is a group of assumptions to explain the behavior of gases. Among these assumptions are that gases are individual molecules moving in straight lines, that they do not react chemically and occupy essentially no volume compared to the volume between molecules. In 1805 English chemist and physicist John Dalton (1766–1844) proposed that atoms form compounds by joining together in simple, whole numbers. In 1811 Italian chemist Amedeo Avogadro (1776–1856) solidified the distinction between molecules and atoms by proposing that, at constant temperature and pressure, equal volumes of all

kinetic theory: theory of molecular motion

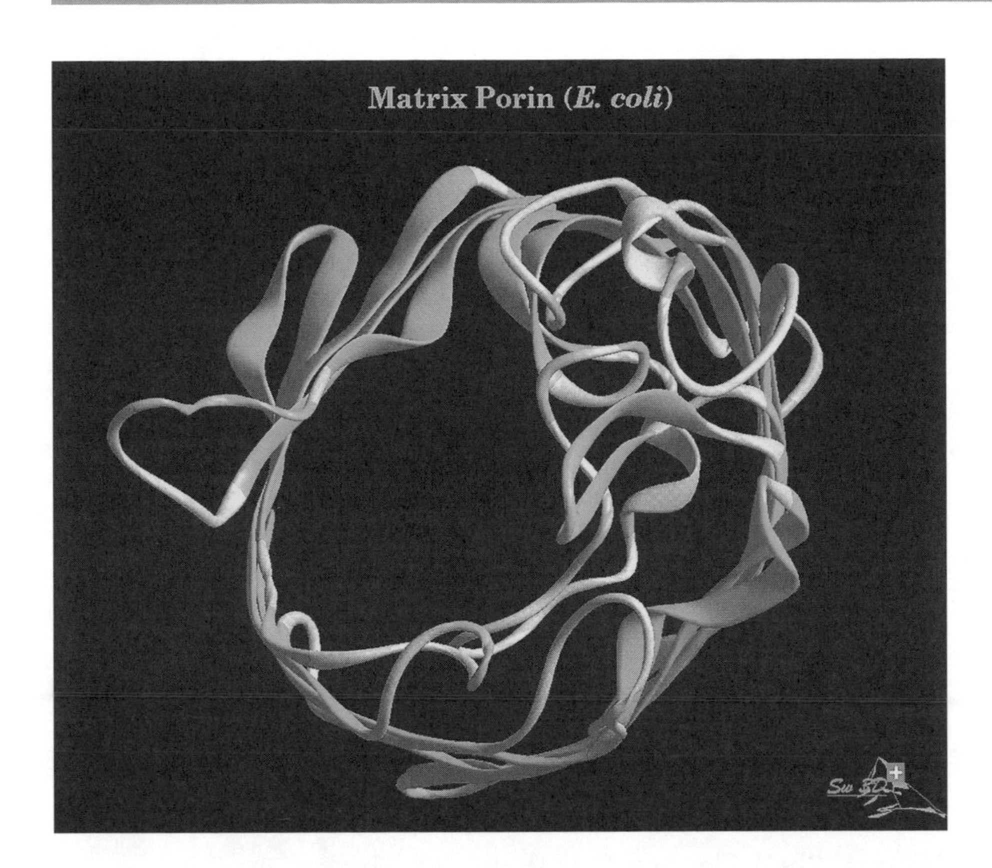

A three-dimensional computer model of a protein molecule of matrix porin found in the *E. coli* bacteria.

gases contain equal numbers of molecules. While Avogadro's theory was published, it was ignored by the scientific community until 1858, when it was revived by Italian chemist Stanislao Cannizzaro (1826–1910), thereby reconciling many inconsistencies chemists were observing. During this same time, valency (the combining capacity of an atom) was defined as the number of hydrogens an atom can combine with.

Initially the structure of molecules was studied using chemical methods, thereby identifying composition, chemical reactions, and the existence of **isomers**. It was understood that bonds had direction, rigidity, and a certain degree of independence from molecule to molecule. The discovery of the electron in 1897 by English physicist Joseph John Thomson (1856–1940) immediately linked electrons with covalent bonding. Though attacked vigorously for his views, Dutch physical chemist Jacobus Hendricus van't Hoff (1852–1911) discarded the flat-molecule model in favor of geometric relations within each molecule. His brilliant postulate of the tetrahedral arrangement of carbon (proposed simultaneously, but independently, by French chemist Joseph-Achille Le Bel [1847–1930]) was a major breakthrough for chemistry. Later in the nineteenth century the advent of physical methods of investigation led to a great deal of additional information regarding atomic configuration.

isomer: molecules with identical compositions but different structural formulas

Danish physicist Niels Bohr (1885–1962) proposed a quantum theory of the hydrogen atom by suggesting that the electron moves about its nucleus in discrete **quanta** (the energies of electrons are restricted to having only certain values, quanta, much as stairs do as opposed to a ramp), establishing a balance between the electron's centrifugal force and its **attraction** for the nucleus. It was not until 1927 that covalent bonding was properly

quantum: smallest amount of a physical quantity that can exist independently, especially a discrete amount of electromagnetic energy

attraction: force that brings two bodies together, such as two oppositely charged bodies

understood, thanks to the contributions of American physical chemist Gilbert N. Lewis (1875–1946), American physicist Edward Uhler Condon (1902–1974), German physicist Walter Heitler (1904–1981), and German physicist Fritz London (1900–1954).

In his 1916 paper *The Atom and the Molecule*, Lewis proposed that a chemical (covalent) bond between two atoms involves the sharing of electrons between the nuclei. Thus a single bond (for hydrogen, H-H) results when an electron from each atom forms an electron pair that is shared between the two nuclei (H:H); a double bond involves two electrons from each atom (e.g., the carbon-carbon bond in $(H:)_2C::C(:H)_2$); and a triple bond involves three electrons from each atom (e.g., the carbon-carbon bond in H:C:::C:H). Such representations are referred to as Lewis dot structures. Lewis further postulated that an electron octet (and in a few cases an electron pair) forms a complete shell of electrons with spatial rigidity and chemical inertness—hence a stable arrangement.

American chemist Irving Langmuir (1881–1957) proposed that many chemical facts could be coordinated by applying these new ideas. Others followed by suggesting that a bond is a balance between nucleus-nucleus and electron-electron repulsions and electron-nuclei attractions. American chemist Linus Pauling (1901–1994) assembled these ideas in his seminal book, *The Nature of the Chemical Bond.*

Valence Shell Electron Pair Repulsion Theory

Molecular geometries are determined by the number and locations of valence electrons around the atoms. Both bonded and lone pair electrons repel each other, staying as far apart as possible, thereby causing the molecule to occupy specific shapes (much as balloons assume fixed arrangements when tied together). These geometries are important in determining chemical properties. One method for determining the structure of covalent molecules is the valence shell electron pair repulsion (VSEPR) method, proposed in 1957 by Canadian chemist Ronald Gillespie and Australian chemist Ronald Nyholm in a classic paper titled "Inorganic Stereochemistry." The theory states that the geometry around a given atom is a function primarily of minimizing the electron pair repulsions. The key postulates of the VSEPR theory are:

- All electrons are negatively charged.
- Bonds are electron groups.
- Lone pair and bonded electrons (and therefore bonds) repel each other.

Geometries of most covalent molecules may be determined by following these steps:

- Determine the central atom. This may be the atom present singly (e.g., B in BF_3), the larger atom (e.g., P in $POCl_3$), the atom written in the center (e.g., C in HCN), or the atom with the largest number of bonds (e.g., C in Cl_2CO).
- Determine the number of bonds needed for each atom to be bonded to the central atom and write the corresponding Lewis dot structure. Thus, for Cl_2CO, each chlorine needs a single bond and oxygen needs

GEOMETRIES OF MOLECULES WITH VARIOUS BONDED ELECTRON GROUPS

Electron Groups About the Central Atom	Example	Shape
2	$BeCl_2$	linear
3	BCl_3	trigonal planar
4	CH_4	tetrahedral
5	PF_5	trigonal bipyramidal
6	SF_6	octahedral

Table 1. Geometries of molecules with various bonded electron groups

two bonds; the Lewis dot structure would be $(Cl:)_2C::O$. Note that a single bond needs a pair of electrons (one group), a double bond needs two pairs (also one group), and a triple bond needs three pairs (still just one group, since it points in one direction only).

- Count the total number of bonded and lone pair electron groups about the central atom. For Cl_2CO it would be three (all bonded) groups. In the case of $:NH_3$ it would be one lone pair group and three bonded groups for a total of four groups.
- Establish the best electronic (counting all electron groups) and molecular (counting only bonded groups) geometries. Table 1 summarizes this information for bonded groups.

trigonal bipyramidal: geometric arrangement of five ligands around a central Lewis acid, with ligands occupying the vertices of two trigonal pyramids that share a common face; three ligands share an equatorial plane with the central atom, two ligands occupy an axial position

axial bond: covalent bond pointing along a molecular axis

equatorial bond: covalent bond perpendicular to a molecular axis

The **trigonal bipyramidal** shape merits a special note. Contrary to the other shapes, it possesses two types of bonds: the two **axial bonds** located at 180° from each other, and the three **equatorial bonds** located perpendicularly to the axis and at 120° from each other.

Each of the examples given in Table 1 has only bonded electrons around its central atom. The existence of lone pair electrons has an effect on the geometry, as seen in Table 2.

For example, water (H_2O) has two bonded and two lone pair valence electrons about the central atom, oxygen. Its electronic geometry, determined by four total groups, is tetrahedral, and its molecular geometry (meaning the H-O-H shape) is bent. Similarly, the $:NH_3$ molecule has three

Table 2. Electronic and molecular geometries of covalent molecules with bonded and lone pair electrons about the central atom

ELECTRONIC AND MOLECULAR GEOMETRIES OF COVALENT MOLECULES WITH BONDED AND LONE PAIR ELECTRONS ABOUT THE CENTRAL ATOM

Electron Groups About the Central Atom			Shape	
Bonded	Lone Pair	Example	Electronic	Molecular
1	1	TlCl	linear	linear
1	2	BiCl	trigonal	linear
1	3	HCl	tetrahedral	linear
2	1	$SnCl_2$	trigonal	bent
2	2	H_2O	tetrahedral	bent
2	3	XeF_2	trigonal bipyramidal	linear
3	1	PCl_3	tetrahedral	pyramidal
3	2	BrF_3	trigonal bipyramidal	T-shaped
4	1	SF_4	trigonal bipyramidal	distorted tetrahedral
4	2	XeF_4	octahedral	square planar
5	1	$XeOF_4$	octahedral	pyramidal

Models representing the arrangement of atoms in a chemical molecule.

bonded and one lone pair electron groups about nitrogen, giving an electronic geometry that is nearly tetrahedral, and a molecular geometry that is **pyramidal**. Because two bonded pairs repulse less than a bonded pair and a lone pair, which in turn repulse less than two lone pairs, the H-O-H bond angle in water is not 109.5° as expected for a tetrahedron, but 104.5°, with the H-O bonds having been pushed by the lone pairs toward each other. For the trigonal bipyramidal shape, lone pairs always occupy equatorial planar positions. Thus, the molecular geometry of BrF_3 is T-shaped, rather than trigonal planar.

pyramidal: relating to a geometric arrangement of four electron-donating groups at the four vertices of a pyramid

Properties

Both physical and chemical properties are affected by the geometry of a molecule. For instance, the polarity of a molecule is determined by the electronegativity differences of its atoms (electronegativity is the ability of an atom in a molecule to draw electrons toward itself), and the relative geometries of the atoms within the molecule. The molecule BCl_3, for example, displays a flat triangle (120°) with each Cl atom pulling electrons symmetrically, making the molecule **nonpolar**. In the pyramidal molecule PCl_3, however, all chlorines are pulling electrons more or less to one side, making the molecule polar. Since polarity goes hand in hand with solubility, CF_4 is a nonpolar tetrahedral molecule not soluble in water, whereas SF_4, a distorted tetrahedron, is instantly hydrolyzed by water.

nonpolar: molecule, or portion of a molecule, that does not have a permanent, electric dipole

Chemical properties are also very dependent on geometries. For example, in the square planar molecule $Pt(NH_3)_2Cl_2$, the chloro (and hence the ammonia) **ligands** may be placed adjacent to each other (cis isomer), or they may be opposite each other (trans isomer). In addition to having different physical properties, their chemical reactivities are also quite remarkable. The cis isomer is an effective treatment of testicular, ovarian,

ligand: molecule or ion capable of donating one or more electron pairs to a Lewis acid

and certain other cancers, whereas the trans isomer is ineffective. Similarly, the linear, nonpolar CO_2 molecule is inert, whereas the polar CO molecule is a poison.

Other Theories

The VSEPR theory allows chemists to successfully predict the approximate shapes of molecules; it does not, however, say why bonds exist. The **quantum mechanical** valence bond theory, with its overlap of **atomic orbitals**, overcomes this difficulty. The resulting hybrid orbitals predict the geometries of molecules. A quantum mechanical graph of radial electron density (the fraction of electron distribution found in each successive thin spherical shell from the nucleus out) versus the distance from the nucleus shows maxima at certain distances from the nucleus—distances at which there are higher probabilities of finding electrons. These maxima correspond to Lewis's idea of shells of electrons.

quantum mechanical: theoretical model to describe atoms and molecules by wave functions

atomic orbital: mathematical description of the probability of finding an electron around an atom

This theory, however, treats electrons as localized, does not account for unpaired electrons, and does not give information on bond energies. The molecular orbital theory attempts to solve these shortcomings by considering nuclei arranged as in a molecule and determining the resulting molecular orbitals when electrons are fed in one by one.

The electronic and molecular geometries of covalent molecules, and hence their resulting polarities, can thus be predicted fairly accurately. Armed with these tools, one can predict whether or not a molecule should be soluble, reactive, or even toxic. SEE ALSO BONDING; AVOGADRO, AMEDEO; BOHR, NIELS; CANNIZZARO, STANISLAO; DALTON, JOHN; LE BEL, JOSEPH-ACHILLE; LEWIS, GILBERT N.; LEWIS STRUCTURES; PAULING, LINUS; THOMSON, JOSEPH JOHN; VAN'T HOFF, JACOBUS.

Erwin Boschmann

Bibliography

Atkins, Peter W. (1996). *Molecules.* New York: W. H. Freeman and Company.

Gillespie, R. J., and Nyholm, R. S. (1957). "Inorganic Stereochemistry." *Quarterly Reviews* (London) 11:339–380.

Lewis, Gilbert N. (1916). "The Atom and the Molecule." *Journal of the American Chemical Society* 38:762–786.

Pauling, Linus (1960). *The Nature of the Chemical Bond.* Ithaca, NY: Cornell University Press.

Pfennig, Brian W., and Frock, Richard L. (1999). "The Use of Molecular Modeling and VSEPR Theory in the Undergraduate Curriculum to Predict the Three-Dimensional Structure of Molecules." *Journal of Chemical Education* 76(7):1018–1022.

Molybdenum

MELTING POINT: 2,623°C
BOILING POINT: 4,639°C
DENSITY: 10.22 g/cm^3
MOST COMMON IONS: Mo^{3+}, $Mo_2(OH)_2^{4+}$, $M_2O_4^{2+}$

Molybdenum is a hard, silver-white **metal** discovered by Swedish chemist Carl Wilhelm Scheele in 1778. Scheele had been researching a mineral called molybdenite, which many suspected of containing lead (the Greek word *molybdos* means "lead"). He instead found that it contained a new element

metal: element or other substance the solid phase of which is characterized by high thermal and electrical conductivities

which he named "molybdenum" after the mineral. Molybdenum was first isolated by Swedish mineralogist Peter Jacob Hjelm in 1782.

Molybdenum has an abundance in Earth's crust of approximately 1.1 parts per million (ppm) or 1.2 milligrams per kilogram. Its chief source is the mineral molybdenite (MoS_2), but it is also found in the ores wulfenite ($PbMoO_4$) and powellite ($CaMoO_4$) or obtained as a by-product of copper mining. The leading producers of molybdenum are the United States, Canada, Chile, Mexico, Peru, China, Russia, and Mongolia.

isotope: form of an atom that differs by the number of neutrons in the nucleus

There are seven known **isotopes** of molybdenum that occur naturally: ^{92}Mo, ^{94}Mo, ^{95}Mo, ^{96}Mo, ^{97}Mo, ^{98}Mo, and ^{100}Mo. Their natural abundances range from 9.25 percent (^{94}Mo) to 24.13 percent (^{98}Mo). Common compounds of molybdenum include molybdenum disulfide (MoS_2), molybdenum trioxide (MoO_3), molybdic acid (H_2MoO_4), molybdenum hexafluoride (MoF_6), and molybdenum phosphide (MoP_2).

alloy: mixture of two or more elements, at least one of which is a metal

Molybdenum's melting point (2,623°C, or 4,753.4°F) exceeds that of steel by 1,000°C (1,832°F) and that of most rocks by 500°C (932°F). For this reason, the element is used in various **alloys** to improve strength, particularly at high temperatures. Approximately 75 percent of molybdenum produced is used by the iron and steel industries. The element is also utilized to make parts for furnaces, light bulbs, missiles, aircraft, and guns. Molybdenum disulfide is used as a high temperature lubricant. SEE ALSO COORDINATION COMPOUNDS; INORGANIC CHEMISTRY; SCHEELE, CARL.

Stephanie Dionne Sherk

Bibliography

Lide, David R., ed. (2003). *The CRC Handbook of Chemistry and Physics*, 84th edition. Boca Raton, FL: CRC Press.

Other Resource

Powell, Darryl. "Molybdenum." Mineral Information Institute. Available from <http://www.mii.org/Minerals/photomoly.html>.

Monosaccharides ***See Carbohydrates.***

Morgan, Agnes Fay

AMERICAN CHEMIST AND NUTRITIONIST
1884–1968

Born in Peoria, Illinois, on May 4, 1884, Agnes Fay Morgan excelled in high school and studied chemistry at the University of Chicago. After receiving a master of science in chemistry in 1905, she spent the next several years teaching at various colleges across the United States. She returned to the University of Chicago in 1914 to obtain a Ph.D. in chemistry.

In 1915 Morgan was appointed assistant professor of nutrition in the Department of Home Economics at the University of California at Berkeley. Her shift in focus from chemistry to nutrition was a result of the limited professional opportunities available to female chemists at the time. In 1919 Morgan became associate professor of household science, and in 1923 she was promoted to full professor.

As chairperson of the department, Morgan worked vehemently to change the prestige of home economics. She strove to establish a scientific basis for the field, which was generally perceived as the course of study in which young women learned how to become proficient wives and mothers. Under her leadership, the home economics curriculum at Berkeley became largely science-based, with strict requirements. It was not until 1960, however, that her efforts to change the name of the department to better define its work were successful. At that time, six years after her retirement, the department was renamed the Department of Nutritional Sciences.

Throughout the course of her career, Morgan published more than 250 papers on topics with far-reaching effects. Her research would become the foundation for understanding the nutritional effects of many **vitamins**. It also established that certain vitamin deficiencies can lead to health problems. She was the first to determine that a deficiency in pantothenic acid, a B vitamin, can lead to damage to the adrenal glands and abnormal skin and hair pigmentation. Morgan also showed that high doses of vitamin D can have a toxic effect on the body. Her studies on vitamins A and C led to their discovery in a wide variety of foods. She demonstrated that proteins become denatured (i.e., their physical structure becomes changed) when heated, reducing their nutritional value. Other areas of Morgan's research included the analysis of processed foods, the association between vitamins and hormones, the effects of food preservation on vitamin content, and the basis for low weight gain in children.

vitamins: organic molecules needed in small amounts for the normal function of the body; often used as part of an enzyme catalyzed reaction

Despite the importance of Morgan's research, much of her efforts remained unrecognized until late in her career. In 1949 she was awarded the prestigious Garvan Medal by the American Chemical Society for her groundbreaking research in nutrition. In 1950 she became the first woman to receive the status of faculty research lecturer at the University of California. Other honors imparted on Morgan included the 1954 Borden Award from the American Institute of Nutrition and the Phoebe Apperson Hearst Gold Medal, which recognized her as one of ten outstanding women in San Francisco in 1963. In 1961 the home economics building at Berkeley was renamed Agnes Fay Morgan Hall.

Morgan maintained her research efforts well after her official retirement in 1954. She continued to frequent her office until her death in 1968, two weeks after a heart attack. As a former staff member was quoted as saying in a 1969 memorial, "We can only feel that her going marked the end of an era in the education of women." SEE ALSO DENATURATION; FOOD PRESERVATIVES.

Stephanie Dionne Sherk

Bibliography

Nerad, Maresi (1999). *The Academic Kitchen: A Social History of Gender Stratification at the University of California, Berkeley.* New York: State University of New York Press.

Internet Resources

American Chemical Society. "Agnes Fay Morgan." *Journal of Chemical Education.* Available from <http://jchemed.chem.wisc.edu/JCEWWW/Features/eChemists/Bios/Morgan.html>.

Okey, Ruth; Johnson, Barbara Kennedy; and Mackinney, Gordon (1969). "Agnes Fay Morgan, Home Economics: Berkeley." Available from <http://dynaweb.oac.cdlib.org:8088/dynaweb/uchist/public/inmemoriam/inmemoriam1969/@Generic__BookTextView/1214>.

English physicist Henry Moseley, who arranged the Periodic Table in order of the atomic numbers of the elements.

Moseley, Henry

ENGLISH PHYSICIST
1887–1915

Henry Moseley's research career lasted only forty months before tragically ending with his death on a Gallipoli battlefield in World War I. But in his classic study of the x-ray spectra of elements, he established the truly scientific basis of the Periodic Table by arranging chemical elements in the order of their atomic numbers.

Henry Gwyn Jeffreys Moseley, who was always called "Harry" by his family, was born in Weymouth, England, on November 23, 1887. His family was wealthy, aristocratic, and scientifically accomplished, and young Henry showed an early interest in **zoology**. He attended Eton on a King's scholarship, where he excelled in mathematics, and was introduced to the study of x rays by his physics teacher. He entered Trinity College, Oxford, in 1906. At that time, Oxford did not have a particularly notable science curriculum, but Moseley chose the school in order to be near his widowed mother. He graduated in 1910 with high honors in mathematics and science, and secured a position in the laboratory of Ernest Rutherford at the University of Manchester.

zoology: branch of biology concerned with the animal kingdom

It was a time of great excitement and ferment in science, and Rutherford's laboratory was one of the epicenters of discovery in atomic physics. The first coherent theory of the structure of the atom was just then being developed by Rutherford and his research group, which, besides Moseley, included Niels Bohr, Hans Geiger, Kasimir Fajans, and others.

The nature of x rays was also receiving new interest because of the discovery by the German physicist Max von Laue in 1912 that they were diffracted by their passage through crystals and therefore possessed a wave nature. Succeeding experiments by William L. Bragg the same year showed that similar results could be obtained by the reflection of x rays from the face of a crystal. Moseley persuaded Rutherford to allow him and a colleague, C. S. Darwin, to further study the nature of x rays. Their work demonstrated that the **spectral line** of platinum, which they were using as the anticathode in their x-ray tube, was characteristic of that element alone. Moseley returned to Oxford, and despite the experimental deficiencies of his laboratories, measured the x-ray spectral lines of nearly all the elements from aluminum to gold. The results of his study showed a clear and simple progression of the elements that was based on the number of protons in the atomic nucleus, rather than the order based on atomic weights that was then the basis of the Periodic Table.

spectral line: line in a spectrum representing radiation of a single wavelength

His work, called a "classical example of the scientific method," was the second and last of his independent publications. Despite the pleadings of his colleagues, he enlisted in the British Army at the outbreak of World War I, and was killed in battle on August 10, 1915. SEE ALSO Bohr, Niels; Bragg, William Lawrence; Lanthanides; Radiation; Rutherford, Ernest.

Bartow Culp

Bibliography

Heilbron, J. L. (1966). "The Work of H. G. J. Moseley." *Isis* 57:336–366.

Heilbron, J. L. (1974). *H. G. J. Moseley: The Life and Letters of an English Physicist, 1887–1915.* Berkeley: University of California Press.

Jaffe, Bernard (1971). *Moseley and the Numbering of the Elements.* Garden City, NY: Doubleday.

Mutagen

Mutagens are chemical agents that cause changes in the genetic **code** which are then passed on to future generations of an organism. Mutations are usually chemical in nature and often carcinogenic, but may also be caused by physical damage produced by x rays or other causes. A mutation changes the activity of a gene. Mutations are frequent in lower forms of life and may help these organisms adapt to changes in their environments.

Proteins are composed of chains of amino acids. In the genetic code of deoxyribonucleic acid (**DNA**), a codon or three-base sequence codes for the placement of each amino acid; for example, the codon UUU places phenylalanine at that location in the protein and replacement of the third base with **adenine** results in the placement of leucine instead of phenylalanine. If a portion of the original code read . . . UUUACG . . . , deleting one of the uridine bases would cause that portion of the code to read . . . UUACG . . . ; the sequence UUA would then specify leucine. A point mutation changing one base might result in the formation of a different protein.

Mutations can occur by several mechanisms, such as replacing one nucleotide base with another or by adding or removing a base; they can also develop when a carcinogenic agent such as an aromatic hydrocarbon molecule is inserted between the strands of DNA, causing the code to be misread. Some chemical mutagens such as nitrites change one base into another, resulting in a new sequence of amino acids and the **synthesis** of a new protein. The modified protein might function normally or might not be useful at all, but it could be dangerous.

Many mutagenic agents are also carcinogenic, and the Ames test provides a quick method for screening foods and other substances for potential cancer-causing agents. SEE ALSO Carcinogen; Codon.

Dan M. Sullivan

HOW DOES THE AMES TEST WORK?

The Ames test is a method for screening potential mutagens. The test uses auxotrophs (strains that have lost the ability to synthesize a needed substance) of *Salmonella typhimurium* that carry mutant genes, making them unable to synthesize histidine. They can live on media containing histidine, but die when the amino acid is depleted. The bacteria are especially sensitive to back mutations that reactivate the gene for the synthesis of histidine; exposure to mutagenic substances allows the bacteria to grow rapidly, developing large and numerous colonies.

code: mechanism to convey information on genes and genetic sequence

DNA: deoxyribonucleic acid—the natural polymer that stores genetic information in the nucleus of a cell

adenine: one of the purine bases found in nucleic acids, $C_5H_5N_5$

synthesis: combination of starting materials to form a desired product

Bibliography

Devlin, Thomas M., ed. (2002). *Textbook of Biochemistry: With Clinical Correlations*, 5th edition. New York: Wiley-Liss.

Voet, D.; Voet, J. G.; and Pratt, C. W. (2003). *Biochemistry*, 3rd edition. New York: Wiley.

Mutation

Any heritable change in the genetic information or **DNA** is called a mutation. A change in the base sequence of DNA that is then replicated and transmitted to future generations of cells becomes a permanent change in the **genome**. Mutations, all of which appear to occur as random events, can range from a single replacement of a base (substitution) to larger changes that result from the deletion or addition of more than one base (often large stretches of a DNA molecule).

genome: total genetic material in a cell or organism

Most mutations are thought to be harmful to the life of the cell. These harmful mutations occur during the development of a cancer cell, for example. In these cases (cancerous transformation), numerous point mutations or deletion mutations are well-established as causative agents. A point mutation

A six-legged frog, a result of a genetic disorder.

occurs when a single base is changed in a DNA sequence. This can be either: (1) a transition, in which a **purine base** is replaced by another purine base, or pyrimidine by pyrimidine (e.g., base pair AT becomes base pair GC); or (2) a transversion, in which a purine is replaced by a pyrimidine, or vice versa (e.g., base pair AT becomes base pair CG). A point mutation that changes a codon with the result that it codes for a different amino acid is called a missense mutation. Such a mutation can change the nature of the protein being formed. It can change the amino acid composition and the protein sequence and, therefore, the structure of that protein. This process may have a deleterious effect on protein activity in essential metabolic functions in the cell. In contrast, there are cases in which a mutation can change the protein sequence but have little or no consequence on the protein function. These are silent mutations. In these cases, the change is a conservative one (a single amino acid is substituted for another of similar type, such as lysine for an arginine, or the **amino acid residue** may reside on the outside surface of the protein where it will have little effect on protein structure). Such silent mutations exhibit no phenotypic (observable) changes. Alternatively, a mutation can occur in intergenic or noncoding regions and thus have no direct effect on the protein product. There can also be rare changes in DNA sequence that may provide a selective advantage to an organism.

purine base: one of two types of nitrogen bases found in nucleic acids

amino acid residue: in a protein, which is a polymer composed of many amino acids, that portion of the amino acid that remains to become part of the protein

Mutations may occur spontaneously, or as a result of external physical agents (radiation) or chemical agents (mutagens). The most common spontaneous mutations result from errors in DNA replication that are not corrected. Virtually all forms of life are exposed to ultraviolet light from the Sun, which can react with adjacent **thymine** bases in DNA in such a way as to link them together to produce an intrastrand thymine dimer. A number of chemicals, including dimethylsulfate, nitrous acid, and nitrogen mustards, react with bases in DNA so as to modify them. As a result, the subsequent replication cycle changes the complementary base or bases and leads to a permanent change in the form of a transition or transversion. In the case of the thymine dimers or the loss of a base, repair enzymes exist that scan the DNA in an attempt to correct the problem. There are a number of inher-

thymine: one of the four bases that make up a DNA molecule

ited disease conditions, such as xeroderma pigmentosum and Cockayne syndrome, that result from defects in genes associated with DNA repair.

In a number of cancers, a deletion of much or all of a gene that completely inactivates the gene has occurred. It is claimed that about 80 percent of human cancers may be caused by carcinogens that damage DNA or interfere with its replication and/or repair. Bruce Ames, a microbiologist at the University of California at Berkeley, developed a simple experimental procedure using bacterial cells that can detect mutagenic chemicals. It has been shown that about 80 percent of carcinogenic compounds are also mutagenic using the Ames test. SEE ALSO DNA; MUTAGEN; TERATOGEN.

William M. Scovell

Bibliography

Fairbanks, Daniel J., and Andersen, W. Ralph (1999). *Genetics: The Continuity of Life.* Pacific Grove, CA: Brooks/Cole Publishing.

Nelson, David L., and Cox, Michael M. (2000). *Lehninger Principles of Biochemistry,* 3rd edition. New York: Worth Publishers.

NAD/NADH *See Nicotinamide Adenine Dinucleotide.*

Nanochemistry

In recent years nanoscale science and technology have grown rapidly. Nanochemistry, in particular, presents a unique approach to building devices with a molecular-scale precision. One can envision the advantages of nanodevices in medicine, computing, scientific exploration, and electronics, where nanochemistry offers the promise of building objects atom by atom. The main challenges to full utilization of nanochemistry center on understanding new rules of behavior, because nanoscale systems lie at the threshold between classical and quantum behavior and exhibit behaviors that do not exist in larger devices.

Although nanochemical control was proposed decades ago, it was only recently that many of the tools necessary for studying the nanoworld were developed. These include the scanning tunneling microscope (STM), atomic force microscope (AFM), high resolution scanning and transmission electron microscopies, x rays, ion and electron beam probes, and new methods for nanofabrication and lithography.

Studies of nanochemical systems span many areas, from the study of the interactions of individual atoms and how to manipulate them, how to control chemical reactions at an atomic level, to the study of larger molecular assemblies, such as dendrimers, clusters, and polymers. From studies of assemblies, significant new structures—such as nanotubes, nanowires, three-dimensional molecular assemblies, and lab-on-a-chip devices for separations and biological research—have been developed.

Single Atoms

The ultimate frontier of nanochemistry is the chemical manipulation of individual atoms. Using the STM, single atoms have been assembled into larger structures, and researchers have observed chemical reactions between two atoms on a surface. The use of atoms as building blocks opens new routes to novel materials and offers the ability to create the smallest features possible in integrated circuits (IC) and to explore areas like quantum computing. Until now the ever-decreasing size of IC circuitry has been well described by Moore's law, but further shrinkage of circuit size will halt by 2012 because of **quantum mechanical** effects. Quantum computing provides a way to circumvent this apparent roadblock and use these quantum effects to advantage. Atomic-scale devices, although promising, present major challenges in how to achieve spatial control and stability.

quantum mechanical: theoretical model to describe atoms and molecules by wave functions

Dendrimers

Dendrimers are highly branched three-dimensional nanoscale molecular objects of the same size and weight as traditional polymers. However, dendrimers are synthesized in a stepwise fashion, allowing for extremely precise control of their size and geometry (see Figure 1, a molecular model of a dendrimer). In addition, the chemical reactivity and properties of their periphery and core can be controlled easily and independently. Dendrimers are already being used in molecular recognition, nanosensing, light harvesting, and optoelectrochemical devices. Because they are built up layer by layer and the properties of any individual layer can be controlled through selection of the monomer, they are ideal building blocks in nanochemistry for the creation of more complex three-dimensional structures.

Nanocrystals and Clusters

Nanocrystals are crystals of nanometer dimensions, usually consisting of aggregates of a few hundred to tens of thousands of atoms combined into a cluster. Nanocrystals have typical dimensions of 1 to 50 nanometers (nm),

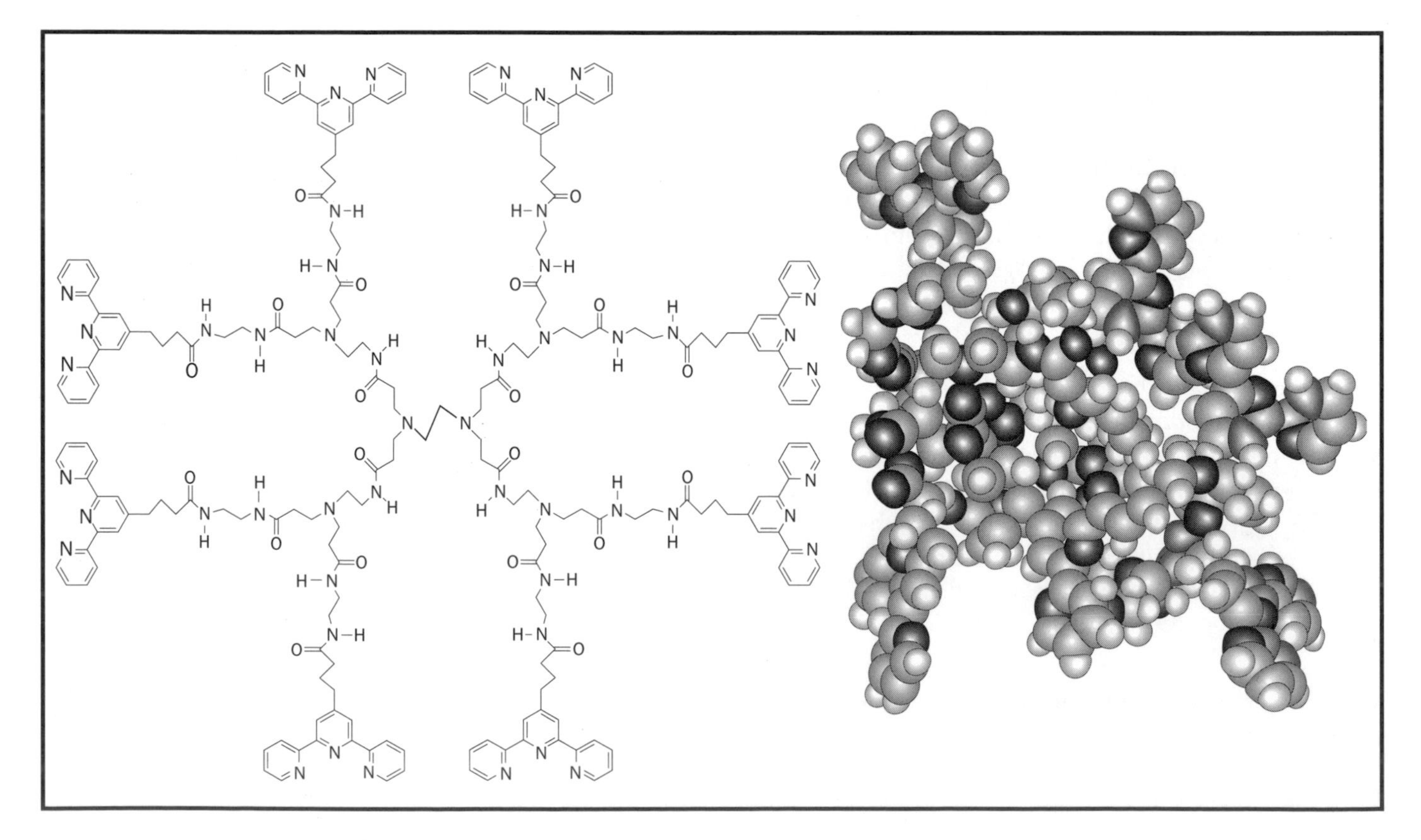

Figure 1. Molecular representation of a dendrimer (left) and a 3-D molecular model of the same dendrimer (right).

and thus they are intermediate in size between molecules and bulk materials and exhibit properties that are also **intermediate**. For example, the small size of semiconductor quantum "dots" leads to a shifted light emission spectrum through quantum confinement effects—with the magnitude of the shift being determined by the size of the nanocrystal. Nanocrystals are of great interest because of their promise in high density data storage and in optoelectronic applications, as they can be efficient light emitters. Nanocrystals have also found applications as biochemical tags, as laser and optical components, for the preparation of display devices, and for chemical **catalysis**.

intermediate: molecule, often short-lived, that occurs while a chemical reaction progresses but is not present when the reaction is complete

catalysis: the action or effect of a substance in increasing the rate of a reaction without itself being converted

Nanotubes

Recently, hollow carbon tubes of nanometer dimensions have been prepared and studied. These nanotubes constitute a new form of carbon, configurationally equivalent to a graphite sheet rolled into a hollow tube (see Figure 2, a molecular model of a carbon nanotube). Carbon nanotubes may be synthesized, with sizes ranging from a few microns to a few nanometers and with thicknesses of many carbon layers down to single-walled structures. The unique structure of these nanotubes gives them advantageous behavior relative to properties such as electrical and thermal conductivity, strength, stiffness, and toughness. Carbon nanotubes can also be functionalized with molecular recognition agents so that they may bind specifically to discrete molecular targets, allowing them to be used as high resolution AFM probes, as channels for materials separation, and as selective gates for molecular sensing.

Nanowires

Like nanotubes, nanowires are very small rods of atoms, but nanowires are solid, dense structures, much like a conventional wire. Controlling the atom (material) used for building the wire, as well as its impurity **doping**, allows for control of its electrical conduction properties. Ultimately, chemists wish to fabricate and control nanowires that are a single atom or molecule in diameter, thus creating an unprecedented laboratory for studying how small structures affect electron transfer within the wire and between the wire and external agents. Clearly, nanowires offer the potential for creating very small IC components.

dope: to add a controlled amount of an impurity to a very pure sample of a substance, which can radically change the properties of a substance

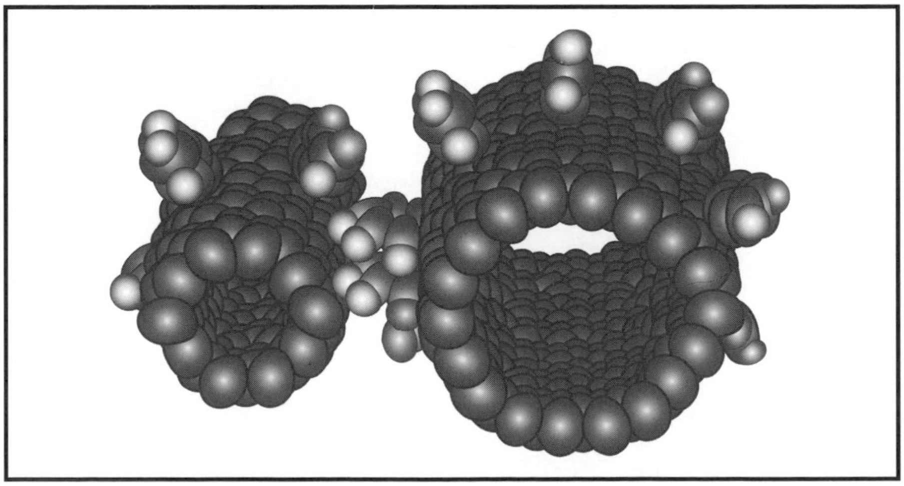

Figure 2. Molecular diagram of a carbon nanotube.

Nanocomposites

Nanocomposites encompass a large variety of systems composed of dissimilar components that are mixed at the nanometer scale. These systems can be one-, two-, or three-dimensional; organic or inorganic; crystalline or amorphous. A critical issue in nanocomposite research centers on the ability to control their nanoscale structure via their **synthesis**. The behavior of nanocomposites is dependent on not only the properties of the components, but also morphology and interactions between the individual components, which can give rise to novel properties not exhibited by the parent materials. Most important, the size reduction from microcomposites to nanocomposites yields an increase in surface area that is important in applications such as mechanically reinforced components, nonlinear optics, batteries, sensors, and catalysts.

THE SCANNING TUNNELING MICROSCOPE AND ATOMIC FORCE MICROSCOPE

The scanning tunneling microscope (STM) and the atomic force microscope (AFM) are very high resolution microscopes that allow scientists to obtain high resolution images of surfaces with atomic or molecular resolution. Both microscopes work by scanning a very sharp tip on a surface and measuring current (STM) or intermolecular forces (AFM) between the tip and the surface.

synthesis: combination of starting materials to form a desired product

Lab on a Chip

Lab-on-a-chip devices are designed to carry out complex chemical processes at an ultrasmall scale, for example, synthesizing chemicals efficiently; carrying out biological, chemical, and clinical analyses; performing combinatorial chemistry; and conducting separations and analysis on a single, miniaturized device. When the amount of material in a sample is small or when it is highly toxic or dangerous, lab-on-a-chip devices offer an ideal way to complete complex chemical manipulations with extremely small sample sizes. Further, because the volumes used to carry solutions are extremely small, even very small sample amounts can be present in reasonable concentrations. Lab-on-a-chip technology has been aggressively pursued in biotechnology, where better ways to separate and analyze **DNA** and proteins are of great interest. It has also sparked great interest in the analysis of dangerous materials where it can be used, for example, by law enforcement or the military to analyze explosives and biological or chemical agents, while maintaining low risks.

DNA: deoxyribonucleic acid—the natural polymer that stores genetic information in the nucleus of a cell

Nano-Electro-Mechanical Systems

Nano-electro-mechanical systems have also generated significant interest in the creation of tiny devices that can use electrochemical energy to carry out mechanical tasks, for example, nanomotors. One can envision that the coupling of chemical energy to mechanical transducers will enable the construction of devices that may be applied in medicine to treat illnesses, explore dangerous areas, or just reach places that larger-scale devices cannot. Research in this area focuses on understanding the preparation of nanoscale components to build such devices as well as the interactions between the components, especially the coupling between the electrochemical and mechanical components. In addition, a new understanding of effects such as friction and wear is required as the nanoscale components obey a different set of rules than their macroscopic counterparts. SEE ALSO Computational Chemistry; Fullerenes; Materials Science.

MOORE'S LAW

In 1965 Gordon Moore, cofounder of Intel, predicted the trend that the number of transistors on integrated circuits (IC) was going to follow. The number of transistors in IC has grown exponentially, with the implication that the size of such transistors has decreased in a similar fashion. His law stills hold, but it is predicted that the law will not be applicable by the year 2012 as further shrinking of transistors will not be possible after that time.

Diego J. Díaz
Paul W. Bohn

Bibliography

Reed, Mark A., and Tour, James M. (2000). "Computing with Molecules." *Scientific American* 282(6):86–93.

Service, Robert F. (2000). "Atom-Scale Research Gets Real." *Science* 290:1524–1531.

Internet Resources

Amato, Ivan. "Nanotechnology: Shaping the World Atom by Atom." Report of the Interagency Working Group on Nanoscience, Engineering and Technology (IWGN) National Nanotechnology Initiative. Available from <http://itri.loyola.edu/nano/IWGN.Public.Brochure>.

Drexler, K. Eric. (1987). *Engines of Creation: The Coming Era of Nanotechnology.* Reprinted and adapted by Russell Whitaker, with permission. Available from <http://www.foresight.org/EOC/index.html>.

Feynman, Richard. "There's Plenty of Room at the Bottom." Available from <http://www.zyvex.com/nanotech/feynman.html>.

Moore, Gordon E. "Cramming More Components onto Integrated Circuits." Available from <http://www.intel.com/research/silicon/mooreslaw.htm>.

Whitesides, G., and Alivisatos, P. (1999). "Fundamental Scientific Issues for Nanotechnology." In *Nanotechnology Research Directions IWGN Workshop Report Vision for Nanotechnology Research and Development in the Next Decade*, ed. by M. C. Roco, S. Williams, and P. Alivisatos. Available from <http://itri.loyola.edu/nano/IWGN.Research.Directions>.

Natural Gas *See Fossil Fuels.*

Neodymium

MELTING POINT: **1,021°C**
BOILING POINT: **3,127°C**
DENSITY: **7.0 g/cm^3**
MOST COMMON IONS: **Nd^{2+}, Nd^{3+}, Nd^{4+}**

Neodymium oxide was first isolated from a mixture of oxides called didymia. The element neodymium is the second most abundant **lanthanide** element in the igneous rocks of Earth's crust. Hydrated neodymium(III) salts are reddish and anhydrous neodymium compounds are blue. The compounds neodymium(III) chloride, bromide, iodide, nitrate, perchlorate, and acetate are very soluble; neodymium sulfate is somewhat soluble; the fluoride, hydroxide, oxide, carbonate, oxalate, and phosphate compounds are insoluble.

lanthanides: a family of elements (atomic number 57 through 70) from lanthanum to lutetium having from 1 to 14 4f electrons

Neodymium is used to color special glasses, giving these glasses a blue-violet shade. It is also used to color television faceplates, to reduce the reflectivity of television screens. $Nd_2Fe_{14}B$ magnets are among the most powerful. Neodymium compounds are used as laser materials, specifically as optically pulsed solid-state laser materials. One of the most important of these is Nd–YAG garnet (YAG = $Y_3A_{l5}O_{12}$), which generates light having wavelengths of 1.06 micrometers (4.17×10^{-5} inches). This garnet laser has potential use in dental caries prevention. Finally, neodymium is used in the making of photographic filters (Nd_2O_3), magnets used in headphones, and ceramic capacitors. SEE ALSO Cerium; Dysprosium; Erbium; Europium; Gadolinium; Holmium; Lanthanum; Lutetium; Praseodymium; Promethium; Samarium; Terbium; Ytterbium.

Lea B. Zinner
Geraldo Vicentini

Bibliography

American Chemical Society. Division of Chemical Education (2000) *Chemistry Come Alive.* Washington, DC: American Chemical Society.

Maestro, Patrick (1998). "From Properties to Industrial Applications." In *Rare Earths*, ed. R. S. Puche and P. Caro. Madrid: Editorial Complutense.

Weber, M. S. (1984). "Rare Earth Lasers." In *Handbook on the Physics and Chemistry of Rare Earths*, Vol. 4, ed. K. A. Gschneidner Jr. and L. R. Eyring. Amsterdam: North-Holland Physics Publishing.

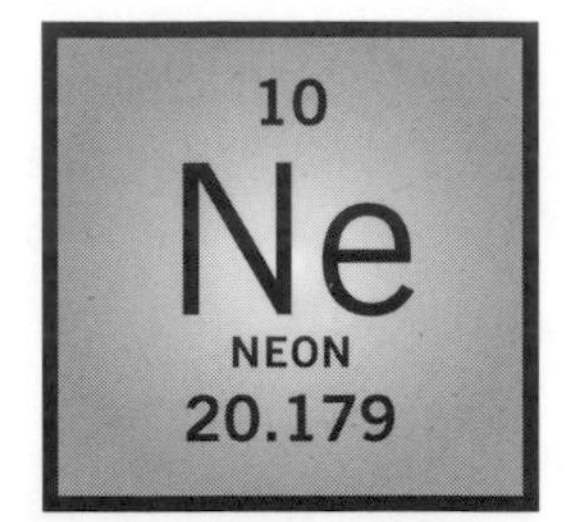

Neon

MELTING POINT: −248.59°C
BOILING POINT: −246.08°C
DENSITY: 0.8999 kg/m^3
MOST COMMON IONS: None

Neon (from the Greek word *neos*, meaning "new") is the second lightest of the **noble gases**. It forms no stable compounds with other elements. Discovered in 1898 by Sir William Ramsay and Morris Travers during their experiments with liquid air, neon accounts for 18 millionths (18 ppm) of the volume of Earth's atmosphere. Trace amounts are also found in the oceans. It is present in the Sun and in the atmosphere of Mars.

Neon in a vacuum discharge tube is commonly used in electric signage.

At room temperature neon is a colorless, odorless gas. Upon freezing it forms a crystal with a face-centered cubic structure. In a vacuum discharge tube, neon emits its famous red-orange light, which has long been used in advertising signs and discharge display tubes. More recently neon (mixed with helium) has been used in common lasers and laser pointers, which produce a characteristic red neon **spectral line** corresponding to light of wavelength 632.8 nanometers (2.49 × 10^{-5} inches).

Neon is produced within stars during **nucleosynthesis.** It has three stable **isotopes**, of which neon-20 is the most abundant (90.5%). SEE ALSO GASES; NOBLE GASES; RAMSAY, WILLIAM; TRAVERS, MORRIS.

Richard Mowat

Bibliography

Lide, David R., ed. (2002). *The CRC Handbook of Chemistry and Physics*, 83rd edition. Boca Raton, FL: CRC Press.

Nobel Lecures in Chemistry 1901-1921 (1966). New York: Elsevier.

Neptunium

MELTING POINT: 640°C
BOILING POINT: 3,930°C
DENSITY: 20.45 g/cm^3
MOST COMMON IONS: Np^{3+}, Np^{4+}, or NpO_2^{+}, NpO_2^{2+}, NpO_3^{+}

Neptunium was discovered by the U.S. physicists Edwin M. McMillan and Philip Abelson, in 1940, via the bombardment of ^{238}U with neutrons. The name of the element is related to the planet Neptune. Neptunium-237 occurs as a product of ^{238}U **fission**, and appears in uranium fuel elements.

$$^{238}_{92}U + ^{1}_{0}n \rightarrow ^{239}_{92}U \xrightarrow{-\beta^{-}} ^{239}_{93}Np(t_{1/2} = 2.35 \text{ days})$$

Neptunium is used to produce plutonium (^{238}Pu), via the irradiation of NpO_2 with neutrons. The **isotope** $^{238}{}_{94}$Pu is used as a power source for satellites.

$$^{237}_{93}Np + ^{1}_{0}n \rightarrow ^{238}_{93}Np \xrightarrow{-\beta^{-}} ^{238}_{94}Pu$$

noble gas: element characterized by inert nature; located in the rightmost column in the Periodic Table

spectral line: line in a spectrum representing radiation of a single wavelength

nucleosynthesis: creation of heavier elements from lighter elements via fusion reactions in stars

Neptunium has several **valence** states: Np^{3+} (purplish in solution), stable in water, is easily oxidized in air to Np^{4+} (yellow-green), and then slowly oxidized in air to the stable ion, NpO_2^{2+} (pink); NpO_2^{+} (green) is obtained by the **oxidation** of Np^{4+} with hot nitric acid. NpO_2^{2+} can also be obtained by the oxidation of lower state ions with Ce^{4+}, MnO^{4-}, O_3, and BrO_3^{-}. The production of ^{237}Np (as NpO_2^{+}) involves the oxidation of Np^{3+} by nitric acid, followed by the extraction of the NpO_2^{+} compound with tributylphosphate in kerosene.

isotope: form of an atom that differs by the number of neutrons in the nucleus

fission: process of splitting an atom into smaller pieces

valence: combining capacity

oxidation: process that involves the loss of electrons (or the addition of an oxygen atom)

The oxide NpO_2, **isostructural** with UO_2, is obtained by heating neptunium nitrates or hydroxides in air. The oxide Np_3O_8, isomorphous with U_3O_8, is also obtained by heating such compounds in air at specific temperatures. The oxidation of Np^{4+} hydroxide compounds with ozone gives the hydrated trioxides $NpO_3 \cdot 2H_2O$ (brown) and $NpO_3 \cdot H_2O$ (red-gold). The fluorides NpF_3 and NpF_4 are precipitated from **aqueous solutions**.

isostructural: relating to an arrangement of atomic constituents that is geometrically the same although different atoms are present

aqueous solution: homogenous mixture in which water is the solvent (primary component)

$$NpO_2 + 1/2H_2 + 3HF(g) \xrightarrow{500°C} NpF_3 + 2H_2O$$

Other neptunium halides are: NpF_3 (purple-black); $NpCl_3$ (white); $NpBr_3$ (green); NpI_3 (brown); NpF_4 (green); $NpCl_4$ (red-brown); $NpBr_4$ (red-brown); and NpF_6 (orange). The removal of highly radioactive neptunium solids or solutions must be performed via remote control. Neptunium is one of the actinides that is found in **nuclear** waste (in oxidate states +3, +4, +5, +6, and maybe +7) and must be kept out of the environment. SEE ALSO ACTINIUM; BERKELIUM; EINSTEINIUM; FERMIUM; LAWRENCIUM; MENDELEVIUM; NOBELIUM; PLUTONIUM; PROTACTINIUM; RUTHERFORDIUM; THORIUM; URANIUM.

nuclear: having to do with the nucleus of an atom

Lea B. Zinner

Geraldo Vicentini

Bibliography

Ball, M. C., and Norbury, A. H. (1974). *Physical Data for Inorganic Chemists.* London: Longman.

Cotton, Frank A., and Wilkinson, Geoffrey (1972). *Advanced Inorganic Chemistry: A Comprehensive Text*, 3rd edition. New York: Wiley Interscience.

Greenwood, Norman N., and Earnshaw, A. (1997). *Chemistry of the Elements*, 2nd edition. Boston: Butterworth-Heinemann.

Nernst, Walther Hermann

GERMAN CHEMIST
1864–1941

Walther Hermann Nernst, born in Briesen, Prussia (now Wabrzezno, Poland), was a pioneer in the field of chemical thermodynamics in a wide range of areas. His most outstanding contributions were his laws for electrochemical cells and his heat theorem, also known as the third law of thermodynamics, for which he was awarded the Nobel Prize in chemistry in 1920.

Nernst first studied physics before he became an assistant in 1887 to German physical chemist Friedrich Wilhelm Ostwald at the University of Leipzig, then the only institute for physical chemistry in Germany. In 1891

German chemist and physicist Walther Hermann Nernst (front holding vial), recipient of the 1920 Nobel Prize in chemistry, "in recognition of his work in thermochemistry."

he was appointed associate professor at the university in Göttingen and, three years later, convinced officials there to create an institute for physical chemistry modeled on the Leipzig center. He served as its director until his move in 1905 to Berlin, where he once again established an institute renowned worldwide.

During his Leipzig period, Nernst performed a series of electrochemical studies from which, at the age of twenty-five, he arrived at his well-known equations. These equations described the concentration dependence of the potential difference of **galvanic** cells, such as batteries, and were of both great theoretical and practical importance. Nernst started with the investigation of the diffusion of electrolytes in one solution. Then he turned to the diffusion at the boundary between two solutions with different electrolyte concentrations; he determined that the osmotic pressure difference would result in an electric potential difference or electromotive force (emf). Next he divided both solutions into two concentration half-cells, connected to each other by a liquid junction, and measured the emf via electrodes dipped into both solutions. The data supported his first equation where the

galvanic: relating to direct current electricity especially when produced chemically emf was proportional to the logarithm of the concentration ratio. Finally, he investigated galvanic cells where a redox reaction (e.g., $Zn + 2Hg^{+} \rightarrow Zn^{2+} + 2Hg$) was divided such that **oxidation** ($Zn \rightarrow Zn^{2+} + 2e^{-}$) and reduction ($2Hg^{+} + 2e^{-} \rightarrow 2Hg$) occurred at the electrodes in two half-cells. By combining this with Helmholtz's law, which related thermodynamics to the emf of electrochemical cells, and van't Hoff's equation, which related chemical equilibria to thermodynamics, Nernst derived his second equation for galvanic cells. Supported by many measurements, the equation described the emf of galvanic cells as a function of the concentration of all substances involved in the reaction.

oxidation: process that involves the loss of electrons (or the addition of an oxygen atom)

$$E = E^{\circ} - \frac{RT}{NF}\left(\frac{[\text{Products}]}{[\text{Reactants}]}\right)$$

Nernst's formulation of the third law of thermodynamics was originally an ingenious solution to a crucial practical problem in chemical thermodynamics, namely, the calculation of chemical equilibria and the course of chemical reactions from thermal data alone, such as reaction heats and heat capacities. Based on the first two laws of thermodynamics and van't Hoff's equation, chemical equilibria depended on the free reaction enthalpy ΔG, which was a function of both the reaction enthalpy ΔH and the reaction entropy ΔS according to the Gibbs-Helmholtz equation:

$$E = \frac{E^{\circ} - RT}{NF}$$

The problem was that, although enthalpy values could be calculated from thermal measurements, entropy values required data at the absolute zero of temperature, which was practically inaccessible. Guided by theoretical reasoning and then supported by a huge measurement program at very low temperatures, Nernst in 1906 suggested his heat theorem. According to a later formulation, it stated that all entropy changes approach zero at the absolute zero.

The theorem not only allowed the calculation of chemical equilibria, it was also soon recognized as an independent third law of general thermodynamics with many important consequences. One such consequence was that it is impossible to reach the absolute zero. Another consequence was that one could *define* a reference point for entropy functions, such that the entropies of all elements and all perfect crystalline compounds were taken as zero at the absolute zero.

Nernst made numerous other important contributions to physical chemistry. For example, his distribution law described the concentration distribution of a solute in two immiscible liquids and allowed the calculation of extraction processes. He also formulated several significant theories, such as those on the electrostriction of ions, the diffusion layer at electrodes, and the solubility product. In addition, he established new methods to measure dielectric constants and to synthesize ammonia, on which the German chemist Fritz Haber later successfully followed up. SEE ALSO ELECTROCHEMISTRY; HABER, FRITZ; OSTWALD, FRIEDRICH WILHELM; PHYSICAL CHEMISTRY.

Joachim Schummer

Bibliography

Barkan, Diana (1999). *Walther Nernst and the Transition to Modern Physical Science.* Cambridge, U.K.: Cambridge University Press.

Hiebert, Erwin N. (1978). "Nernst, Hermann Walther." In *Dictionary of Scientific Biography,* Vol. XV, Supplement I, ed. Charles C. Gillispie. New York: Scribners.

Mendelssohn, Kurt (1973). *The World of Walther Nernst: The Rise and Fall of German Science.* London: Macmillan.

Neurochemistry

Neurochemistry refers to the chemical processes that occur in the brain and nervous system. The fact that one can read this text, remember what has been read, and even breathe during the entire time that these events take place relies on the amazing chemistry that occurs in the human brain and the nerve cells with which it communicates.

There are two broad categories of chemistry in nerve systems that are important. The first is the chemistry that generates electrical signals which propagate along nerve cells. The key chemicals involved in these signals are sodium and potassium ions. To see how they give rise to a signal, one must first look at a nerve cell that is at rest.

Like any other cell, a nerve cell has a membrane as its outer "wall." On the outside of the membrane, the concentration of sodium ions will be relatively high and that of potassium ions will be relatively low. The membrane maintains this concentration gradient by using channels and enzymes.

The channels are pores that may be opened or closed by enzymes which are associated with them. Some ion channels allow the movement of sodium ions and others allow potassium ions to cross the membrane. They are also called "gated" channels because they can open and close much like a gate in a fence. The **voltage** they experience dictates whether the gate is open or closed. Thus, for example, a gated sodium ion channel in a membrane opens at certain voltages to allow sodium ions to pass from regions of high concentration to regions of low concentration.

voltage: potential difference expressed in volts

Active transport mechanisms are also present. Enzymes that span the membrane can actively pump sodium and potassium ions from one side of the membrane to another. When the nerve cell is at rest, these mechanisms maintain a high potassium and low sodium environment inside the cell.

Even when it is at rest, a nerve cell is in contact with many other nerve cells. When a neighboring cell passes on a signal to the resting cell (by a mechanism to be discussed shortly), a dramatic change occurs in the ion concentrations. Once the nerve cell at rest has received a sufficient signal from a neighbor to surpass a threshold level, some of the sodium ion channels near the connection point open and sodium ions flow into the cell. This flow of charge results in an electrical potential that is called the action potential. The action potential does not stay localized, however. Farther down the nerve cell, more sodium ion channels surpass their threshold and open so that the sodium ions flow into them as well. Thus, the action potential moves down the nerve. After the sodium ion gates open, the potassium ion gates also open and potassium ions flow out of the cell. This flow of ions offsets the charge from sodium ions flowing into the cell and the signal has receded in that region (and has moved on).

Once the cell propagates a signal, how does that cell send its signal to a neighbor? This question leads to the second broad category of neurochemistry: the chemistry at the synapse. Nerve cells do not actually touch their neighbors, but rather form a small gap called the synapse. The signal is transferred across this gap by chemicals called neurotransmitters.

The communication that occurs across the synapse may either excite or inhibit the action of the neighboring nerve cell. Thus, synapses are further categorized as either **excitatory** synapses or **inhibitory** synapses. The cell that is **propagating** the signal is called the presynaptic cell, and the cell that receives the signal is the postsynaptic cell.

excitatory: causing cells to become active

inhibitory: relating to the prevention of an action that would normally occur

propagating: reproducing; disseminating; increasing; extending

The end of the presynaptic cell contains small **vesicles**, spherical collections of the same **lipid** molecules that make up the cell membrane. Inside these vesicles, neurotransmitters exist in high concentrations. When the action potential reaches the end of the presynaptic cell, some of the vesicles merge with the cell membrane and release their contents (a process called exocytosis). The released neurotransmitters experience an immediate concentration gradient. They diffuse away from the release point to counteract the gradient, and in doing this, they cross the synapse and arrive at the neighboring cell.

vesicle: small compartment in a cell that contains a minimum of one lipid bilayer

lipid: a nonpolar organic molecule; fatlike; one of a large variety of nonpolar hydrophobic (water-hating) molecules that are insoluble in water

On the postsynaptic cell, there are **receptors** that are capable of interacting with the neurotransmitters. Once these messenger molecules cross the synapse, they connect with the receptors and the two cells have successfully communicated. The proteins of the receptors are capable of opening sodium gated ion channels, and a new action potential is engaged in the postsynaptic cell.

receptor: area on or near a cell wall that accepts another molecule to allow a change in the cell

The remaining step in the process is also a critical one. Somehow the action of the neurotransmitters must cease. If they continue to cross the synapse, or are not removed from the receptors of the postsynaptic cell, they will continue to activate that cell. An overexcited or inhibited nerve cell is not capable of proper function. For example, schizophrenia is a mental disease that is caused by the brain's inability to eliminate excitatory neurotransmitters. The nerve cells continue firing, even when they need not, and the incorrect brain chemistry results in debilitating symptoms such as auditory hallucinations—hearing voices that are not actually there. SEE ALSO ENZYMES; NEUROTOXINS; NEUROTRANSMITTERS; STIMULANTS.

Thomas A. Holme

Bibliography

Bloch, K. (1999). *Blondes in Venetian Paintings, the Nine Banded Armadillo and Other Essays in Biochemistry.* New Haven, CT: Yale University Press.

Bradford, H. F. (1986). *Chemical Neurobiology.* New York: W.H. Freeman.

Darnell, J.; Lodish, H.; and Baltimore, D. (1990). *Molecular Cell Biology.* New York: Scientific American Books.

Neurotoxins

Many chemical compounds, some natural and some made by humans, show toxic effects in humans or other animals. Every **toxin** is harmful, but toxins that target the nervous system have been developed into chemical warfare agents, so the public concern about them is enhanced.

toxin: poisonous substance produced during bacterial growth

Despite the connection with weapons of mass destruction, the most common neurotoxin in society is ethanol, found in alcoholic beverages. Neurons convey signals by manipulating ion concentrations, and neurotoxins reduce their ability to do so. Alcohol does this by essentially overloading the entire cell and hindering its ability to function. Many of the characteristics of alcohol intoxication, such as slurred speech and erratic motion, are the result of improper function of neurons in the brain. As the body metabolizes the alcohol and removes it from the blood, the neurotoxic effects wear off. With large overdoses of alcohol, however, the effects do not wear off, and death due to alcohol poisoning is a dramatic and unfortunately too common manifestation of neurotoxins.

The neurotoxins that are associated with chemical warfare typically operate in a different fashion. A neuron carries a signal as a miniature electric current. Ions carry charges, and when they move across the cell membrane in a specific region of a neuron at a rapid rate they change the electrical potential in that region. The rapid movement of ions migrates along the neuron and propagates an electrical signal (called an *action potential*). When this signal reaches the end of the neuron, it must somehow trigger a response in the next neuron. In a few cases, neurons are packed closely enough so that the charge associated with the moving action potential directly excites the next neuron. In most cases, the first neuron releases small molecules called *neurotransmitters* that diffuse across a small gap (the **synaptic cleft**) and interact with the next neuron, triggering its response. Many neurotoxins, including both human-made agents of chemical warfare and natural agents found in venoms and other natural toxins, work by disrupting this communication process.

synaptic cleft: tiny space between the terminal button of one neuron and the dendrite or soma of another

There are two common mechanisms by which nerve signaling is disrupted. The cell that receives the signal does so when **receptors** within its membrane interact with the neurotransmitters. Some neurotoxins act by blocking these receptors, making it impossible for them to receive signals. When signaling stops, nerve function is impaired or eliminated and, the neurotoxin has caused its damage.

receptor: area on or near a cell wall that accepts another molecule to allow a change in the cell

The other key component of interneuron communication is that the neurotransmitters, once they have carried a signal across a synaptic cleft, must be removed. If a "receiving" neuron is continually stimulated because neurotransmitters continue to activate it, the neuron's function will be impaired, and the neuron may even be killed. There are special enzymes in the synaptic cleft that break down certain neurotransmitters, such as **acetylcholine**, to end the signaling. Some neurotoxins block the actions of these hydrolytic enzymes, thereby preventing the removal of acetylcholine (or other neurotransmitters), leading to continuous stimulation of the neurons and, ultimately, cell death. SEE ALSO ACETYLCHOLINE; INHIBITORS; NEUROTRANSMITTERS.

acetylcholine: neurotransmitter with the chemical forumula $C_7H_{17}NO_3$; it assists in communication between nerve cells in the brain and central nervous system

Thomas A. Holme

Bibliography

Changeux, Jean-Pierre; Devillers-Thiery, Anne; and Chemeuilli, Phillippe (1984). "The Acetylcholine Receptor: An Allosteric Protein." *Science* 25:1335–1345.

Crosby, Donald G. (1998). *Environmental Toxicology and Chemistry*. New York: Oxford University Press.

Simpson, Lance L. (1971). *Neuropoisons: Their Pathophysiological Actions*. New York: Plenum Press.

Neurotransmitters

Neurotransmitters are chemical messengers produced by the nervous systems of higher organisms in order to relay a nerve impulse from one cell to another cell. The two cells may be nerve cells, also called neurons, or one of the cells may be a different type, such as a muscle or gland cell. A chemical messenger is necessary for rapid communication between cells if there are small gaps of 20 to 50 nanometers (7.874×10^{-7}–19.69×10^{-7} inches), called synapses or synaptic clefts, between the two cells. The two cells are referred to as either presynaptic or postsynaptic. The term "presynaptic" refers to the neuron that produces and releases the neurotransmitter, whereas "postsynaptic" refers to the cell that receives this chemical message.

Neurotransmitters include small molecules with amine **functional groups** such as **acetylcholine**, certain amino acids, amino acid derivatives, and peptides. Through a series of chemical reactions, the amino acid **tyrosine** is converted into the catecholamine neurotransmitters dopamine and norepinephrine or into the hormone epinephrine. Other neurotransmitters that are amino acid derivatives include γ-aminobutyric acid, made from glutamate, and serotonin, made from the amino acid tryptophan.

functional group: portion of a compound with characteristic atoms acting as a group

acetylcholine: neurotransmitter with the chemical forumula $C_7H_{17}NO_3$; it assists in communication between nerve cells in the brain and central nervous system

tyrosine: one of the common amino acids

Peptide neurotransmitters include the enkephalins, the endorphins, oxytocin, substance P, vasoactive intestinal peptide, and many others. The gaseous free radical **nitric oxide** is one of the more recent molecules to be added to the list of possible neurotransmitters. It is commonly believed that there may be fifty or more neurotransmitters. Although there are many

nitric oxide: compound, NO, which is involved in many biological processes; the drug Viagra enhances NO-stimulation of pathways to counteract impotence; may be involved in killing tumors

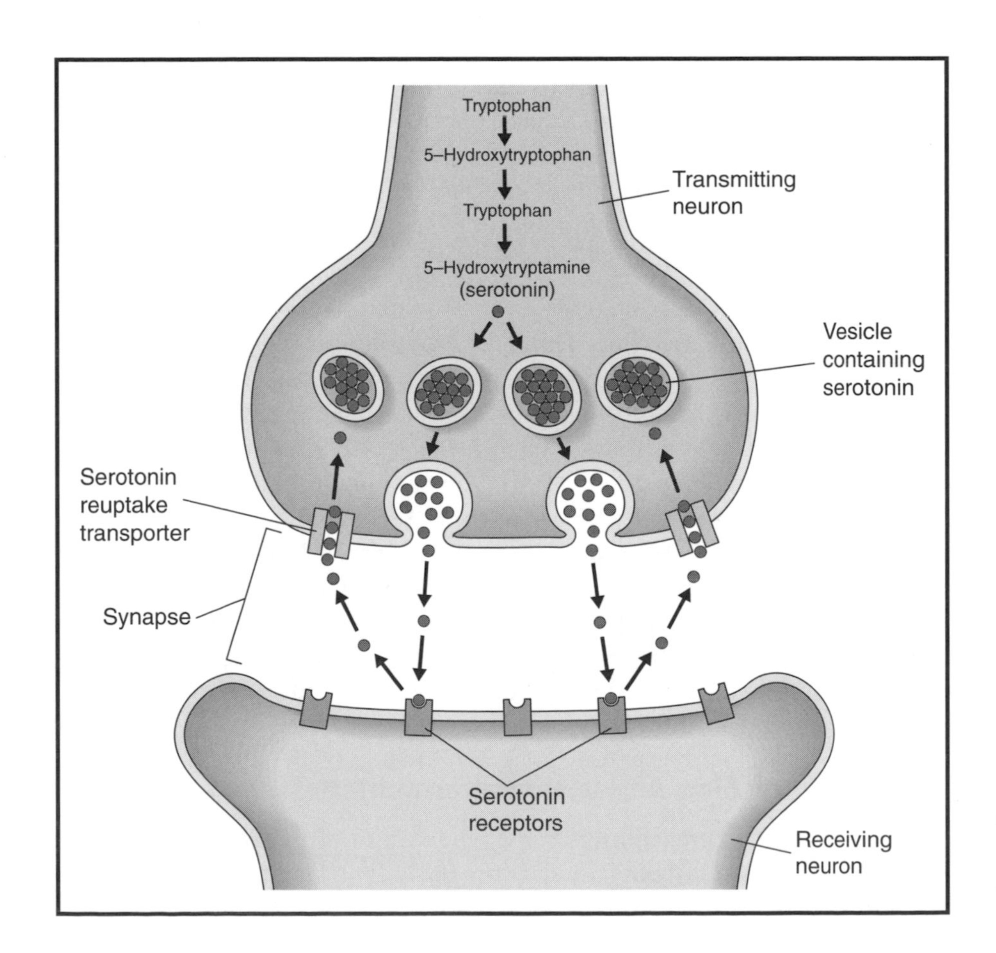

This diagram shows the transmission and reception of neurons and the role of serotonin in communication between neurons.

different neurotransmitters, there is a common theme by which they are released and exert their actions. In addition, there is always a mechanism for termination of the chemical message.

General Mechanism of Action

vesicle: small compartment in a cell that contains a minimum of one lipid bilayer

Neurotransmitters are formed in a presynaptic neuron and stored in small membrane-bound sacks, called **vesicles**, inside this neuron. When this neuron is activated, these intracellular vesicles fuse with the cell membrane and release their contents into the synapse, a process called exocytosis.

receptor: area on or near a cell wall that accepts another molecule to allow a change in the cell

Once the neurotransmitter is in the synapse, several events may occur. It may (1) diffuse across the synapse and bind to a **receptor** on the postsynaptic membrane, (2) diffuse back to the presynaptic neuron and bind to a presynaptic receptor causing modulation of neurotransmitter release, (3) be chemically altered by an enzyme in the synapse, or (4) be transported into a nearby cell. For the chemical message to be passed to another cell, however, the neurotransmitter must bind to its protein receptor on the postsynaptic side. The binding of a neurotransmitter to its receptor is a key event in the action of all neurotransmitters.

Mechanism of Fast-Acting Neurotransmitters

ligand: molecule or ion capable of donating one or more electron pairs to a Lewis acid

depolarization: process of decreasing the separation of charge in nerve cells; the opposite of hyperpolarization

hyperpolarization: process of causing an increase in charge separation in nerve cells; opposite of depolarization

Some neurotransmitters are referred to as fast-acting since their cellular effects occur milliseconds after the neurotransmitter binds to its receptor. These neurotransmitters exert direct control of ion channels by inducing a conformational change in the receptor, creating a passage through which ions can flow. These receptors are often called **ligand**-gated ion channels since the channel opens only when the ligand is bound correctly. When the channel opens, it allows for ions to pass through from their side of highest concentration to their side of lowest concentration. The net result is **depolarization** if there is a net influx of positively charged ions or **hyperpolarization** if there is a net inward movement of negatively charged ions. Depolarization results in a continuation of the nerve impulse, whereas hyperpolarization makes it less likely that the nerve impulse will continue to be transmitted.

excitatory: causing cells to become active

amino acid sequence: twenty of the more than five hundred amino acids that are known to occur in nature are incorporated into a variety of proteins that are required for life processes; the sequence or order of the amino acids present determines the nature of the protein

inhibitory: relating to the prevention of an action that would normally occur

The first ligand-gated ion channel whose structure and mechanism were studied in detail was the nicotinic acetylcholine receptor of the neuromuscular junction. This receptor contains five protein subunits, each of which spans the membrane four times. When two acetylcholine molecules bind to this receptor, a channel opens, resulting in sodium and potassium ions being transported at a rate of 10^7 per second. Acetylcholine's action at these receptors is said to be **excitatory** due to the resulting depolarization. Other receptors for fast transmitters have a similar **amino acid sequence** and are believed to have a similar protein structure. Glycine and γ-aminobutyric acid (GABA) also act on ligand-gated ion channels and are fast-acting. However, they cause a net influx of chloride ions, resulting in hyperpolarization; thus, their action is **inhibitory**.

Mechanism of Slow-Acting Neurotransmitters

Slower-acting neurotransmitters act by binding to proteins that are sometimes called G-protein-coupled receptors (GPCRs). These receptors do not form ion channels upon activation and have a very different architecture than the ion channels. However, the timescale for activation is often relatively fast, on the order of seconds. The slightly longer time frame than that for fast-acting neurotransmitters is necessary due to additional molecular interactions that must occur for the postsynaptic cell to become depolarized or hyperpolarized. The protein structure of a GPCR is one protein subunit folded so that it transverses the membrane seven times. These receptors are referred to as G-coupled protein receptors because they function through an interaction with a **GTP**-binding protein, called G-protein for short.

GTP: guanosine triphosphate, a nucleotide consisting of ribose, guanine, and three linked phosphate groups

The conformational change produced when a neurotransmitter binds to a GPCR causes the G-protein to become activated. Once it becomes activated, the protein subunits dissociate and diffuse along the intracellular membrane surface to open or close an ion channel or to activate or inhibit an enzyme that will, in turn, produce a molecule called a second messenger. Second messengers include cyclic **AMP**, cyclic **GMP**, and calcium ions and phosphatidyl inositol. They serve to activate enzymes known as protein kinases. Protein kinases in turn act to phosphorylate a variety of proteins within a cell, possibly including ion channels. Protein **phosphorylation** is a common mechanism used within a cell to activate or inhibit the function of various proteins.

AMP: adenosine monophosphate, a form of ATP after removal of two phosphate groups

GMP: guanosine monophosphate, a nucleotide consisting of ribose, guanine, and one phosphate group

phosphorylation: the process of addition of phosphates into biological molecules

Termination of Transmission

For proper control of neuronal signaling, there must be a means of terminating the nerve impulse. In all cases, once the neurotransmitter dissociates from the receptor, the signal ends. For a few neurotransmitters, there are enzymes in the synapse that serve to chemically alter the neurotransmitter, making it nonfunctional. For instance, the enzyme acetylcholinesterase hydrolyzes acetylcholine. Other neurotransmitters, such as catecholamines and glutamate, undergo a process called reuptake. In this process, the neurotransmitter is removed from the synapse via a transporter protein. These proteins are located in presynaptic neurons or other nearby cells.

Drugs of Abuse

The actions of neurotransmitters are important for many different physiological effects. Many drugs of abuse either mimic neurotransmitters or otherwise alter the function of the nervous system. Barbiturates act as depressants with effects similar to those of anesthetics. They seem to act mainly by enhancing the activity of the neurotransmitter GABA, an inhibitory neurotransmitter. In other words, when barbiturates bind to a GABA receptor, the inhibitory effect of GABA is greater than before. Opiates such as heroin bind to a particular type of opiate receptor, resulting in effects similar to those of naturally occurring endorphins. **Amphetamines** can displace catecholamines from synaptic vesicles and block reuptake of catecholamines in the synapse, prolonging the action of catecholamine neurotransmitters. SEE ALSO Acetylcholine; Dopamine; Hydrolysis; Ion Channels; Norepinephrine.

amphetamine: class of compounds used to stimulate the central nervous system

Jennifer L. Powers

Bibliography

Changeux, Jean-Pierre (1993). "Chemical Signaling in the Brain." *Scientific American* 269(5):58.

Garrett, Reginald H., and Grisham, Charles M. (1995). "Excitable Membranes, Neurotransmission, and Sensory Systems." In *Molecular Aspects of Cell Biology*. Philadelphia: Saunders.

Powledge, Tabitha M. (2002). "Beating Abuse." *Scientific American* 286(1):20.

Internet Resources

"Hallucinogens." Available from <http://www.pharmcentral.com>.

"Narcotics." Available from <http://www.pharmcentral.com>.

"Neurotransmitters." Available from <http://www.pharmcentral.com>.

"Sedatives." Available from <http://www.pharmcentral.com>.

"Stimulants." Available from <http://www.pharmcentral.com>.

Neutrons ***See Atomic Nucleus; Atomic Structure.***

New Battery Technology

The need for better batteries is a recurring theme in the effort to reduce energy consumption and in the effort to make electricity increasingly portable. As the world depends more and more on portable devices and turns to electric vehicles to reduce pollution, it becomes important that lightweight, long-lived batteries be developed. Additional desirable features of batteries include safety, dependability, environmental friendliness, and cost. This article considers battery technology currently used in transportation, including in electric vehicles and spacecraft.

Battery Basics

A battery is a collection of one or more electrochemical cells that convert chemical energy into electrical energy via electrochemical reactions (**oxidation-reduction reactions**). These reactions take place at the battery's anode and cathode. The electrochemical cells are connected in series or in parallel depending on the desired **voltage** and capacity. Series connections provide a higher voltage, whereas parallel connections provide a higher capacity, compared with one cell.

oxidation-reduction reaction: reaction, sometimes called redox, that involves the movement of electrons between reactants to form products

voltage: potential difference expressed in volts

A cell typically consists of a negative electrode, a positive electrode, and an electrolyte. A cell discharges when a load such as a motor is connected between the negative and positive electrodes. The negative electrode, the anode, produces electrons that flow in an external circuit. The positive electrode, the cathode, consumes the electrons from the external circuit. The uniform flow of electrons around the circuit results in an electric current. Within the cell, the electrons received at the positive electrode react with the active material of this electrode, in reduction reactions that continue the flow of charge by sending ions through the electrolyte to the negative electrode. At the negative electrode, oxidation reactions between the active material of this electrode and the ions flowing through the electrolyte results in a surplus of electrons that are donated to the external circuit. For every electron generated in an oxidation reaction at the negative electrode, there is an electron consumed in a reduction reaction at the positive electrode. As the electrode reactions continue spontaneously, the active materials become depleted and the reactions slow down until the battery is no longer capable of supplying electrons; the battery is said to be fully discharged.

A battery is either a primary or a secondary battery. Primary batteries, such as those used in a flashlight, are used once and replaced. The chemical reactions producing the current in such batteries are too difficult to make it worth trying to reverse them. Secondary batteries, such as car batteries, can be recharged and reused because the chemical reactions are easily reversed. By reversing the flow of electricity (i.e., putting current in rather than taking it out), the chemical reactions are reversed to restore active material that had been depleted. Secondary batteries are also known as rechargeable or storage batteries and are used in transportation applications.

Battery performance is measured in terms of voltage and capacity. The voltage is determined by the chemistry of the **metals** and electrolytes used in the battery. The capacity is the number of electrons that can be obtained from a battery. Since current is the number of electrons released per unit time, cell capacity is the current supplied by a cell over time and is normally measure in ampere-hours. Battery specialists experiment with many different redox combinations and try to balance the energy output with the costs of manufacturing the battery. Other factors, such as battery weight, shelf life, and environmental impact also factor into the battery's design.

metal: element or other substance the solid phase of which is characterized by high thermal and electrical conductivities

Present-Day Battery Technology

Lead-acid batteries are used in gasoline-driven automobiles and in electric and hybrid vehicles. They have the best discharge rate of secondary battery technology, they are the cheapest to produce, and they are rechargeable. The chemical reactions are:

Cathode (+):

$$PbO_2 + H_2SO_4 \xrightarrow{discharge} PbSO_4 + 2H_2O$$

Anode (–):

$$Pb + H_2SO_4 \xrightarrow{discharge} PbSO_4 + 2H^+ + 2e$$

The positive electrode is made of lead dioxide (PbO_2) and is reduced to lead sulfate ($PbSO_4$), while sponge metallic lead (Pb) is oxidized to lead sulfate at the negative electrode. The electrolyte is sulfuric acid (H_2SO_4), which provides the sulfate ion (SO_4^{2-}) for the discharge reactions.

The nickel-cadmium battery (Ni-Cd) is the most common battery used in communication satellites, in Earth orbiters, and in space probes. The chemical reactions are:

Cathode (+):

$$2NiO(OH) + 2H_2O + 2e \xrightarrow{discharge} 2Ni(OH)_2 + 2OH^-$$

Anode (–):

$$Cd + 2OH^- \xrightarrow{discharge} Cd(OH)_2 + 2e$$

Nickel hydroxide, NiO(OH), is the active cathode material, cadmium, Cd, is the active anode material, and aqueous potassium hydroxide, KOH, is the electrolyte.

There is considerable interest in the development of nickel-metal hybrid (Ni/MH) batteries for electric and hybrid vehicles. These batteries operate in concentrated KOH electrolyte. The electrode reactions are:

Cathode (+):

$$NiO(OH) + H_2O + e \xrightarrow{discharge} Ni(OH)_2 + OH^-$$

Anode (–):

$$MH + OH^- \xrightarrow{discharge} M + H_2O + e$$

Ni/MH batteries use nickel hydoxide, NiO(OH), as the active material for the cathode, a metal hydride, MH, as the anode, and a potassium hydroxide, KOH, solution as the electrolyte. The metal hydride is a type of **alloy** (hydrogen absorption alloy) that is capable of undergoing a reversible hydrogen absorbing-desorbing process while the battery is discharged and charged. Current research is directed at improving the performance of the metal hydride anode and making the battery rechargeable.

alloy: mixture of two or more elements, at least one of which is a metal

Lithium ion (Li-ion) batteries are environmentally friendly batteries that offer more energy in smaller, lighter packages and thus are promising candidates for electric and hybrid vehicle applications. The electrode reactions are:

Cathode (+):

$$xLi^+ + xe + Li_{1-x}AB \xrightarrow{discharge} LiAB$$

Anode (–):

$$Li_xC_6 \xrightarrow{discharge} xLi^+ + C_6 + xe$$

Li-ion batteries use various forms of carbon (C) as anode material because carbon can reversibly accept and donate significant amounts of lithium (as Li_xC_6. Li-intercalation compounds (such as $LiCoO_2$, $LiMn_2O_4$, and $LiNiO_2$) are used as cathode materials. Electrolyte mixtures include a lithiated salt ($LiPF_6$ or $LiClO_4$) dissolved into a nonaqueous solvent (ethylene carbonate, propylene carbonate, or dimethyl carbonate). Because Li is a highly reactive metal in **aqueous solution**, Li-ion batteries are constructed to keep Li in its ionic state, and nonaqueous solvents are used. The next step in lithium-ion battery technology is believed to be the lithium polymer battery, in which a gelled or solid electrolyte will replace the liquid electrolyte.

aqueous solution: homogenous mixture in which water is the solvent (primary component)

Fuel Cells

Unlike the batteries described in the previous section, a fuel cell does not run down or require recharging; it will produce energy in the form of electricity and heat as long as fuel is supplied. Additionally, the electrode materials (usually platinum) serve only as a site for the reactions to occur (i.e., as a **catalyst**) and are not involved in the chemical reactions. As hydrogen flows over the anode, it is oxidized to hydrogen ions and electrons in a proton-exchange membrane or PEM fuel cell. The hydrogen ions pass through the membrane to the cathode, where they combine with oxygen from the air and with the electrons flowing in the external circuit from the anode to form water, which is expelled from the cell. A fuel cell system that includes a "fuel reformer" utilizes hydrogen from any hydrocarbon fuel, such as natural gas or methanol. This also makes a fuel cell quiet, dependable, and very fuel-efficient. Fuel cell reactions include:

catalyst: substance that aids in a reaction while retaining its own chemical identity

Cathode (+):

$$O_2 + 4H^+ + 4e \rightarrow 2H_2O$$

Anode (–):

$$2H_2 \rightarrow 4H^+ + 4e$$

Fuel cells are lighter and more compact, compared with batteries that make available the same amount of energy.

Solar Cells

Solar cells (photovoltaic cells) convert sunlight to electricity. Photovoltaic cells are made of semiconductor materials such as silicon and gallium arsenide. When light strikes the cell, photons are absorbed within the semiconductor and create electron-hole pairs that move within the cell. This generates the energy that is used to power space vehicles.

Electric and Hybrid Vehicles

Electric vehicles have an electric motor rather than a gasoline engine. The electric motor is usually powered by two banks of twenty-five 12-volt lead-acid rechargeable batteries, providing a total of 300 volts for each battery bank. Problems with lead-acid battery technology include battery weight (a typical lead-acid battery pack weighs 1,000 pounds or more), limited capacity (a typical battery pack holds about 15 kilowatt-hours of electricity, giving the car a range of approximately 80 kilometers [50 miles]), long recharging times (typically between four and ten hours for full charge), short life (three to four years) and cost (about $2,000 for each battery pack). The hybrid vehicle, in which a small gas engine is combined with an electric motor, is a compromise between gas-powered and electric vehicles.

The car of the future will likely be an electric or hybrid vehicle that gets its electricity from a fuel cell. It is unlikely that these vehicles will ever be solar powered, because solar cells produce too little power to make using them to run a full-size car practical.

Space Power

Spacecraft and space stations are powered by solar cells or collections of solar cells called solar panels. To get the most power, solar panels must be pointed directly at the Sun. Spacecraft are built so that the solar panels can be pivoted as the spacecraft moves, so that they can always stay in the direct path of the rays of light.

A desalination system on this roof in Jeddah, Saudi Arabia, is using 210 photovoltaic modules to supply operating power.

Solar cells generate electricity in the sunshine but not in the dark. Thus space stations and spacecraft run on power from batteries during dark periods. As of June 2003, solar power has been practical for spacecraft operating no farther from the Sun than the orbit of Mars. For example, Magellan, Mars Global Surveyor, Mars Observer, and the Earth-orbiting Hubble Space Telescope operate on solar power.

A Look to the Future

Exciting research is underway to improve the performance and longevity of batteries, fuel cells, and solar cells. Much of this research is directed at enhancing the chemistry in these systems through the use of polymer electrolytes, nanoparticle catalysts, and various membrane supports. Additionally, considerable effort is being put into the construction of three-dimensional microbatteries. SEE ALSO Electrochemistry; Materials Science; Solar Cells.

Cynthia G. Zoski

Bibliography

Hamann, Carl H.; Hamnett, Andrew; and Vielstich, Wolf (1998). *Electrochemistry.* New York: Wiley-VCH.

Hart, Ryan W.; White, Henry S.; Dunn, Bruce; and Rolison, Debra R. (2003). *Electrochemistry Communications* 5:120–123.

Linden, David, and Reddy, Thomas B., eds. (2002). *Handbook of Batteries*, 3rd edition. New York: McGraw-Hill.

McFarland, Eric W., and Tang, Jing (2003). *Nature* 421:616–618.

Tarascon, Jean-Marie, and Armand, Michel (2001). *Nature* 414:359–367.

Royal Society: The U.K. National Academy of Science, founded in 1660

alchemy: medieval chemical philosophy having among its asserted aims the transmutation of base metals into gold

Newton, Isaac

ENGLISH PHYSICIST AND MATHEMATICIAN
1642–1727

Sir Isaac Newton was born on December 25, 1642, in Woolsthorpe, Lincolnshire, England. His father died shortly before he was born. Newton attended Trinity College, starting in 1661, and remained there for the early part of his career. During the year of the plague (1665 to 1666), Trinity College was closed and Newton returned to his family home in the country. It was during this one incredibly productive year that much of Newton's most important work began. In 1703 Newton was knighted and elected president of the **Royal Society**, a post he held until his death in 1727.

English physicist Sir Isaac Newton, author of *Philosophiae Naturalis Principia Mathematica.*

Newton's best-known contributions to science were his three laws of motion and law of universal gravitation. These were first published in his *Principia* of 1687. Newton's other seminal work was *Opticks*, initially published in 1704. Newton also developed differential and integral calculus (although with different terminology and notation than used today), fluid mechanics, equations describing heat transfer, and an experimental scientific method. His other major intellectual interests were **alchemy**, theology, history, and biblical chronology. While working at the Royal Mint, Newton successfully oversaw the recoinage of the nation's currency to control coin clipping (the illicit trimming of gold or silver from the edges of coins) and its related inflation.

It is historically known that Newton owned one of Europe's largest book collections on alchemy. Unfortunately, this part of his library was dispersed at the time of his death without an adequate inventory. Although Newton did not publish any large work on alchemy, the subject did continue to preoccupy him during the course of his life. Alchemy had obvious relevance to his work at the mint and its associated work on **metallurgy**. It also interested him because of its relevance to questions about the ultimate structure of matter. Much of Newton's published work on chemistry or alchemy appears in the form of "Queries" placed at the end of *Opticks*. These are rhetorical questions with postulated answers, some of which are quite extensive. Together, the Queries cover some sixty-seven pages. Query 31 alone is thirty text pages long.

metallurgy: the science and technology of metals

Newton's postulated answers concerning the ultimate structure of matter by advancing the idea of atoms with some level of internal structure, a notion anticipating the modern concept of molecules. Newton also postulated on the existence of a nonmaterial substance, an imponderable (unweighable) fluid called ether, which might work at very small distances to repel atoms from one another. Heat, light, electricity, or the reactions of chemistry might be used, Newton suggested, to probe this subtle, imponderable fluid. SEE ALSO Alchemy; Atoms.

David A. Bassett

Bibliography

Andrade, Edward Neville daCosta (1979). *Sir Isaac Newton.* Westport, CT: Greenwood Press.

Cohen, I. B., and Westfall, Richard S. (1995). *Newton: Texts, Background, and Commentaries.* New York: W. W. Norton.

Manuel, Frank E. (1990). *A Portrait of Isaac Newton.* New York: Da Capo Press.

Newton, Sir Isaac (1979). *Opticks.* New York: Dover. Based on the 4th edition. London, 1730.

Newton, Sir Isaac (1999). *The Principia, Mathematical Principles of Natural Philosophy.* Translated by I. Bernard Cohen and Anne Whitman. Berkeley: University of California Press.

Thackray, Arnold (1970). *Atoms and Powers: An Essay on Newtonian Matter-Theory and the Development of Chemistry.* Cambridge, MA: Harvard University Press.

Niacin ***See Nicotinamide.***

Nickel

MELTING POINT: 1,455°C
BOILING POINT: 2,913°C
DENSITY: 8.9 g/cm^3
MOST COMMON IONS: Ni^{2+}, Ni^{3+}, Ni^{4+}

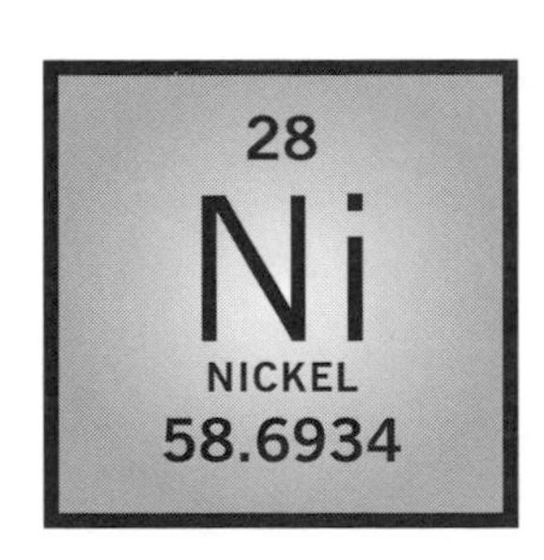

Nickel is a silver-white, lustrous **metal**. It was first isolated by Swedish chemist Axel Fredrik Cronstedt in 1751. Cronstedt had been attempting to **isolate** copper from a mineral called niccolite (the German word *kupfernickel* means "Devil's copper" or "Old Nick's copper"). He instead found nickel, which he named after the mineral.

metal: element or other substance the solid phase of which is characterized by high thermal and electrical conductivities

isolate: part of a reaction mixture that is separated and contains the material of interest

The abundance of nickel in Earth's crust is 90 parts per million (ppm); in ocean water, its abundance is 2 parts per billion (ppb). In meteorites, however, its abundance approaches 13,000 ppm. Much of the world's supply of nickel is found in Ontario, Canada, where it is isolated from the ores pentlandite and pyrrhotite. Other large deposits are found in Australia, New Caledonia, Cuba, Indonesia, and Greenland.

isotope: form of an atom that differs by the number of neutrons in the nucleus

The most common **isotope** of nickel is ^{58}Ni, which has a natural abundance of 68.1 percent. Other stable isotopes include ^{60}Ni (26.2%), ^{61}Ni (1.1%), ^{62}Ni (3.6%), and ^{64}Ni (0.9%). Important nickel compounds include nickel oxides (NiO and Ni_2O_3), nickel sulfides (NiS, NiS_2, Ni_3S_2), and nickel chloride ($NiCl_2$).

ductile: property of a substance that permits it to be drawn into wires

alloy: mixture of two or more elements, at least one of which is a metal

Nickel metal is malleable, **ductile**, and a fairly good conductor of electricity and heat. Its most common use is in stainless steels, where it may be combined with various other metals (such as iron, chromium, chromium, copper, etc.) to form **alloys** that are highly resistant to corrosion. Nickel is also used to make coins (U.S. five-cent pieces contain 25 percent nickel), batteries, magnets, and jewelry; to protectively coat other metals; and to color glass and ceramics green. SEE ALSO COORDINATION COMPOUNDS; INORGANIC CHEMISTRY.

Stephanie Dionne Sherk

Bibliography

Lide, David R., ed. (2003). *The CRC Handbook of Chemistry and Physics*, 84th edition. Boca Raton, FL: CRC Press.

Other Resources

Winter, Mark "Nickel." The University of Sheffield and WebElements Ltd., U.K. Available from <http://www.webelements.com>.

Nicotinamide

Nicotinamide is the most common form of the vitamin niacin. Nicotinamide is found in the body as part of nicotinamide adenine dinucleotide (NAD), an important cofactor of many enzymes involved in metabolism and the production of energy from sugars and fats. The structure of nicotinamide, shown in Figure 1, incorporates a six-atom ring, with one nitrogen atom in the ring and another in the amide group side chain. Nicotinic acid is the other common form of niacin. It has the same ring structure but, as shown in Figure 2, oxygen atoms replace the nitrogen atom in the side chain. The nicotinic acid form of the vitamin produces severe side effects when taken in large doses; however, it is sometimes used as a medication to reduce high cholesterol levels in blood. Nicotinic acid was first produced from nicotine long before it was known to be a nutrient. Despite the similarities between nicotine and nicotinic acid, their functions are very different.

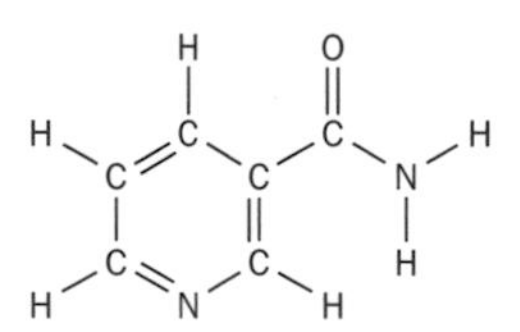

Figure 1. Structure of nicotinamide.

Figure 2. Structure of nicotinic acid.

Niacin and Pellagra

Pellagra is a disease characterized by skin rashes, diarrhea, mental deterioration, and death. Early in the last century it was a serious health problem. Alan Kraut in "Dr. Joseph Goldberger & the War on Pellagra" describes the situation as follows. In 1912 South Carolina alone reported 30,000 cases of pellagra and a mortality rate of 40 percent. In 1914 Dr. Joseph Goldberger (1874–1929) was assigned to study the disease. Goldberger had extensive prior experience treating yellow fever, dengue fever, and typhus. He noted that unlike these other diseases, pellagra was never transmitted from patients to doctors or hospital staff. He also determined that those patients likely to exhibit symptoms of pellagra shared a diet of refined corn flour, molasses, and pork fat. Such observations led Goldberger to deduce that poor nutrition might be the cause of the disease. In 1915 he tested his hypothesis with volunteers from a Mississippi prison. They were fed only the suspect diet and half developed the signs of pellagra within a few months. The symptoms disappeared when meat and vegetables were added to the volunteers' diet. Despite the fact that his study clearly indicated poor nutrition was the cause of pellagra, Goldberger spent the rest of his career attempting to convince political and medical authorities that germs were not the root cause of this dreaded disease. He was perhaps hampered in his efforts by the inability to determine exactly what was missing in the diet. Not until 1937 did Conrad Elvehjem identify the chemical nicotinamide as the cure for pellagra in dogs, followed almost immediately by the work of Thomas Spies, who demonstrated that niacin also cures human pellagra.

Sources of Niacin

The Federal Enrichment Act of 1942 required the millers of flour to restore iron, niacin, thiamin and riboflavin lost in the milling process. Enriched flours and baked goods made from them are now excellent sources of niacin. Niacin may also be found in meat, poultry, fish, whole grains, and peanut butter. Besides direct niacin intake, humans can convert the amino acid tryptophan to niacin. Many people take daily vitamin supplements to ensure they get enough niacin and other essential nutrients. SEE ALSO COENZYME; NICOTINAMIDE ADENINE DINUCLEOTIDE.

David Speckhard

Bibliography

Anderson, Jean, and Barbara Deskins. *The Nutrition Bible.* New York: William Morrow and Co., 1995.

Internet Resources

Kraut, Alan. "Dr. Joseph Goldberger & the War on Pellagra." Available from <http://www.nih.gov/od/museum/exhibits/goldberger/full-text.html>.

Nicotinamide Adenine Dinucleotide

Nicotinamide **adenine** dinucleotide (NAD) is the coenzyme form of the **vitamin** niacin. Most biochemical reactions require protein catalysts (enzymes). Some enzymes, lysozyme or trypsin, for example, catalyze reactions by themselves, but many require helper substances such as coenzymes, **metal** ions, and **ribonucleic acid** (RNA). Niacin is a component of two coenzymes: NAD, and nicotinamide adenine dinucleotide phosphate (NADP). NAD^+ (the oxidized form of the NAD coenzyme) is important in **catabolism** and in the production of metabolic energy. $NADP^+$ (the oxidized form of NADP) is important in the **biosynthesis** of fats and sugars.

Hans von Euler is generally recognized as the first to establish the chemical structure of NAD (Metzler, p. 468). Von Euler and Arthur Harden shared the 1929 Nobel Prize in physiology or medicine for the discovery of coenzymes (including NAD). Later von Euler showed that NAD contains two units of the sugar ribose, two phosphate groups, one adenine unit, and

adenine: one of the purine bases found in nucleic acids, $C_5H_5N_5$

vitamins: organic molecules needed in small amounts for the normal function of the body; often used as part of an enzyme catalyzed reaction

metal: element or other substance the solid phase of which is characterized by high thermal and electrical conductivities

ribonucleic acid: a natural polymer used to translate genetic information in the nucleus into a template for the construction of proteins

catabolism: metabolic process involving breakdown of a molecule into smaller ones resulting in a release of energy

biosynthesis: formation of a chemical substance by a living organism

Figure 1a. Structure of NAD+

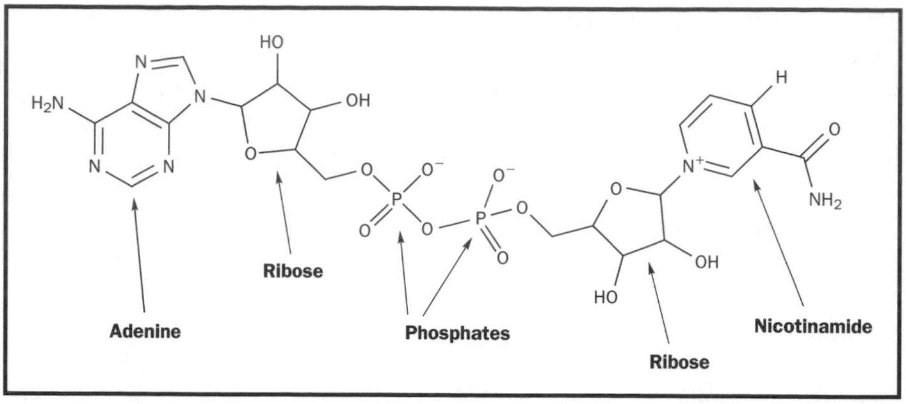

one nicotinamide unit (derived from niacin). (See Figure 1a.) The adenine, ribose, and phosphate compounds are linked exactly as in the nucleotide molecule adenosine diphosphate (ADP). In the case of NAD^+, the nicotinamide ring has a positive charge on its nitrogen atom: This is the + indicated in the designation NAD^+. This is often confusing, because the molecule as a whole is negatively charged due to the presence of the phosphate groups, as shown in the figure. In 1934 Otto Warburg and William Christian discovered a variant of NAD^+ in human red blood cell extracts (Metzler, p. 466). This form, called $NADP^+$, contains a third phosphate group attached to one of the ribose rings (see Figure 1b).

NAD^+ and $NADP^+$ play an essential role in many biochemical reactions, especially redox reactions in which oxidoreductase enzymes transfer hydrogen. (See Table 1 for a partial list of enzymes that require NAD^+ and $NADP^+$.) The redox reaction shown in Figure 2 is catalyzed by the oxidoreductase enzyme alcohol dehydrogenase. In this reaction, two hydrogen atoms and two electrons (the two electrons of the C-H bond) are removed from the ethanol molecule. One hydrogen atom and both electrons, shown in red, are transferred to NAD^+, generating NADH. (A molecule's acquisition of electrons is called reduction, thus NADH is the reduced form of NAD^+. Conversely, NAD^+ is the oxidized form, the form with fewer electrons.) $NADP^+$ can be reduced to NADPH, just as NAD^+ can be reduced to NADH, although different enzymes will be involved. Enzymes that use NAD^+ rarely use $NADP^+$ and vice versa, making it possible to separate the

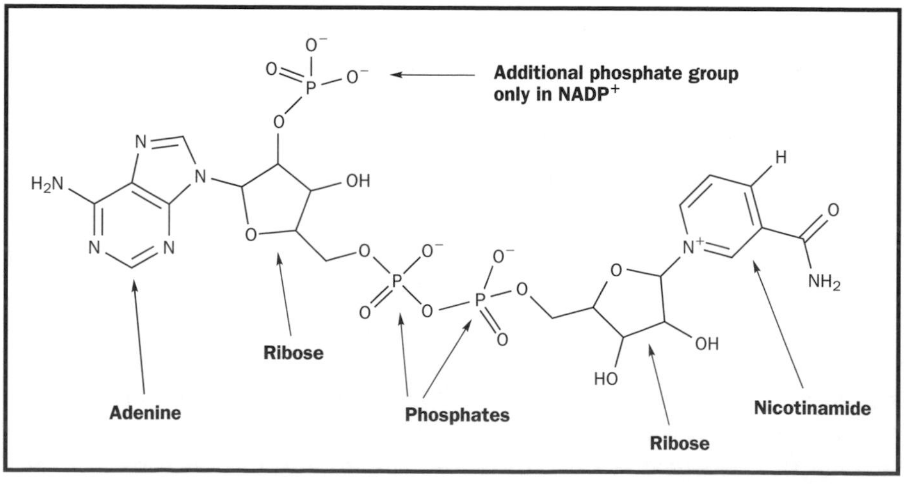

Figure 1b. Structure of NADP+

ENZYMES THAT REQUIRE NAD AND NADP

	Enzyme	Function
Enzymes that use NAD^+/NADH	Alcohol dehydrogenase	Metabolizes alcohol
	Glyceraldehyde phosphate dehydrogenase	Catalyzes important step in glycolysis
	Lactate dehydrogenase	Catalyzes reactions in muscle and liver cells
	Pyruvate dehydrogenase	Catalyzes reactions connecting glycolysis to the Krebs cycle
	α-keto-glutarate dehydrogenase, isocitrate dehydrogenase, malate dehydrogenase	Catalyzes reactions in the Krebs cycle, aerobic metabolism
	NADH dehydrogenase	Catalyzes oxidative phosphorylation reactions
	Hydroxy-acyl-SCoA dehydrogenase	Important in fat catabolism
Enzymes that use $NADP^+$/NADPH	Glucose 6-phosphate dehydrogenase	Catalyzes reactions in the pentose phosphate pathway
	β-ketoacyl-ACP reductase β-enoyl-ACP reductase	Catalyzes reactions in fatty acid synthesis
	Chloroplast glyceraldehyde Phosphate dehydrogenase	Catalyzes reactions in the Calvin cycle, glucose synthesis

Table 1. Enzymes that use NAD^+/NADH and enzymes that use $NADP^+$/NADPH

biosynthetic and energy-producing functions of $NADP^+$ and NAD^+. NAD^+ and $NADP^+$ act as electron acceptors in oxidoreductase catalyzed reactions; NADH and NADPH act as electron donors.

The transfer of hydrogen to NAD^+ is **stereospecific**, and dehydrogenases are now classified as H_A side or H_B side enzymes, according to the "side" of the NAD molecule they act on. Alcohol dehydrogenase, which catalyzes the reaction shown in Figure 2, is an H_A side enzyme, as it promotes the transfer of hydrogen from ethanol to the "A position" of NADH. This specificity was somewhat unexpected, as the flat nicotinamide group can be approached (by a dehydrogenase enzyme) equally well from either side in solution, when NAD^+ is bound to the enzyme; however, one side or the other of NAD is more approachable and therefore preferred.

stereospecific: yielding one product when reacted with a given compound but the opposite product with its stereoisomer

Both NADH and NADPH have a distinctive signal in ultraviolet **spectroscopy**. This signal is lost when NADH or NADPH is oxidized (to NAD^+ or $NADP^+$). This phenomenon has been employed by thousands of scientists to monitor a wide variety of enzyme-catalyzed reactions. SEE ALSO Coenzyme; Enzymes; Nicotinamide.

spectroscopy: use of electromagnetic radiation to analyze the chemical composition of materials

David Speckhard

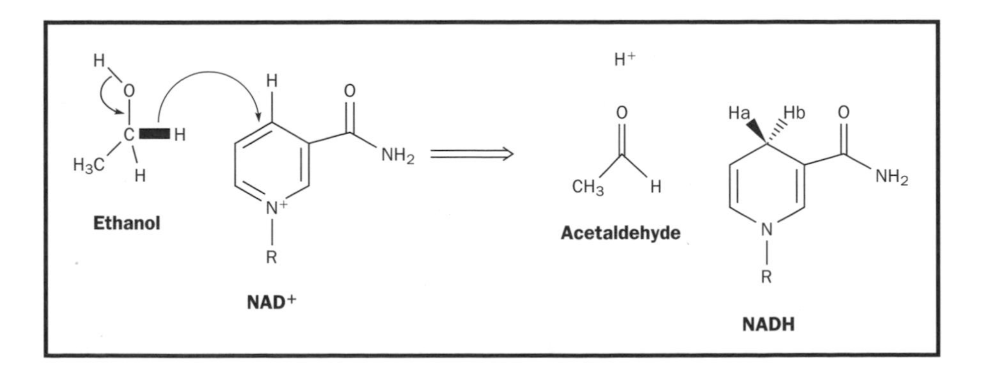

Figure 2. Reduction of NAD to NADH by alcohol dehydrogenase. Note R stands for the remainder of the NAD+ and NADH molecules not shown in the figure or changed in the reaction.

Bibliography

Magill, Frank (1991). *The Nobel Prize Winners: Physiology or Medicine,* Vol. 1. Pasadena, CA: Salem Press.

Metzler, David E. (1977). *Biochemistry: The Chemical Reactions of Living Cells.* New York: Academic Press.

Nicotine

alkaloid: alkaline nitrogen-based compound extracted from plants

Nicotine, $C_{10}H_{14}N_2$, is a highly toxic, pale yellow **alkaloid** produced in tobacco plants in response to leaf damage. Nicotine is synthesized in the roots of tobacco plants in response to hormones released by damaged tissue, and it is then carried to the leaves, where it is stored in concentrations of between 2 percent and 8 percent by weight. Nicotine is used commercially as an insecticide (it is one of the few poisons to which insects have not become resistant). Tobacco smoke contains nicotine, believed to be the active (and addictive) ingredient.

Mayan peoples of South America used tobacco for recreational and ceremonial, as well as medicinal, purposes. Mayan sculptures depict high-ranking persons smoking cigars and priests blowing tobacco smoke over human sacrifices. By the time of the arrival of Christopher Columbus in the New World, tobacco use had spread throughout both North America and South America. Early accounts by European explorers describe Native Americans carrying glowing sticks from which they inhaled, and many pipes are found among Native American artifacts. Tobacco was often chewed by Native Americans; the juice was dropped into eyes to improve night vision and applied to skin as an agent having antiseptic properties.

The men who accompanied Columbus encountered many users of tobacco, but early European explorers showed little interest in the plant until they acquired an awareness that it might be used to treat diseases. Europeans at first forbade tobacco use, but tobacco gradually gained a reputation among court physicians as a medicine. For many Europeans, tobacco was suddenly a valuable New World commodity.

nonpolar: molecule, or portion of a molecule, that does not have a permanent, electric dipole

receptor: area on or near a cell wall that accepts another molecule to allow a change in the cell

Nicotine is the active ingredient of tobacco. Nicotine is soluble in water and in **nonpolar** solvents. It can be absorbed by the body from smoke that has been taken into the lungs, or through the skin. It rapidly crosses the blood-brain barrier, appearing in brain tissue minutes after its absorption into capillaries lining the alveoli of the lungs. The presence of nicotine in the body stimulates nicotinic-cholinergic **receptors** of the nervous system, resulting in increased attention span, increased heart rate and blood pressure, and increases in the concentrations of some hormones. Habitual users have a feeling of well-being after intake of nicotine, ascribed to the increased concentrations of dopamine in the brain. The increased metabolic rate that is associated with nicotine use may be what is in back of the common belief that it is easier to lose weight when using nicotine.

Nicotinic-cholinergic receptors that are part of the autonomic nervous system may be stimulated at low concentrations of nicotine, but blocked at higher concentrations. The repeated use of nicotine-containing products (which includes chewing tobacco, chewing nicotine-containing gum, or the use of therapeutic patches that release nicotine for skin absorption) promotes the formation of (new) nicotinic-cholinergic receptors. The tolerance and eventual addiction that go along with repeated use may result in increased craving for nicotine.

Many environmentally hazardous substances, such as asbestos and radon, are much more hazardous when they become mixed with cigarette smoke, probably because the particulate matter in smoke in the atmosphere may adsorb these dangerous substances and carry them into the alveoli of lungs. Many cancers may be caused by substances or materials associated with nicotine use, such as tobacco smoke or the tobacco plant itself (as in chewing tobacco). Nicotine itself, although not known to cause cancer directly, causes proliferation of both healthy and neoplastic cells, and may further the development of cancer by stimulating angiogenesis (the growth of new blood vessels) and thus providing cancerous tissues with increased blood supplies. The effect of nicotine on cell growth is especially strong in tissue environments having low concentrations of carbon dioxide, for example, in damaged lungs; thus, the effect would be greater in persons whose breathing was already impaired. Nicotine's stimulation of cell growth may account for the observation that atherosclerotic plaques (which are intracellular accumulations of **lipids**) grow more rapidly in the presence of this alkaloid substance. This effect may actually become the basis of medical treatments intended to improve blood flow to tissues damaged by atherosclerosis.

lipid: a nonpolar organic molecule; fatlike; one of a large variety of nonpolar hydrophobic (water-hating) molecules that are insoluble in water

Single exposure to nicotine in quantities as small as 50 mg (0.0018 oz) may result in vomiting and seizures; the average cigarette yields about 3 mg (0.00011 oz). As nicotine can be absorbed through skin, accidental exposures in persons working with nicotine-containing pesticide preparations may be fatal. Extracts of chewing tobacco are effective insecticides; commercial insecticide products contain much higher amounts of nicotine than products intended for human consumption. SEE ALSO DOPAMINE; RADON; TOXICITY.

Dan M. Sullivan

Bibliography

Brautbar, N. (1995). "Direct Effects of Nicotine on the Brain: Evidence for Chemical Addiction." *Archives of Environmental Health* (July 1):263.

Nicotine and Tobacco Research. Various issues.

Internet Resources

"A Brief History of Tobacco." Available from <http://www.cnn.com/US/9705/tobacco/history/index/html>.

Niobium

MELTING POINT: 2,475°C ±10°C
BOILING POINT: ~4,740°C
DENSITY: 8.57g/cm^3 at room temperature
MOST COMMON IONS: Nb^{3+}

Niobium **metal** is typically gray or dull silver in color. It is one of the refractory metals along with tantalum, tungsten, molybdenum, and rhenium, due to its very high melting point. It is estimated that niobium has a natural occurrence in Earth's crust of approximately 20 parts per million (ppm). The largest niobium-containing mineral reserves are located in Brazil and Canada.

metal: element or other substance the solid phase of which is characterized by high thermal and electrical conductivities

English chemist Charles Hatchett originally discovered niobium in 1801 while examining an ore sample obtained in Connecticut. Since the ore sample came from the United States, he named the unknown material columbium (at the time, Columbia was another name for America). In the 1840s German chemist Heinrich Rose rediscovered the element and named it niobium. Chemically, niobium and the element tantalum are very similar, so niobium was named for Niobe, a daughter of Tantulus (root name for the element tantalum) in ancient mythology. It was not until 1950, at a meeting of the International Union of Pure and Applied Chemistry, that it was finally settled that the element would be called niobium. Many metallurgists and engineers, especially in the United States, still refer to the element as columbium.

alloy: mixture of two or more elements, at least one of which is a metal

nuclear: having to do with the nucleus of an atom

Pure niobium has relatively poor mechanical properties and readily oxidizes in air to niobium pentoxide (Nb_2O_5) at elevated temperatures. Various niobium-containing **alloys** such as Nb-1Zr and C-103 have been successfully used in specific liquid-metal based **nuclear** applications and in the fabrication of various rocket components. SEE ALSO INORGANIC CHEMISTRY.

Daniel P. Kramer

Bibliography

American Society for Metals (1998). *Metals Handbook*, Vol. 2: *Properties and Selection: Nonferrous Alloys and Pure Metals*. Metals Park, OH: American Society for Metals.

Nitrogen

MELTING POINT: **−210°C**
BOILING POINT: **−196°C**
DENSITY: **0.0012506 g/cm^3**
MOST COMMON IONS: **NH_4^+, N^{3-}, NO_2^-, NO_3^-**

Nitrogen is a gaseous element that is abundant in the atmosphere as the molecule dinitrogen (N_2). Scottish chemist Daniel Rutherford, Swedish chemist Carl Wilhelm Scheele, and English chemist Henry Cavendish independently discovered the element in 1772. Nitrogen received its name in 1790 from French chemist Jean-Antoine Chaptal, who realized that it was present in nitrate (NO_3^-) and nitric acid (HNO_3).

isotope: form of an atom that differs by the number of neutrons in the nucleus

spectroscopy: use of electromagnetic radiation to analyze the chemical composition of materials

RNA: natural polymer used to translate genetic information in the nucleus into a template for the construction of proteins

DNA: deoxyribonucleic acid—the natural polymer that stores genetic information in the nucleus of a cell

saltpeter: potassium nitrate; chile saltpeter is sodium nitrate

Nitrogen is the most abundant terrestrial element in an uncombined state, as it makes up 78 percent of Earth's atmosphere as N_2, but it is a minor component (19 parts per million) of Earth's crust. Nitrogen exists as two **isotopes**: ^{14}N (99.63% relative abundance) and ^{15}N (0.4% abundance). Both isotopes are nuclear magnetic resonance (NMR) active, with the rarer ^{15}N isotope being utilized more commonly in NMR **spectroscopy** because of its nuclear spin of one-half.

In its reduced state nitrogen is essential for life because it is a constituent of the nucleotides of deoxyribonucleic acid (**DNA**) and ribonucleic acid (**RNA**) molecules that encode genetic information) and of the amino acids of proteins. The nitrogen-containing minerals **saltpeter** (KNO_3) and sodium nitrate ($NaNO_3$) are found in Chile, India, Bolivia, the former Soviet Union, Spain, and Italy; they were significant as fertilizers and explo-

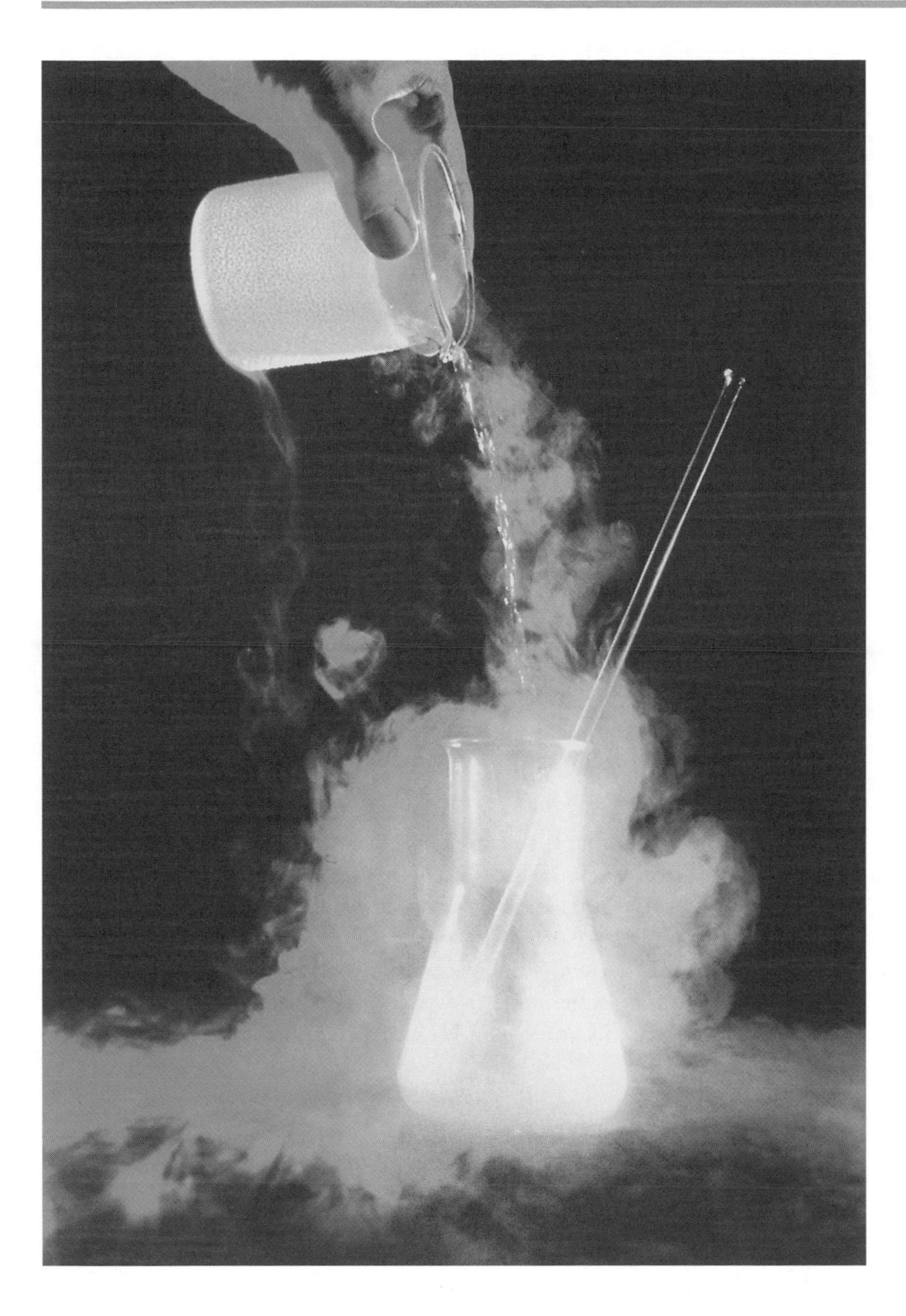

Liquid nitrogen is being poured into a beaker.

sives precursors prior to modern industrial nitrogen fixation. The global nitrogen cycle between the atmosphere and the biosphere is based on continuous exchanges whereby dinitrogen is fixed by the enzyme nitrogenase in symbiotic bacteria associated with some plant roots, by the Haber-Bosch industrial process for the reduction of N_2 with H_2 to ammonia, and by atmospheric **oxidation** during electrical discharges such as lightning.

oxidation: process that involves the loss of electrons (or the addition of an oxygen atom)

Dinitrogen possesses the strongest known chemical bond, with a high bond dissociation energy of 945 kJ mol^{-1} and a short N-N triple bond length of 109.8 picometers. This colorless, tasteless, odorless gas is relatively unreactive because of its strong N-N triple bond, a stable electronic configuration, and the lack of a dipole moment. Dinitrogen is reduced by

metal: element or other substance the solid phase of which is characterized by high thermal and electrical conductivities

liquefaction: process of changing to a liquid form

fractional distillation: separation of liquid mixtures by collecting separately the distillates at certain temperatures

lithium **metal** at room temperature to give the saline (saltlike) lithium nitride (Li_3N).

Dinitrogen is obtained from the atmosphere by either membrane separations or repetitive cycles of compression and cooling (termed **liquefaction**), followed by **fractional distillation** to separate it from other gases. The major uses of dinitrogen are as blanketing atmospheres for chemical processing and metallurgical production, in glove boxes for handling of dioxygen- and water-sensitive compounds, in electronic materials manufacturing, and in food packaging. Liquid dinitrogen is used as a refrigerant in the laboratory and food industry and in the preservation of biological samples.

The major industrial applications of nitrogen-containing compounds are in fertilizers and explosives. The most important nitrogenous compounds are ammonia (NH_3), which is used as a fertilizer, refrigerant, nonaqueous solvent, and precursor for many nitrogen compounds including nylon and plastics; nitric acid (HNO_3); ammonium nitrate (NH_4NO_3), a fertilizer and explosive; fertilizers ammonium phosphate and urea ($H_2NC[O]NH_2$). Other important oxides include nitrous oxide (N_2O), used as a dental anesthetic and aerosol propellant, and **nitric oxide** (NO), the simplest stable odd-electron molecule and a short-lived, biologically active neurotransmitter, cytotoxic agent in immunology, vasoconstrictor for blood pressure control, and major component along with NO_2 in **acid rain** and smog. The strong **reducing agent** hydrazine (N_2H_4) is used in controlling the attitude of spacecraft and in rocket fuels.

nitric oxide: compound, NO, which is involved in many biological processes; the drug Viagra enhances NO-stimulation of pathways to counteract impotence; may be involved in killing tumors

acid rain: precipitation that has a pH lower than 5.6; term coined by R. A. Smith during the 1870s

reducing agent: substance that causes reduction, a process during which electrons are lost (or hydrogen atoms gained)

Covalent, intermetallic metal nitrides are among the most stable compounds and are hard, refractory materials that can possess useful properties. For example, titanium nitride (TiN) is used as a gold-colored coating on costume jewelry and as a wear-resistant coating on tool bits; silicon nitride (Si_3N_4) is a strong, thermally stable ceramic material; and gallium nitride (GaN) is a compound semiconductor with optoelectronic applications (e.g., lasers, LEDs). SEE ALSO CAVENDISH, HENRY; GASES; INORGANIC CHEMISTRY; SCHEELE, CARL.

Louis Messerle

Bibliography

Greenwood, Norman N., and Earnshaw, A. (1997). *Chemistry of the Elements*, 2nd edition. Oxford, U.K.: Butterworth-Heinemann.

Nitrogen Fixation ***See Haber, Fritz.***

Nobel, Alfred Bernhard

SWEDISH MANUFACTURER, INVENTOR, AND PHILANTHROPIST
1833–1896

Alfred Bernhard Nobel was born in Stockholm, Sweden, on October 21, 1833, as the third of four sons to Immanuel and Andriette (Ahlsell) Nobel. That same year, his father, an engineer and builder, went bankrupt when barges full of building materials were lost at sea. In 1837 Immanuel left Stockholm and moved to St. Petersburg, Russia, where he started manufacturing equipment for the Russian army. His factory flourished, especially with the manufacture and sale of naval mines of his own construction.

Immanuel was eventually able to bring his family to Russia, where his sons were given a private education. Alfred Nobel's interests ranged from literature and poetry to physics and chemistry. Nobel's command of foreign languages was excellent; by the age of seventeen he was fluent in Swedish, Russian, French, English, and German, which aided him in his future business transactions.

In 1863 Nobel began trying to master the process of the **synthesis** of nitroglycerine. His first partial success was a mixture of nitroglycerine with black gunpowder, called "blasting oil." The danger of working with such an unstable material was a problem. After an explosion in Nobel's Stockholm factory that claimed five lives, including that of his brother Emil (1843–1864), the municipal authorities forbade him to carry out further experiments in the town. He then continued his work on a ship anchored on Lake Mälaren.

Swedish manufacturer Alfred Nobel, the inventor of dynamite.

synthesis: combination of starting materials to form a desired product

Nobel began to realize that, for handling purposes, nitroglycerine would have to be absorbed in some kind of stabilizing carrier. After many unsuccessful trials using sawdust, charcoal, paper, and brick-dust, he finally succeeded with Kieselguhr, a diatomaceous earth found in Germany. Even when saturated with nitroglycerine, this earth was quite safe to handle, a blasting cap and detonator being required to force it to explode. Originally called "Kieselguhr-dynamite," its name was later abridged to "dynamite" (the Greek *dynamis* meaning "power"). Nobel was granted a patent for dynamite in England on May 7, 1867, and on September 13 of the same year in Sweden.

In 1868 Nobel and his father were awarded the Letterstedt Prize by the Royal Swedish Academy of Sciences. Nobel highly valued this award, which was the only prize he ever received, and which was perhaps the source of his idea for a similar prize he later established.

Nobel, one of the wealthiest men of his time, constantly moved between his factories and his houses equipped with laboratories. He was both an industrialist and an administrator, handling his business without a secretary. As he wrote in a letter: "My home is where I work, and I work everywhere." His prodigious activities had a negative effect on his health, which had been poor since his youth. After 1890 he preferred to stay at his home in San Remo, Italy. By that time he had 350 patents and ninety-three factories in several countries.

On November 27, 1895, Nobel wrote his last will, in which he generously bequeathed his wealth to his relatives and friends. The second part of his will, however, is more famous, for it is here that he established the Nobel Prizes. Nobel's property that was designated for the fund was worth seventy million Swedish crowns, and has continued to grow since then. Nobel Prizes are awarded in physics, chemistry, medicine, literature, and peace. Since 1969 a Nobel Prize, funded by the Swedish Bank, has also been awarded for outstanding achievements in economy.

Nobel died on December 10, 1896, in San Remo. Shortly before his death he wrote: "It sounds like the irony of fate that I should be ordered to take nitroglycerin internally," which had been prescribed to him as a treatment for angina pectoris. The Nobel Foundation, established in accordance with his will, awarded the first Nobel Prizes in 1901. SEE ALSO EXPLOSIONS.

Vladimir Karpenko

Bibliography

Fant, Kenne (1993). *Alfred Nobel: A Biography*, tr. Marianne Ruuth. New York: Arcade.

Hellberg, Thomas, and Jansson, Lars Magnus (1986). *Alfred Nobel.* Karlshamn, Sweden: Lagerblads Förlag AB.

Ihde, Aaron J. (1984). *Development of Modern Chemistry.* New York: Dover.

Roberts, Royston M. (1989). *Seredipity: Accidental Discoveries in Science.* New York: Wiley.

Internet Resources

"Alfred Nobel: Biographical." The Nobel Foundation. Available from <http://www.nobel.se/nobel/alfred-nobel/biographical/index.html>.

Nobelium

MELTING POINT: Unknown
BOILING POINT: Unknown
DENSITY: Unknown
MOST COMMON IONS: No^{2+}, No^{3+}

The first claim for the discovery of the element nobelium was made in Sweden in 1957. However, neither American nor Soviet researchers could duplicate the original results, which are now known to have been interpreted incorrectly. The actual discovery of nobelium is credited to researchers in Berkeley, California, who in 1958 bombarded a curium target (95% ^{244}Cm and 4.5% ^{246}Cm) plated on a nickel foil with 60 to 100 MeV ^{12}C ions, and detected both the 8.4 MeV α-particles created by the **radioactive decay** of ^{252}No and the ^{250}Fm created from the α-decay of ^{254}No. Known **isotopes** of nobelium possess 148 to 160 neutrons and 102 protons; all are radioactive, with half-lives ranging between 2.5 milliseconds and 58 minutes, and decay by spontaneous **fission**, α-particle emission, or electron capture. ^{259}No has the longest half-life: 58 minutes.

radioactive decay: process involving emission of subatomic particles from the nucleus, typically accompanied by emission of very short wavelength electromagnetic radiation

isotope: form of an atom that differs by the number of neutrons in the nucleus

fission: process of splitting an atom into smaller pieces

lanthanides: a family of elements (atomic number 57 through 70) from lanthanum to lutetium having from 1 to 14 4f electrons

rare earth elements: older name for the lanthanide series of elements, from lanthanum to lutetium

Nobelium is a member of the actinide series of elements. The ground state electron configuration is assumed to be (Rn)5f147s2, by analogy with the equivalent **lanthanide** element ytterbium ([Kr]4f146s2); there has never been enough nobelium made to experimentally verify the electronic configuration. Unlike the other actinide elements and the lanthanide elements, nobelium is most stable in solution as the dipositive cation No^{2+}. Consequently its chemistry resembles that of the much less chemically stable dipositive lanthanide cations or the common chemistry of the alkaline earth elements. When oxidized to No^{3+}, nobelium follows the well-established chemistry of the stable, tripositive **rare earth elements** and of the other tripositive actinide elements (e.g., americium and curium). SEE ALSO ACTINIUM; BERKELIUM; EINSTEINIUM; FERMIUM; LAWRENCIUM; MENDELEVIUM; NEPTUNIUM; PLUTONIUM; PROTACTINIUM; RUTHERFORDIUM; THORIUM; URANIUM.

Mark Jensen

Bibliography

Hoffman, Darleane C.; Ghiorso, Albert; and Seaborg, Glenn Theodore (2000). *The Transuranium People: The Inside Story.* London: Imperial College Press.

Nobel Metals ***See Gold; Palladium; Platinum; Silver.***

Noble Gases

The noble gases, also known as rare or **inert** gases, form Group 18 of the Periodic Table, embedded between the alkali **metals** and the **halogens**. The elements helium, neon, argon, krypton, xenon, and radon are the members of this group.

inert: incapable of reacting with another substance

metal: element or other substance the solid phase of which is characterized by high thermal and electrical conductivities

halogen: element in the periodic family numbered VIIA (or 17 in the modern nomenclature) that includes fluorine, chlorine, bromine, iodine, and astatine

Discovery

In 1785 English physicist and chemist Henry Cavendish performed an experiment in which he passed electric sparks through an air bubble enclosed by a soap solution (NaOH). While nitrogen and oxygen were absorbed by the solution, about 1/120th of the volume of the original bubble remained—it is now known that the residual gas was mainly argon. However, it was a century later before argon was finally recognized as a new element. In 1894 English physicist John William Strutt noticed that nitrogen produced from air had a slightly higher density than that from nitrogen compounds. Sir William Ramsay, together with Strutt, repeated the Cavendish experiment and identified argon as the unreactive species. The **liquefaction** of air in 1895 by Carl von Linde allowed Ramsay the further discovery of neon, krypton, and xenon. Extraterrestrial helium had been discovered earlier (in 1868), based on its **spectral lines** in the Sun. Ramsay realized that the new elements did not fit into the contemporary periodic system of the elements and suggested that they form a new group, bridging the alkali metals and the halogens. The last member of the family, radon, was discovered in 1900 by Ernest Rutherford and Frederick Soddy as a decay product of radium.

liquefaction: process of changing to a liquid form

spectral line: line in a spectrum representing radiation of a single wavelength

Physical and Chemical Properties

The chemical inertness of the noble gases is based on their electronic structure. Each element has a completely filled **valence** shell. In fact their inertness helped to develop the key idea of a stable octet.

valence: combining capacity

The atomic sizes of the noble gas elements increase from top to bottom in the Periodic Table, and the amount of energy needed to remove an electron from their outermost shell, the **ionization** energy, decreases in the same order. Within each period, however, the noble gases have the largest ionization energies, reflecting their chemical inertness. Based on increasing atomic size, the electron clouds of the spherical, **nonpolar**, atoms become increasingly polarizable, leading to stronger interactions among the atoms (van der Waals forces). Thus, the formation of solids and liquids is more easily attained for the heavier elements, as reflected in their higher melting points and boiling points. As their name implies, all members of the family are gases at room temperature and can, with the exception of helium, be liquefied at atmospheric pressure.

ionization: dissociation of a molecule into ions carrying + or − charges

nonpolar: molecule, or portion of a molecule, that does not have a permanent, electric dipole

Compounds

Until 1962 only physical inclusion compounds were known. Argon, krypton, and xenon form cage or clathrate compounds with water (clathrate hydrates) and with some organics such as quinol. The host molecules are arranged in such a way that they form cavities that can physically trap the noble gas atoms, referred to as "guests." The noble gas will be released upon dissolution or melting of the host **lattice**.

lattice: systematic geometrical arrangement of atomic-sized units that constitute the structure of a solid

A scientist studying helium being released from a hot spring in Yellowstone National Park.

In 1962 the first chemical noble gas compound, formulated as $XePtF_6$, was synthesized by Neil Bartlett. This result spurred intense research activity and led to the discovery of numerous xenon and krypton compounds. In 2000 the formation of the first argon compound, argon fluorohydride (HArF), was reported by Leonid Khriachtchev and colleagues. SEE ALSO ARGON; CAVENDISH, HENRY; HELIUM; KRYPTON; NEON; RAMSAY, WILLIAM; RUTHERFORD, ERNEST; SODDY, FREDERICK; STRUTT, JOHN; XENON.

Tanja Pietraß

Bibliography

Greenwood, Norman N., and Earnshaw, A. (1984). *Chemistry of the Elements.* New York: Pergamon Press.

Khriachtchev, Leonid; Pettersson, Mika; Runeberg, Nino; Lundell, Jan; and Räsänen, Markku (2000). "A Stable Argon Compound." *Nature* 406(6798):874–876.

Nomenclature of Inorganic Chemistry

The purpose of nomenclature in chemistry is to convey information about the material being described. The designation chosen should be unequivocal, at least within the limitations of the type of nomenclature adopted. The type adopted will depend in part on the total amount of information to be conveyed, the kind of compound to be described, and the whim of the person describing the compound.

Nomenclaturists use the terms "trivial" and "systematic" to describe two major divisions of nomenclature. Systematic nomenclature is based on established principles so that it can be extended in a logical way to describe known, new, and hypothetical compounds. A trivial nomenclature is one established by rule of thumb and includes many of the older names (spirit of salt, aqua regia, etc.) and lab nomenclatures (the green chelate, etc.).

Actual usage is often a mixture of the two types, and the fundamental bases of all chemical names, those of the elements, are essentially trivial. Note that the name methane is trivial, but that the name pentane is not. For this discussion, a formula representing a compound can be regarded simply as a kind of name. The principal general (but by no means the only) types of nomenclature used in inorganic chemistry are substitutive and additive (coordination).

Substitutive Nomenclature

Substitutive nomenclature is essentially an organic invention and follows the historical development of organic chemistry. It starts with the designation of an appropriate parent compound from which the compound under discussion can be developed formally by substitution or replacement processes. In organic chemistry these parents can be the paraffins, and in inorganic chemistry they are generally (and arbitrarily) taken to be the hydrides of the elements of Periodic Groups 14, 15, and 16, plus boron, which also has an additional rather specific nomenclature of its own. Thus the formula SiH_3Cl can be named chlorosilane, as a substituted derivative of the saturated parent SiH_4, silane (compare chloromethane). The generation of a radical by the loss of a hydrogen atom from the parent is indicated by modification of the termination, silane becoming silyl, $SiH_3^{\cdot}$ (the superscript dot indicates an unpaired electron). The name silyl can be used to represent a substituent group in another parent hydride (compare methyl) or for the unbound radical, and the procedure is quite general for all parent hydrides to which the methodology is applied. Silane can also be modified formally by the removal of a proton, yielding the **anion** SiH_3^-. The name then takes the characteristic anion ending -ide: silanide. The formal addition of a proton is indicated by another termination (-ium), giving SiH_5^+, silanium. These terminations are used generally in inorganic nomenclature, as in chloride for Cl^- and ammonium for NH_4^+. Other formal operations recognized in substitutive nomenclature include addition or removal of a hydride from the parent. This can be indicated by the termination -ylium, giving the name silylium for SiH_3^+.

anion: negatively charged chemical unit, like Cl^-, CO_3^{2-}, or NO^{3-}

Other Modifications of Names

The terminations cited above can be used generally in inorganic nomenclature. However, they are sometimes not applicable, especially where parent hydrides are not reasonably definable. Inorganic chemists have tended to assign electropositive and **electronegative** character to elements, though numerical values are not necessarily easy to define. **Metals** are generally assigned electropositive character and nonmetals electronegative character. The names developed on this basis may imply formally a saltlike nature even in compounds that are not really salts at all. Thus common salt is called sodium chloride, which is ionic, but phosphorus trichloride is certainly not saltlike.

electronegative: tending to attract electrons

metal: element or other substance the solid phase of which is characterized by high thermal and electrical conductivities

It is not wise to infer the detailed physical nature of a compound from the name alone. In this system the name of the (electropositive) metal is not modified from that of the element, but the name of the electronegative element is, and in the way described in the Substitutive Nomenclature section, above. Similarly we derive oxide and sulfide, for example, from oxygen and sulfur. The same division between electronegative and electropositive parts is evident in the covalent nonionic compound $SiCl_4$, which can be named silicon **tetrachloride**, though an equally valid substitutive name is tetrachlorosilane.

tetrachloride: term that implies a molecule has four chlorine atoms present

Inorganic chemists also use a further termination to indicate the name of a cation. This is the ending -ate, and it is used as a modification of the name of an oxoacid. Thus sulfuric acid, H_2SO_4, gives rise to sulfate, SO_4^{2-}, phosphoric acid to phosphate, PO_4^{3-}, and nitric acid to nitrate, NO_3^-. The partially deprotonated anions such as HSO_4^- and $H_2PO_4^{2-}$ are rather more complicated to deal with, and are discussed in *Nomenclature of Inorganic Chemistry*, often referred to as the Red Book.

In an older procedure that is no longer recommended, the name of an electropositive element displaying more than one **oxidation** state in its compounds was sometimes modified to indicate the particular oxidation state involved. Thus iron chlorides were often named ferrous chloride and ferric chloride to convey the two oxidation states of II and III (note that, like normal arabic numbers, these Roman numerals are positive unless otherwise shown by a negative sign). However, the use was not consistent. Cuprous and cupric chlorides indicated oxidation states I and II, and phosphorous and phosphoric chlorides indicated oxidation states III and V. Modern nomenclature specifies the oxidation state of the electropositive partner in these compounds directly: iron(II) chloride, iron(III) chloride, copper(I) chloride, copper(II) chloride, phosphorus(III) chloride, and phosphorus(V) chloride. These designations are unequivocal. The number of counter anions, 1, 2, 3, or 5, should immediately be evident. Examples of negative oxidation states include oxide(-II) or oxide(2-), and dioxide(-I) or dioxide(2-). Note that in a multi-atom group, of which PO_4^{3-} may be taken as an example, the charge on any given atom may not be evident, even if the overall charge is known. In contrast, the oxidation states phosphorus(V) and oxide(–II) are much more readily defined. The use of such charges in names and formulae in these circumstances is not recommended.

oxidation: process that involves the loss of electrons (or the addition of an oxygen atom)

Formulae

The rules for formulae for the compounds discussed above are rather elastic. At its simplest, a formula is a list of element symbols accompanied by multiplying subscripts indicating the atomic proportions of each kind of atom. These formulae may be empirical, simply corresponding to the atom ratios, or stoichiometric, representing the totality of the atoms within a molecule. The latter can be used to calculate a molecular weight. Strictly speaking, for a compound that exists as discrete molecules, this latter can also be termed a molecular formula, but this is a misnomer for ionic compounds and for compounds of which the structure changes with temperature. The ordering of these symbols can be adjusted to suit the requirements of the user. At the simplest, an alphabetical order is used, since this is the same in most European languages. Many chemists emphasize the importance of carbon and hydrogen and adopt a sequence C, H, N, and then the remaining element symbols in alphabetical order. Such devices are often employed in indexes. Inorganic chemists often group the atomic symbols in a formula in electropositive and electronegative groups, designated as discussed above. This can be a somewhat arbitrary procedure, and the relative positions of atoms in an electronegativity sequence may be established using the Periodic Table. For simple cases, formulae such as NaCl or $SiCl_4$ are used. Anionic groups are assumed to be electronegative, hence $Ca_3(PO_4)_2$. The parentheses are used to define the associated groups of atoms within the formula.

Formulae can also be used to indicate two- or three-dimensional structures. This is particularly useful for coordination compounds, which are discussed next. However, this use is not restricted to classical coordination compounds, as the following examples show. Special devices are often adopted to indicate bonds or lines that are not in the plane of the paper. Their use is not consistent throughout chemistry, but the meaning in any given case is generally obvious.

Figure 1.

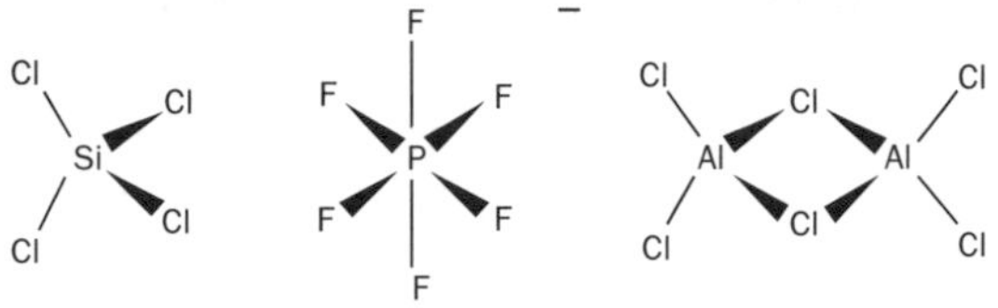

The first example represents a tetrahedral arrangement, because the solid defined by the four chlorine atoms at its apices is a tetrahedron. The second is **octahedral**, and the third represents two edge-fused tetrahedra. The wedge bonds are pointing in front of or behind the plane of the paper; the thin lines designate bonds in the plane of the paper.

octahedral: relating to a geometric arrangement of six ligands equally distributed around a Lewis acid; literally, eight faces

Figure 2.

Inorganic chemists often represent tetrahedra, octahedra, and other shapes in their formulae, to help the reader identify molecular shapes. The broken lines designating these shapes are not intended to represent bonds between atoms.

Oxidation states may also be indicated in formulae where this is helpful, though the need to do so is not common in the simplest cases. The following examples show the formalism employed: $Fe^{II}Cl_2$, $Fe^{III}Cl_3$, $Cu^{I}Cl$, $Cu^{II}Cl_2$, $P^{III}Cl_3$, $P^{V}Cl_5$.

Coordination Nomenclature

This is an additive nomenclature, and just as organic chemists have developed substitutive nomenclature in parallel with the methodology of substitutive chemistry, inorganic chemists have developed a nomenclature for coordination compounds that arises from the formal assembly of a

ligand: molecule or ion capable of donating one or more electron pairs to a Lewis acid

organometallic compound: compound containing both a metal (transition) and one or more organic moieties

coordination entity from its components, a central metal ion (in the simplest cases) and its **ligands**. Such a coordination entity may be neutral or it may carry a charge, positive or negative. Any such charge may be shown in the usual way, using formalisms such as 2- and 3-. Clearly **organometallic compounds**, depending upon their type, may be named either from substituted parent hydrides or as coordination entities.

Formulae in Coordination Nomenclature

The general rule is that the formula of a coordination entity should always appear within square brackets, even when the entity itself is an infinite polymer. The use of enclosing marks (square brackets, curly brackets, and parentheses) is slightly different for the usage that is common in organic chemistry. The usual priority sequence is [( )], [{( )}], [{[( )]}], [{{[( )]}}], and so on. Brackets should always be used if they make the formula clearer. The order of symbol citation within the formula of a coordination entity should begin with the metal ion followed by the ligands, ideally with charged ligands cited in alphabetical order using the first symbol of the ligand formula, and these are then followed as a class by the neutral ligand formulae, similarly ordered. The division into neutral and charged ligands can be somewhat arbitrary. Since a ligand is generally assumed to present a lone pair of electrons to the central metal, groups such as CH_3 are formally regarded as anions rather than as radicals with unpaired electrons, even though they usually carry the names of radicals. Compounds that really do possess unpaired electrons in the free state can cause problems, especially when calculating oxidation states. For coordination nomenclature purposes, NO, nitrogen(II) oxide, is considered to be a neutral ligand. Complicated ligands may be represented by abbreviations rather than formulae, and lists of recommended abbreviations have been published in sources such as *Nomenclature of Inorganic Chemistry*. Some examples of these usages are shown in Table 1. The use of square brackets to indicate the coordination entity is fundamental and is a particularly useful device.

Note the negative oxidation state and the η (hapto) connectivity symbol in the last two examples. Where appropriate, stereochemical descriptors, such as *cis-*, *trans-*, *mer-*, and *fac-*, polyhedral descriptors, and chirality descriptors may be added to give structural information, but these are more often used in names, except for the simplest formulae. Polynuclear species may be described using the appropriate multiplicative suffixes, and bridging ligands can also be shown. The bridging symbol μ_n is useful for this pur-

Compound formulae	Complex ion formulae	Showing oxidation state
$[Co(NH_3)_6]Cl_3$	$[Co(NH_3)_6]^{3+}$	$[Co^{III}(NH_3)_6]^{3+}$
$[CoCl(NH_3)_5]Cl_2$	$[CoCl(NH_3)_5]^{2+}$	$[Co^{III}Cl(NH_3)_5]^{2+}$
$[CoCl(NO_2)(NH_3)_4]Cl$	$[CoCl(NO_2)(NH_3)_4]^{+}$	$[Co^{III}Cl(NO_2)(NH_3)_4]^{+}$
$[PtCl(NH_2CH_3)(NH_3)_2]Cl$	$[PtCl(NH_2CH_3)(NH_3)_2]^{+}$	$[Pt^{II}Cl(NH_2CH_3)(NH_3)_2]^{+}$
$[CuCl_2\{O{=}C(NH_2)_2\}_2]$		$[Cu^{II}Cl_2\{O{=}C(NH_2)_2\}_2]$
$K_2[PdCl_4]$	$[PdCl_4]^{2-}$	$[Pd^{II}Cl_4]^{2-}$
$K_2[OsCl_5N]$	$[OsCl_5N]^{2-}$	$[Os^{VI}Cl_5N]^{2-}$
$Na[PtBrCl(NO_2)(NH_3)]$	$[PtBrCl(NO_2)(NH_3)]^{-}$	$[Pt^{II}BrCl(NO_2)(NH_3)]-$
$[Co(en)_3]Cl_3$	$[Co(en)_3]^{3+}$	$[CoI^{III}(en)_3]^{3+}$
$Na_2[Fe(CO)_4]$	$[Fe(CO)_4]^{2-}$	$[Fe^{-II}(CO)_4]^{2-}$
$[Co(\eta^5-C_5H_5)_2]Cl$	$[Co(\eta^5-C_5H_5)2]^{+}$	$[Co^{II}(\eta^5-C_5H_5)_2]^{+}$

Table 1.

pose. The subscript may be omitted if a ligand bridges only two groups. Polymeric materials can be indicated in an empirical formula using the indeterminate subscript *n*. When there are different central metal ions present in a polynuclear compound, the established priority sequence for metal ions should be used to determine the order of citation.

$$[\{Cr(NH_3)_5\}(OH)\{Cr(NH_3)_5\}]^{5+} \text{ or } [\{Cr(NH_3)_5\}_2(\mu\text{-}OH)]^{5+}$$

$$[Re_2Br_8]^{4-} \text{ or } [(ReBr_4)_2]^{4-}$$

$$[[IrCl_2(CO)\{P(C_6H_5)_3\}_2](HgCl)]$$

$$[\{PdCl_2\}_n] \text{ or } [\{Pd(\mu\text{-}Cl)_2\}_n]$$

Names in Coordination Nomenclature

The names of coordination entities are assembled using principles similar, but not identical, to those used for formulae. The central atom is always cited last. Its name may be modified by an oxidation state symbol. The ligands are presented in the alphabetical order of their initial letters, neglecting for this purpose any multiplicative prefixes. It is not necessary to divide the ligands into neutral and charged groups. However, the names of negatively charged ligands are generally modified by adding the postfix suffix -o in place of the final -e where it occurs, to indicate that they are indeed bound and not free. As an exception, this is not the case with hydrocarbon ligands such as methyl and ethyl, which retain the names of radicals. The names of neutral ligands are not modified. If the coordination entity itself is negatively charged (but not when it is neutral or positively charged), then the name of the central atom is modified by the ending -ate. These practices are illustrated below.

Formula	Name
$[Co(NH_3)_6]Cl_3$	hexaamminecobalt(III) trichloride
$[Co(NH_3)_6]^{3+}$	hexaamminecobalt(3+)
$[CoCl(NH_3)_5]Cl_2$	pentaamminechlorocobalt(III) trichloride
$[CoCl(NH_3)_5]^{2+}$	pentaamminechlorocobalt(2+)
$[CoCl(NO_2)(NH_3)_4]Cl$	tetraamminechloronitritocobalt(III) chloride
$[CoCl(NO_2)(NH_3)_4]^+$	tetraamminechloronitritocobalt(1+)
$[PtCl(NH_2CH_3)(NH_3)_2]Cl$	bisamminechloromethylamineplatinum(II) chloride
$[PtCl(NH_2CH_3)(NH_3)_2]^+$	diamminechloromethylamineplatinum(+)
$[CuCl_2\{O{=}C(NH_2)_2\}_2]$	dichlorobis(urea)copper(II)
$K_2[PdCl_4]$	potassium tetrachloropalladate(II)
$K_2[OsCl_5N]$	potassium pentachloronitrodoosmate(VI)
$[Co(H_2O)_2(NH_3)_4]Cl$	tetraamminediaquacobalt(III) chloride

Note that in some cases it may be useful to introduce additional enclosing marks to ensure clarity: for example, to avoid possible confusion between chloromethylamine, $ClCH_2NH_2$, and (chloro)methylamine, which implies two separate ligands, Cl and CH_3NH_2. It is for the writer to decide whether such a strategy is useful, depending on the particular case under review. Ammonia as a ligand has the name ammine. Similarly, water has the coordination name aqua.

$Na[PtBrCl(NO_2)(NH_3)]$	sodium amminebromochloronitrito-platinate(II)
$[Co(en)_3]Cl_3$	tris(ethane-1,2-diamine)cobalt(II) trichloride
$Na_2[Fe(CO)_4]$	sodium tetracarbonylferrate(-II)
$[Co(\eta^5\text{-}C_5H_5)_2]Cl$	bis(cyclopentadienyl)cobalt(III) chloride or bis(η^5-cyclopentadienyl)cobalt(III) chloride.

coordination chemistry: chemistry involving complexes of metal ions surrounded by covalently bonded ligands

The symbol η is used above and also quite generally throughout organometallic **coordination chemistry** to indicate the number of carbon atoms in a ligand that are coordinated to the metal. Other devices to indicate connectivity are the italicized atomic symbols of the donor atoms (useful for indicating structure in complexes containing chelating and polydentate ligands), and for some complicated cases, the κ symbolism may be useful. Examples follow in Figure 3.

Further devices are used in coordination names to show polymeric structures, which may contain bridging groups and metal-metal bonds.

$[\{Pd(\mu\text{-}Cl)_2\}_n]$	poly(di-μ-chloropalladium)
$[\{Cr(NH_3)_5\}(OH)\{Cr(NH_3)_5\}]^{5+}$	μ-hydroxo-bis[pentaam-minechromium(III)](5+)
$[(ReBr_4)_2]^{4-}$	bis(tetrabromorhenate)(*Re-Re*)(2-)
$[[IrCl_2(CO)\{P(C_6H_5)_3\}_2](HgCl)]$	carbonyl-1κ*C*-trichloro-1κ^2,2κ*Cl*-bis(triphenylphosphine-1κ*P*)iridium mercury(*Hg-Ir*)

Where different metals are present, priority rules must be applied to assign metal locants.

For more information on this and other topics cited above, as well as for descriptions of the use of geometrical and stereo descriptors, polyhedral

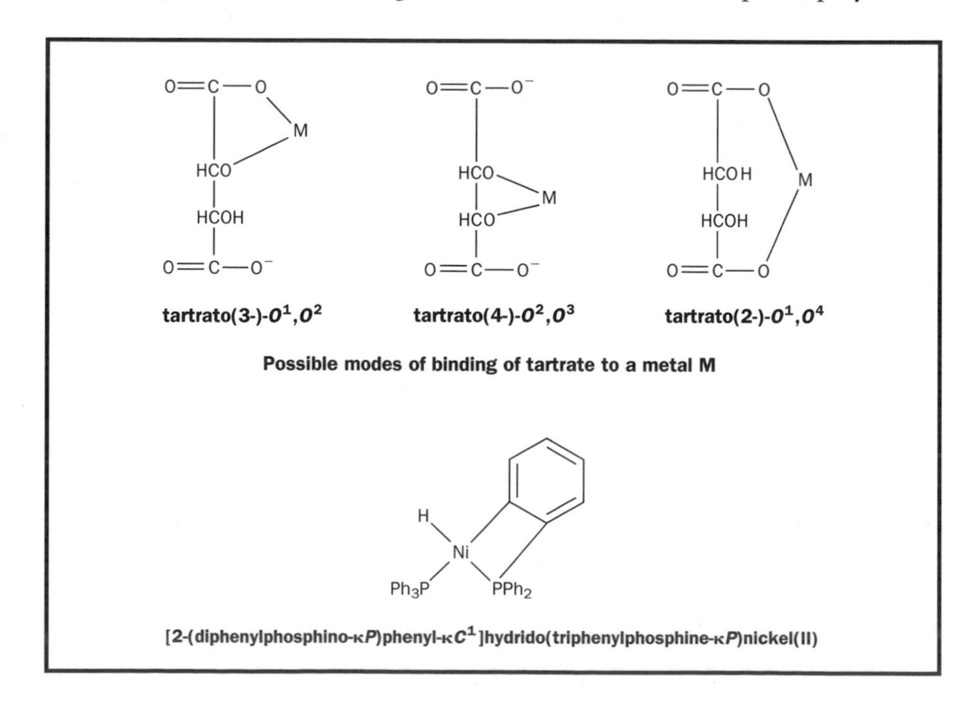

Figure 3.

symbols, and configuration indices in names, the reader is referred to the books cited in the bibliography. The international authority with the task of formalizing nomenclature rules, assigning the names of new elements, etc., is the International Union of Pure and Applied Chemistry (IUPAC). All the publications cited in the bibliography carry the authority of IUPAC. Some more specialized inorganic nomenclatures are described in *Nomenclature of Inorganic Chemistry II. Principles of Chemical Nomenclature* and *A Guide to IUPAC Nomenclature of Organic Compounds* offer more general treatments suitable for those not requiring the most detailed information. SEE ALSO BONDING.

G. J. Leigh

Bibliography

Leigh, G. J. (1990). *Nomenclature of Inorganic Chemistry*. Oxford, U.K.: Blackwell Science. (This title is often referred to as the Red Book, or since 2001, Red Book I.)

Leigh, G. J., ed. (1998). *Principles of Chemical Nomenclature*. Oxford, U.K.: Blackwell Science.

McCleverty, J. A., and Connelly, N. G., eds. (2000). *Nomenclature of Inorganic Chemistry II*. Cambridge, U.K.: The Royal Society of Chemistry. (This is sometimes referred to as Red Book II.)

Richer, J.-C., ed. (1993). *A Guide to IUPAC Nomenclature of Organic Compounds*. Oxford, U.K.: Blackwell Science.

Rigaudy, J., and Klesney, S. P., eds. (1979). *Nomenclature of Organic Chemistry*. Oxford, U.K.: Pergamon Press.

Norepinephrine

Norepinephrine (noradrenaline) belongs to a family of biological compounds called catecholamines. These compounds are synthesized in sympathetic neurons and in the adrenal glands. Norepinephrine is produced from the catecholamine dopamine by the action of the enzyme dopamine β-hydroxylase. This enzyme is responsible for the addition of a hydroxyl (–OH) group at the β carbon. (See Figure 1.) In certain cells of the adrenal glands, norepinephrine is chemically transformed into epinephrine (adrenaline), the hormone responsible for the fight-or-flight response. Epinephrine differs from norepinephrine by the presence of a methyl ($-CH_3$) group on the nitrogen atom.

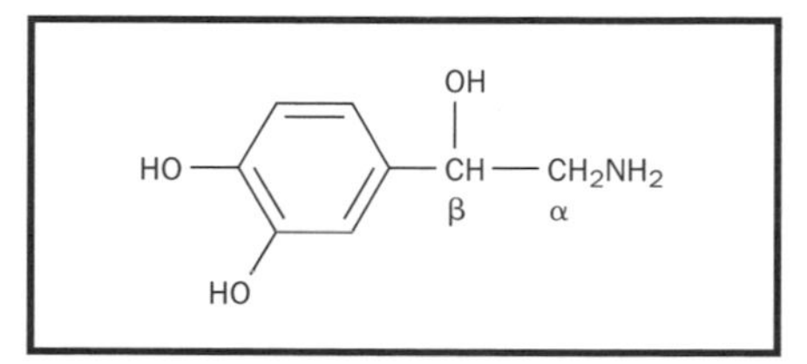

Figure 1. The structure of norepinephrine. Carbon atoms of the side chain are labeled.

Norepinephrine functions biologically as a neurotransmitter, transmitting a signal from one neuron to another neuron or muscle cell. After release from a neuron, norepinephrine diffuses through the tiny space between the cells (the synapse), where it can bind to a **receptor** protein on the surface of a nearby cell. Nerve impulses are typically short-lasting because the neurotransmitter dissociates from its receptor. Once this happens, the neurotransmitter can quickly be chemically altered or transported into another cell, either of these terminating the nerve impulse.

receptor: area on or near a cell wall that accepts another molecule to allow a change in the cell

metabolism: the complete range of biochemical processes that take place within living organisms; comprises processes that produce complex substances from simpler components, with a consequent use of energy (anabolism), and those that break down complex food molecules, thus liberating energy (catabolism)

contraction: the shortening of a normal trend of a quantity

The family of receptors that responds to norepinephrine and related compounds are called adrenergic receptors. Adrenergic receptors in the peripheral nervous system are important in the activity of smooth muscle and cardiac muscle and in **metabolism**. The effect on most smooth muscle is relaxation, whereas the effect on cardiac muscle is to increase the force and rate of **contraction**. Drugs that mimic the action of norepinephrine are often used to treat asthma because they relax bronchial smooth muscle, helping the asthma patient to breathe more easily. Drugs called β-blockers bind to adrenergic receptors but block activation. Because this results in a decrease in blood pressure, β-blockers are commonly prescribed to treat high blood pressure. Adrenergic receptor activity is also important within the central nervous system. Some drugs used to treat depression prolong the adrenergic nerve impulse by allowing norepinephrine to remain in synapses for longer periods. SEE ALSO EPINEPHRINE; INHIBITORS; NEUROTRANSMITTERS.

Jennifer L. Powers

Bibliography

Nemeroff, Charles B. (1998). "The Neurobiology of Depression." *Scientific American* 278:42–49.

Internet Resources

Basic Neuropharmacology. "The Chemistry of the Nervous System." Available from <http://www.ptd.neu.edu/neuroanatomy/>.

King, Michael W. "Biochemistry of Neurotransmitters." Available from <http://web.indstate.edu/thcme/>.

Northrop, John

AMERICAN BIOCHEMIST
1891–1987

John Northrop shared the Nobel Prize in chemistry in 1946 with Wendell Stanley, awarded to them "for their preparation of enzymes and virus proteins in a pure form," and with James Sumner, "for his discovery that enzymes can be crystallized." Although Sumner had been the first, in 1926, to crystallize an enzyme (urease) and to aver that enzymes were proteins, Northrop did more than any other scientist to establish that pure enzymes are indeed proteins.

About 1920 Northrop repeated the earlier claim of Cornelis Pekelharing that he had isolated a protein from gastric juice (the enzyme pepsin). Neither Pekelharing nor Northrop was able to crystallize the protein. However, Sumner's **crystallization** of urease encouraged Northrop to take up the problem again. In 1930 Northrop isolated a crystalline substance from a commercial pepsin preparation, and the crystallized substance appeared to be the enzyme pepsin. Subsequently Northrop, together with Moses Kunitz, isolated and crystallized trypsin, trypsinogen, chymotrypsin, and chymotrypsinogen.

crystallization: process of producing crystals of a substance when a saturated solution in an appropriate solvent is either cooled or some solvent removed by evaporation

Northrop carefully tested his enzyme preparations by means of solubility measurements, ultracentrifuge analysis, and **electrophoresis**, and concluded that they were essentially pure proteins. From measurements of diffusion, denaturation, hydrolysis, and the formation of active enzyme from inactive precursor, he concluded that enzymatic activity was a property of the protein molecule itself and was not due to a nonprotein impurity.

electrophoresis: migration of charged particles under the influence of an electric field, usually in solution; cations, positively charged species, will move toward the negative pole and anions, the negatively charged species, will move toward the positive pole

Northrop had long had an interest in viruses and bacteriophage—things that occupied his attention increasingly in his later years. Though he was more concerned with the protein component of bacteriophage than their nucleic acid component, in 1951 he made the prophetic suggestion: "The nucleic acid may be the essential autocatalytic self-reproducing part of the molecule, and the protein portion may be necessary only to allow the entrance to the host cell."

John Howard Northrop was born into an academic family in Yonkers, New York, in 1891. He entered Columbia University in 1908, from which institution he received his Ph.D. in chemistry in 1915. In 1916 he was appointed to the Rockefeller Institute, and he remained there for the rest of his working life. In 1924 he transferred to the Princeton branch of the institute, where most of his significant work on proteins was performed. In 1949, when the institute closed its Princeton branch, Northrop moved to the University of California at Berkeley as professor of bacteriology and biophysics, while remaining a member of the institute and continuing to receive its support for his work. He retired in 1959. He died in 1987, aged ninety-five. SEE ALSO HYDROLYSIS; PROTEINS.

Keith L. Manchester

Bibliography

Herriott, R. M. (1983). "John H. Northrop: The Nature of Enzymes and Bacteriophage." *Trends in Biochemical Sciences.* 13:296–297.

Robbins, F. C. (1991). "John Howard Northrop (5 July, 1891–May 27, 1987)." *Proceedings of the American Philosophical Society* 135:315–320.

Internet Resources

Northrop, John. Nobel lecture. Available from <http://www.nobel.se/chemistry/laureates>.

Nuclear Chemistry

Nuclear chemistry is the study of the chemical and physical properties of elements as influenced by changes in the structure of the atomic nucleus. Modern nuclear chemistry, sometimes referred to as radiochemistry, has become very interdisciplinary in its applications, ranging from the study of the formation of the elements in the universe to the design of radioactive drugs for diagnostic medicine. In fact, the chemical techniques pioneered by nuclear chemists have become so important that biologists, geologists, and physicists use nuclear chemistry as ordinary tools of their disciplines. While the common perception is that nuclear chemistry involves only the study of radioactive nuclei, advances in modern mass spectrometry instrumentation has made chemical studies using stable, nonradioactive **isotopes** increasingly important.

isotope: form of an atom that differs by the number of neutrons in the nucleus

There are essentially three sources of radioactive elements. Primordial nuclides are radioactive elements whose half-lives are comparable to the age of our solar system and were present at the formation of Earth. These nuclides are generally referred to as naturally occurring radioactivity and are derived from the radioactive decay of thorium and uranium. Cosmogenic nuclides are atoms that are constantly being synthesized from the bombardment of planetary surfaces by cosmic particles (primarily protons ejected from the Sun), and are also considered natural in their origin. The third source of radioactive nuclides is termed anthropogenic and results from human activity in the production of nuclear power, nuclear weapons, or through the use of particle accelerators.

Lasers focus on a small pellet of fuel in attempt to create a nuclear fusion reaction (the combination of two nuclei to produce another nucleus) for the purpose of producing energy.

precipitation: process involving the separation of a solid substance from a solution

Marie Curie was the founder of the field of nuclear chemistry. She was fascinated by Antoine-Henri Becquerel's discovery that uranium minerals can emit rays that are able to expose photographic film, even if the mineral is wrapped in black paper. Using an electrometer invented by her husband Pierre and his brother Jacques that measured the electrical conductivity of air (a precursor to the Geiger counter), she was able to show that thorium also produced these rays—a process that she called radioactivity. Through tedious chemical separation procedures involving **precipitation** of different chemical fractions, Marie was able to show that a separated fraction that had the chemical properties of bismuth and another fraction that had the chemical properties of barium were much more radioactive per unit mass than the original uranium ore. She had separated and discovered the elements polonium and radium, respectively. Further purification of radium from barium produced approximately 100 milligrams of radium from an initial sample of nearly 2,000 kilograms of uranium ore.

In 1911 Ernest Rutherford asked a student, George de Hevesy, to separate a lead impurity from a decay product of uranium, radium-D. De Hevesy did not succeed in this task (we now know that radium-D is the radioactive isotope ^{210}Pb), but this failure gave rise to the idea of using radioactive isotopes as tracers of chemical processes. With Friedrich Paneth in Vienna in 1913, de Hevesy used ^{210}Pb to measure the solubility of lead salts—the first application of an isotopic tracer technique. De Hevesy went on to pioneer the application of isotopic tracers to study biological processes and is generally considered to be the founder of a very important area in which nuclear chemists work today, the field of nuclear medicine. De Hevesy also is credited with discovering the technique of neutron activation analysis, in which samples are bombarded by neutrons in a nuclear reactor or from a neutron generator, and the resulting radioactive isotopes are measured, allowing the analysis of the elemental composition of the sample.

In Germany in 1938, Otto Hahn and Fritz Strassmann, skeptical of claims by Enrico Fermi and Irène Joliot-Curie that bombardment of uranium by neutrons produced new so-called transuranic elements (elements beyond uranium), repeated these experiments and chemically isolated a radioactive isotope of barium. Unable to interpret these findings, Hahn asked Lise Meitner, a physicist and former colleague, to propose an explanation for his observations. Meitner and her nephew, Otto Frisch, showed that it was possible for the uranium nucleus to be split into two smaller nuclei by the neutrons, a process that they termed "**fission.**" The discovery of nuclear fission eventually led to the development of nuclear weapons and, after World War II, the advent of nuclear power to generate electricity. Nuclear chemists were involved in the chemical purification of plutonium obtained from uranium targets that had been irradiated in reactors. They also developed chemical separation techniques to isolate radioactive isotopes for industrial and medical uses from the fission products wastes associated with plutonium production for weapons. Today, many of these same chemical separation techniques are being used by nuclear chemists to clean up radioactive wastes resulting from the fifty-year production of nuclear weapons and to treat wastes derived from the production of nuclear power.

fission: process of splitting an atom into smaller pieces

In 1940, at the University of California in Berkeley, Edwin McMillan and Philip Abelson produced the first manmade element, neptunium (Np), by the bombardment of uranium with low energy neutrons from a nuclear accelerator. Shortly thereafter, Glenn Seaborg, Joseph Kennedy, Arthur Wahl, and McMillan made the element plutonium by bombarding uranium targets with deuterons, particles derived from the heavy isotope of hydrogen, deuterium (^{2}H). Both McMillan and Seaborg recognized that the chemical properties of neptunium and plutonium did not resemble those of rhenium and osmium, as many had predicted, but more closely resembled the chemistry of uranium, a fact that led Seaborg in 1944 to propose that the transuranic elements were part of a new group of elements called the actinide series that should be placed below the lanthanide series on the periodic chart. Seaborg and coworkers went on to discover many more new elements and radioactive isotopes and to study their chemical and physical properties. At the present, nuclear chemists are involved in trying to discover new elements beyond the 112 that are presently confirmed and to study the chemical properties of these new elements, even though they may exist for only a few thousandths of a second.

Nobel laureate Glenn T. Seaborg was among those who discovered many radioactive elements and isotopes.

As early as 1907 Bertram Boltwood had used the discovery of radioactive decay laws by Ernest Rutherford and Frederick Soddy to ascribe an age of over two billion years to a uranium mineral. In 1947 Willard Libby at the University of Chicago used the decay of ^{14}C to measure the age of dead organic matter. The cosmogenic radionuclide, ^{14}C, becomes part of all living matter through photosynthesis and the consumption of plant matter.

Once the living organism dies, the ^{14}C decays at a known rate, enabling a date for the carbon-containing relic to be calculated. Today, scientists ranging from astrophysicists to marine biologists use the principles of radiometric dating to study problems as diverse as determining the age of the universe to defining food chains in the oceans. In addition, newly developed analytical techniques such as accelerator mass spectrometry (AMS) have allowed nuclear chemists to extend the principles of radiometric dating to nonradioactive isotopes in order to study modern and ancient processes that are affected by isotopic fractionation. This isotopic fractionation results from temperature differences in the environment in which the material was formed (at a given temperature, the lighter isotope will be very slightly more reactive than the heavier isotope), or from different chemical reaction sequences.

The newest area in which nuclear chemists play an important role is the field of nuclear medicine. Nuclear medicine is a rapidly expanding branch of health care that uses short-lived radioactive isotopes to diagnose illnesses and to treat specific diseases. Nuclear chemists synthesize drugs from radionuclides produced in nuclear reactors or accelerators that are injected into the patient and will then seek out specific organs or cancerous tumors. Diagnosis involves use of the radiopharmaceutical to generate an image of the tumor or organ to identify problems that may be missed by x rays or physical examinations. Treatment involves using radioactive compounds at carefully controlled doses to destroy tumors. These nuclear medicine techniques hold much promise for the future because they use biological chemistry to specify target cells much more precisely than traditional radiation therapy, which uses radiation from external sources to kill tumor cells, killing nontarget cells

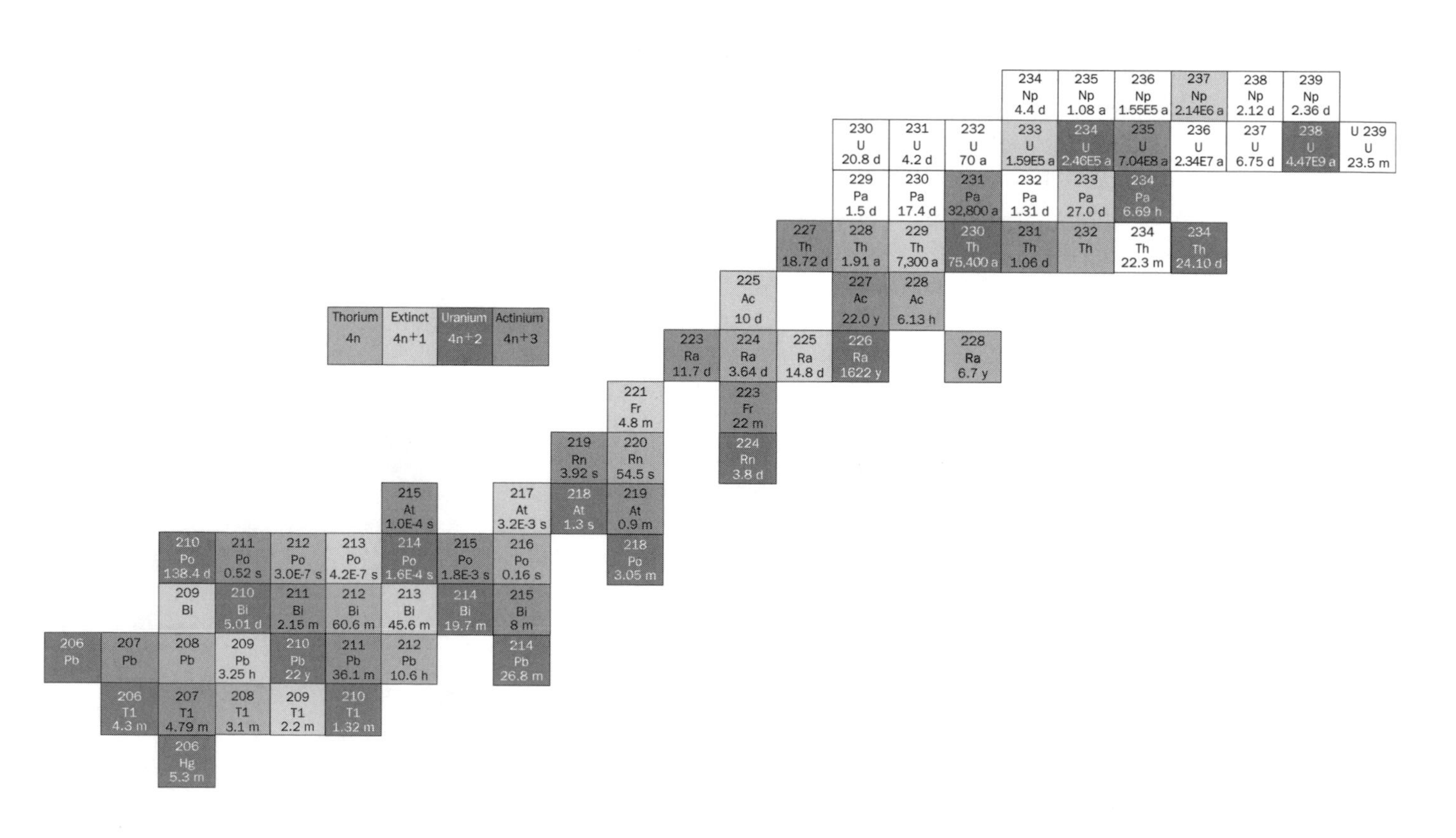

Table of naturally occurring radionuclides.

as well. Additionally, the use of nuclear pharmaceuticals containing the short-lived isotope ^{11}C has allowed nuclear chemists and physicians to probe brain activity to better understand the biochemical basis of illnesses ranging from Parkinson's disease to drug abuse. SEE ALSO Becquerel, Antoine-Henri; Curie, Marie Sklodowska; Fermi, Enrico; Meitner, Lise; Neptunium; Plutonium; Polonium; Radiation; Radioactivity; Radium; Rutherford, Ernest; Seaborg, Glenn Theodore; Soddy, Frederck; Uranium.

W. Frank Kinard

Bibliography

Hoffman, D. C.; Ghiorso, A.; and Seaborg, Glenn T., eds. (2000). *The Transuranium People: An Intimate Glimpse.* London: Imperial College Press.

Morrissey, D.; Loveland, W. T.; and Seaborg, Glenn T. (2001). *Introductory Nuclear Chemistry.* New York: John Wiley & Sons.

Rydberg, J.; Liljenzin, J.-O.; and Choppin, Gregory R. (2001). *Radiochemistry and Nuclear Chemistry,* 3rd edition. Woburn, MA: Butterworth-Heinemann.

Internet Resources

American Institute of Physics History Center. Available from <http://www.aip.org/history/>.

Division of Nuclear Chemistry and Technology of the American Chemical Society. Available from <http://www.cofc.edu/~nuclear/>.

Society of Nuclear Medicine. Available from <http://www.snm.org/>.

Nuclear Fission

Following the discovery of the neutron in the early 1930s, nuclear physicists began bombarding a variety of elements with neutrons. Enrico Fermi in Italy included uranium (atomic number 92) among the elements he bombarded which resulted in formation of nuclei that decayed by emission of negative β-rays. Such decay produces nuclei of higher atomic number, so Fermi assumed that the bombardment of uranium led to a new element with atomic number 93. By 1938 similar research had resulted in reports of the discovery of four new elements with atomic numbers 93, 94, 95, and 96. In 1938 Otto Hahn and Fritz Strassmann in Berlin bombarded uranium with neutrons to study the possibility of production of nuclides with atomic numbers less then 92 due to emission of protons and α-particles. To their surprise, they found that they had made barium ($z = 56$). Hahn informed a former colleague, Lise Meitner, who, with Otto Frisch, reviewed the data and reached the conclusion that the uranium atom was splitting (fissioning) into two new, smaller nuclei with the accompanying release of a large amount of energy. Many laboratories quickly confirmed the occurrence of this process of nuclear fission. Niels Bohr and John Wheeler, within a few months, published a paper explaining many features of fission using a model of nuclear behavior based on an analogy to a droplet of liquid, which, when given extra energy, can elongate from a spherical shape and split into two smaller droplets. Nuclei have two opposing energies: a disruptive energy resulting from the mutual electrostatic repulsion of the positive protons in the nucleus, and an attractive energy due to nuclear forces present between the nuclear particles (both neutrons and protons). The repulsive electrostatic energy of the protons increases as the number of protons increases and decreases as the average distance between them increases. The attractive nu-

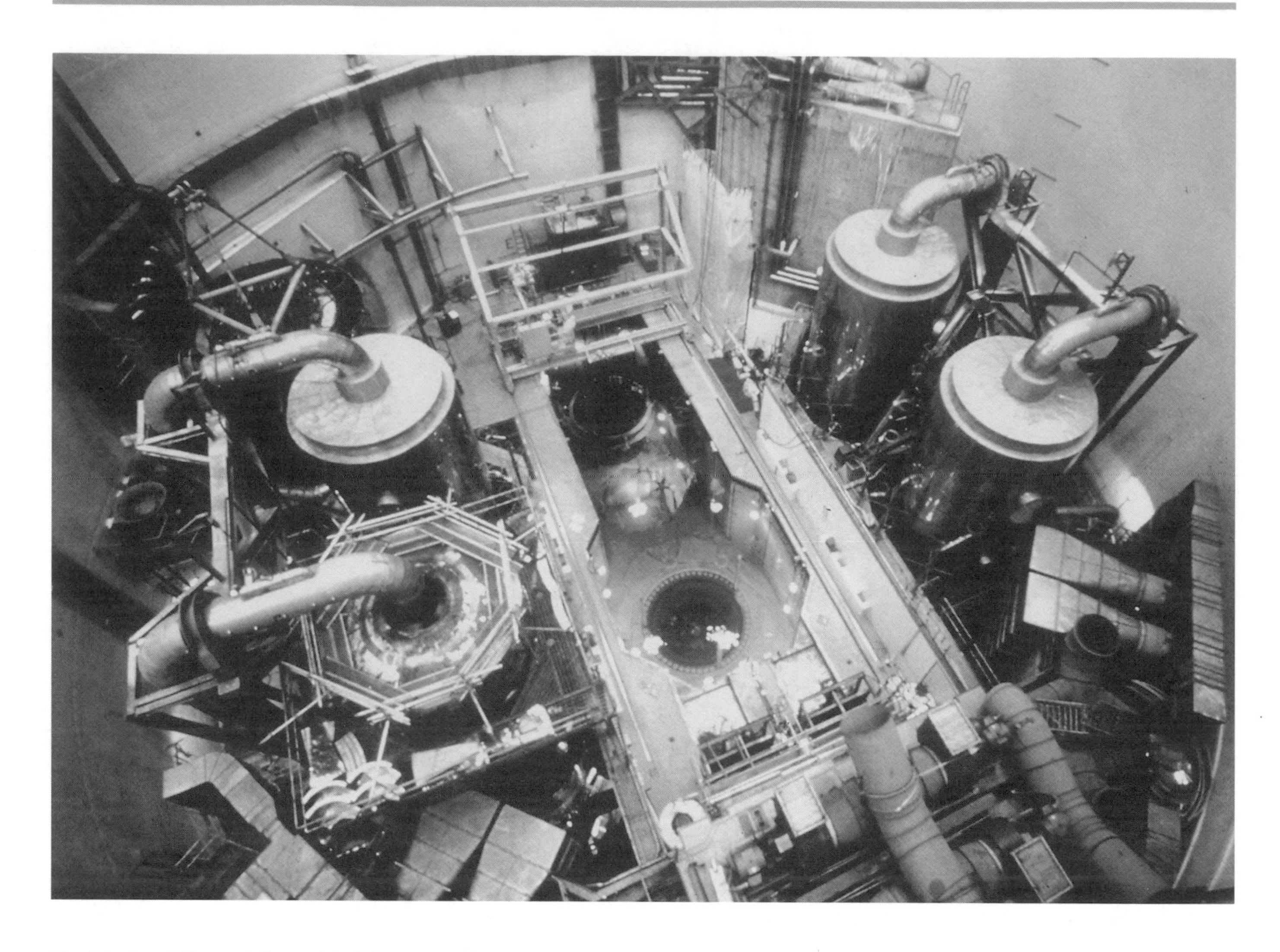

The interior of the containment building at the Trojan Nuclear Plant near Rainier, Oregon. The splitting, or fission, of atoms is a source of energy.

clear force energy increases with the total number of nucleons (protons and neutrons) in the nucleus. The nuclear attractive force is at a maximum when the nucleus has a spherical shape and at a minimum when the nucleus is distorted into two roughly equal fragments.

Nuclei formed in fission, known as fission products, range in atomic number (number of protons) from approximately 30 to 64. The original fissioning nuclide has a neutron to proton ration of about 1.6, whereas stable nuclei having the same range of atomic numbers as the fission products have neutron to proton ratios of 1.3 to 1.4. This means that nuclei formed in fission have too great a number of neutrons for stability and undergo beta (β^-) decay to convert neutrons to protons. In general, fission is restricted to nuclei with atomic numbers above 82 (Pb), and the probability of fission increases as the atomic number increases. Fission produces nuclei of atomic masses from above 60 to about 150.

With very-low-energy neutrons, uranium of mass number 235 emits an average of two to three neutrons per fission event. Because more neutrons are released than absorbed, fission can result in a multiplication of successive fission events. This multiplication can reach very high numbers in about 10^{-14} to 10^{-17} seconds, resulting in the release of a great amount of energy in that time. This was the basis of the development of nuclear weapons. Soon after the discovery of fission it had been calculated that if a sufficient quantity of the fissionable material was assembled under proper conditions, a self-sustaining nuclear explosion could result. The *critical mass* of the fissionable material necessary for explosion is obtained with a spherical shape (minimum surface area per mass). The uranium isotope of mass 235 and the plutonium isotope of mass 239 are incited to fission and release energy in the use of nuclear weapons and in nuclear reactors. Nuclear reactors control the rate of fission and maintain it at a constant level, allowing the released energy to be used for power. Nuclear reactors are used in many nations as a major component of their natural energy. In the United States, approximately 20 percent of the electricity is provided by nuclear reactors, whereas France uses reactors to produce almost 80 percent of its electricity. Reactors used for power have four basic components: (1) fuel, either natural uranium or uranium enriched in ^{235}U or ^{239}Pu; (2) a moderator to reduce neutron energies, which increases the probability of fission; (3) control rods of cadmium and boron to control the rate of fission; and (4) coolants to keep the temperature of the reactor at a reasonable level and to transfer the energy for production of electricity. In power reactors the coolant is commonly H_2O or D_2O, but air, graphite, or a molten mixture of sodium and potassium can be used. Reactors are surrounded by a thick outer shield of concrete to prevent release of radiation.

LEO SZILARD (1898–1964)

Leo Szilard determined that the formation of neutrons occurs during the fission of uranium. This is crucial to sustaining a chain reaction necessary to build an atomic bomb, the first of which he helped to construct in 1942. Shortly thereafter, realizing the destructive power of the atom bomb, Szilard argued for an end to nuclear weapons research.

—*Valerie Borek*

There have been two major accidents (Three Mile Island in the United States and Chernobyl in the former Soviet Union) in which control was lost in nuclear power plants, with subsequent rapid increases in fission rates that resulted in steam explosions and releases of radioactivity. The protective shield of reinforced concrete, which surrounded the Three Mile Island Reactor, prevented release of any radioactivity into the environment. In the Russian accident there had been no containment shield, and, when the steam explosion occurred, fission products plus uranium were released to the environment—in the immediate vicinity and then carried over the Northern Hemisphere, in particular over large areas of Eastern Europe. Much was learned from these accidents and the new generations of reactors are being built to be "passive" safe. In such "passive" reactors, when the power level increases toward an unsafe level, the reactor turns off automatically to prevent the high-energy release that would cause the explosive release of radioactivity. Such a design is assumed to remove a major factor of safety concern in reactor operation. SEE ALSO BOHR, NIELS; FERMI, ENRICO; MANHATTAN PROJECT; PLUTONIUM; RADIOACTIVITY; URANIUM.

Gregory R. Choppin

Nuclear Fusion

Nuclear fusion is a reaction whereby two smaller nuclei are combined to form a larger nucleus. It results in the release of energy for reactions that form nuclei of mass number below 60, with the largest energy release occurring with the lightest nuclides. This stands in contrast to the process of nuclear **fission** in which a heavy nucleus is split into two smaller nuclei with the release of energy. Since light nuclei have smaller repulsion energies (the energy required to bring two like charges together), fusion is much more likely to occur among these nuclei. Two deuterons, ^2H, must have a total kinetic energy of 0.02 million electron volts (MeV) to be able to collide and

fission: process of splitting an atom into smaller pieces

react. Temperatures of greater than 200 million°C (360 million°F) are required for such kinetic energies. Atoms with kinetic energies of 0.02 MeV exist only as gases in which the atoms have lost their electrons. Such gases of ions and electrons are known as plasmas.

Temperatures required for fusion reactions exist in stars where fusion reactions are the principle components of energy release. In the Sun, approximately 90 percent of solar energy is a result of proton–proton interactions in several steps to form helium of mass number 4. These steps all involve binuclear collisions since multinuclei collisions are very improbable events. Initially, two protons interact to form a deuterium nucleus (deuterium is an **isotope** of hydrogen with one proton and one neutron; the nucleus is a **deuteron**) that collides with another proton to form a ^{3}He (tritium) nucleus. This nucleus collides with a neutron or another ^{3}He nucleus (with the emission of two protons) to form ^{4}He. The net reaction can be represented as four protons fusing to form a ^{4}He nucleus releasing 26.7 MeV. When a sufficient number of the ^{3}He and ^{4}He nuclei are formed in the star, they begin fusion reactions to form heavier nuclei such as ^{7}Li and ^{7}Be. The number of proton–proton fusion reactions in the Sun amounts to 1.8×10^{38} s^{-1}. At present the Sun is 73 percent hydrogen, 25 percent helium, 2 percent carbon, nitrogen, oxygen, and all the other elements in the Periodic Table. Approximately 6 percent of the hydrogen originally in the Sun's stellar core has now been burnt.

isotope: form of an atom that differs by the number of neutrons in the nucleus

deuteron: nucleus containing one proton and one neutron, as is found in the isotope deuterium

Since the average mass of elements increases in stars, there is a transition from a proton cycle to a carbon cycle, as was proposed by American physicist Hans Bethe and German physicist Carl F. von Weizsacker in the 1930s. In such stars, the temperature and pressure reach higher values and the consumption of hydrogen accelerates. Since helium has a greater mass then hydrogen, it accumulates in the stellar core, while most of the hydrogen burning fusion of its nuclei moves to a layer outside that core. With an increased average mass in stars, reactions such as ^{8}Be + ^{4}He forming ^{12}C begin to occur. The formation of ^{12}C in sufficient quantities leads to reac-

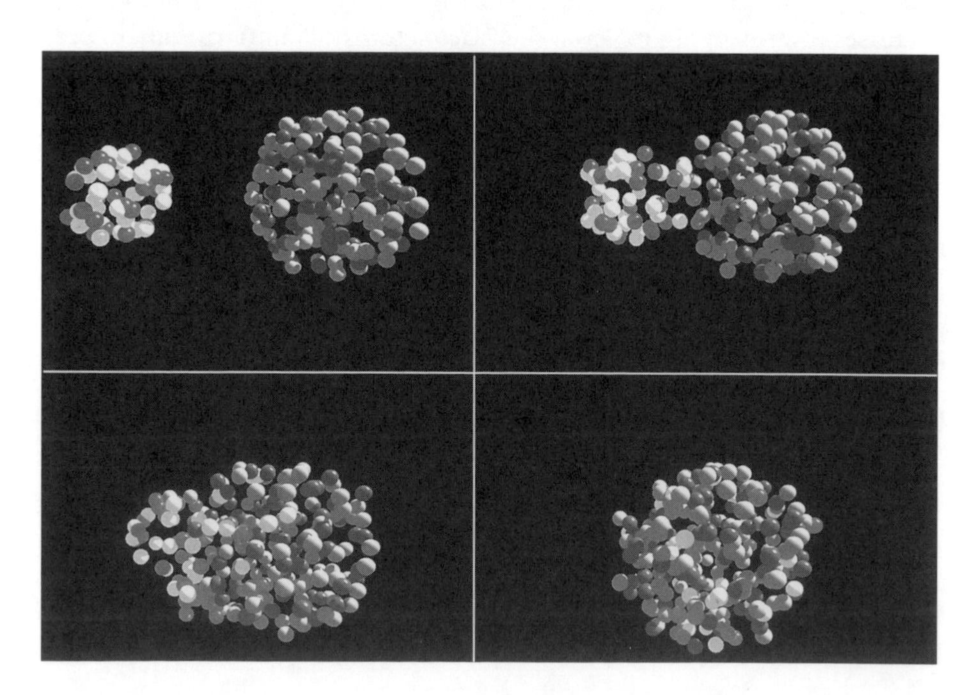

A computer simulation of gold (Au) and nickel (Ni) nuclei fusing.

tions with ^{4}He to form oxygen, neon, and higher elements. Eventually, there is sufficient carbon present in some stars for the fusion of a pair of ^{12}C nuclei to begin.

In stars with very heavy average masses, helium burning may last for only a few million years before it is replaced by carbon fusion. In time this leads to the production of elements such as calcium, titanium, chromium, iron, and nickel fusion partly by helium capture, partly by the direct fusion of heavy nuclides. For example, two ^{28}Si can combine to form ^{56}Ni that can decay to ^{56}Co which then decays to stable ^{56}Fe. These last steps of production may occur rather rapidly in a few thousand years. When the nuclear fuel for fusion is exhausted, the star collapses and a supernova results.

Nuclear fusion became important on Earth with the development of hydrogen bombs. A core of uranium or plutonium is used to initiate a fission reaction that raises the core's temperature to approximately 10^8 K, sufficient to cause fusion reactions between deuterium and tritium. In fusion bombs, LiD is used as ^{6}Li reacts with fission neutrons to form tritium that then undergoes fusion with deuterium. It is estimated that approximately half the energy of a 50 megaton thermonuclear weapon comes from fusion and the other half from fission. Fusion reactions in these weapons also produce secondary fission since the high energy neutrons released in the fusion reactions make them very efficient in causing the fission of ^{238}U.

The deuterium plus tritium and deuterium plus deuterium reactions are of interest in the development of controlled fusion devices for producing energy. A number of designs have been proposed for these fusion reactors, with most attention given to inertial confinement and magnetic confinement systems.

Inertial confinement is a pulsed system in which small pellets of D_2 and T_2 are irradiated by intense beams of photons or electrons. The surface of the pellet rapidly vaporizes, causing a temperature-pressure wave to move through the pellet, increasing its central temperature to greater than 10^8 K and causing fusion. If a fusion rate of approximately 100 pellets per second can be achieved, the result is a power output between 1 and 10 gigawatts.

At temperatures equal or greater than 10^7 K, hydrogen atoms are completely dissociated into protons and free electrons (the plasma state). Since construction material cannot withstand a plasma of this energy, the plasma is kept away from the walls by magnetic fields. The plasma density is limited by heat transfer and other considerations to approximately 10^{20} to 10^{21} particles m^{-3}. For a particle density of 3×10^{20} particles m^{-3}, confined for 0.1 to 1 second, the power density is estimated to be tens of megawatts per cubic meter. Several large machines based on magnetic confinement have been built, and confinement times of 2 seconds with particle densities of 5 $\times 10^{19}$ achieved. However, it seems unlikely that controlled thermonuclear reactors will be in operation for the purpose of power production before the year 2050 as significant technical problems remain to be solved. The availability of hydrogen and deuterium in the sea is so vast that nuclear fusion would outlast other nonrenewable energy sources. For example, a liter of seawater contains deuterium with an energy content equivalent to 300 liters (79.25 gallons) of gasoline. SEE ALSO Explosions; Nuclear Chemistry; Nuclear Fission.

Gregory R. Choppin

Bibliography

Choppin, Gregory R.; Liljenzin, Jan-Olov; and Rydberg, Jan (2001). *Radiochemistry and Nuclear Chemistry*, 3rd edition. Woburn, MA: Butterworth-Heinemann.

Friedlander, Gerhart; Kennedy, Joseph W.; Macias, Edward S.; and Miller, Julian (1981). *Nuclear and Radiochemistry*, 3rd edition. New York: Wiley-Interscience.

Internet Resources

FusEdWeb: Fusion Energy Educational Web Site. Available from <http://fusedweb.pppl.gov>.

General Atomics' Educational Web Site. Available from <http://fusioned.gat.com>.

Nuclear Magnetic Resonance

Nuclear magnetic resonance (NMR) is one of the most useful analytical methods in modern chemistry. It is used to determine the structure of new natural and synthetic compounds, the purity of compounds, and the course of a chemical reaction as well as the association of compounds in solution that might lead to chemical reactions. Although many different kinds of nuclei will produce a spectrum, hydrogen (H) nuclei historically have been the ones most studied. NMR spectroscopy is particularly useful in the study of organic molecules because these usually incorporate a large number of hydrogen atoms.

NMR Spectrometers

While the original NMR spectrometers were built to scan either the frequency or the magnetic field, the usual procedure is to use Fourier transform spectroscopy (FT NMR). The protocol for obtaining a FT NMR spectrum is to place a solution of the compound to be studied in a homogenous magnetic field and irradiate it with a short pulse of the appropriate radio-frequency. The shortness of the radio-frequency pulse results in a band of frequencies that simultaneously radiate all of the nuclei of a particular type to be studied. Each magnetic nucleus that absorbs this radio-frequency energy will then radiate radio-frequency energy at a very specific frequency. The frequencies generated by the various nuclei are then detected, Fourier transformed, and displayed as a plot of frequency versus intensity. This plot is called an NMR spectrum. The frequency at which magnetic resonance occurs depends on the strength of the magnetic field used and on the nucleus to be studied. The stronger the magnetic field used the higher the resonance frequency, the greater the dispersion (separation) of the bands, and the greater the sensitivity of the experiment. Thus, the higher the magnetic field the better the NMR spectrometer. Over the years this has led to the development of spectrometers of ever increasing magnetic fields. Superconducting magnets can be built with much higher fields than the usual electromagnets. Thus most NMR spectrometers used incorporate superconducting magnets. Although most of the elements in the Periodic Table have an isotope that is magnetic, the most common nucleus to be observed by NMR is that of hydrogen. This has led to the common use of the hydrogen resonant frequency for a given NMR spectrometer as a measure of the magnetic field strength of that spectrometer. NMR spectrometers that use permanent or electromagnets range from 60 MHz up to 100 MHz, while spectrometers with superconducting magnets range from 200 MHz to many hundred megacycles.

Data Analysis

In a hydrogen NMR spectrum, the presence of any resonance explains first that the molecule of study contains hydrogen. Second, the number of bands in the spectrum shows how many different positions there are on the molecule to which hydrogen is attached. The frequency of a particular resonance in the NMR spectrum is referred to as the chemical shift. This is the most important measurable part of the NMR spectrum and contains information about the environment of each hydrogen atom and the structure of the compound under study. The third bit of information that an NMR spectrum provides is the ratio of the areas of the different bands, thus explaining the relative number of hydrogen atoms that exist at each position on a given molecule. This ratio is direct evidence of the structure of molecular structure and must correspond absolutely to any proposed structure before that structure may be considered correct.

Finally, the complex structure of the bands may contain information about the distance that separate the various hydrogen atoms through covalent bonds and the spatial arrangement of the hydrogen atoms attached to the molecule, including secondary structure. Secondary structure refers to folding or self-assembly of a molecule due to long-range bonding, such as in the spiral structure of DNA. The complex structure of the NMR bands is due primarily to spin coupling between the various hydrogen atoms. This coupling is, in turn, a function of the distance through the bonds and the geometry of the molecule. In the case of small molecules, the band complexity may be simulated exactly with quantum mechanical calculations or approximated using quantum mechanically derived rules.

Fundamental to the use of NMR is the ready correlation of measurable spectrum quantities with the structure of a molecule under study. Consider the simple molecule ethanol. Ethanol has three bands in its hydrogen NMR spectrum that correlate with the three distinct types of hydrogen atoms present in the molecule. The area ratio of the three bands is 1:2:3 and this reflects the number ratio of the hydrogen atoms seen in the structure of

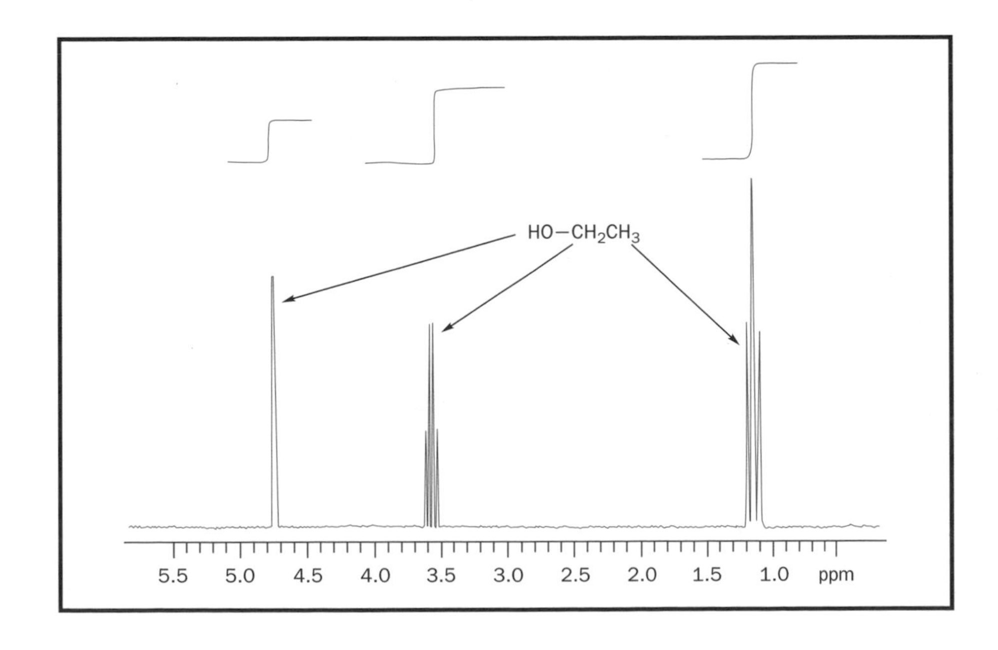

The 300 MHz NMR spectrum of ethanol. This simple alcohol has three types of hydrogen atoms and thus there are three bands. The relative area of the bands is indicated by the vertical displacement of the integral at the top of the spectrum. The relative band area of 1:2:3 is sufficient data to assign all three bands.

ethanol. The fine structure of the bands in this spectrum may also be correlated with the structure of ethanol using simple rules. As more complex molecules are considered, more complex rules must be used until finally two-dimensional NMR (2D NMR) and even more complex spectroscopic procedures must be used for complete analysis.

Discovery

Although NMR was thought to be possible for many years, it was first demonstrated in 1946 simultaneously and independently by two physicists working on the East Coast and the West Coast of the United States: Felix Bloch at Stanford University and Edward Mills Purcell at Harvard University. For their work they shared the Nobel Prize in physics in 1952. The first commercial spectrometers appeared later in the 1950s and quickly became an indispensable tool for research chemists. The first commercial spectrometers were based on conventional electromagnets and permanent magnets, but during the 1960s the superconducting magnetic had already been largely adopted. In 1966 the chemist Richard Ernst demonstrated Fourier transform nuclear magnetic resonance (FT NMR). This procedure quickly replaced the older scanning techniques and earned Ernst the Nobel Prize in chemistry in 1991. He continued to make contributions to many areas of NMR but most notable are his contributions to 2D NMR and magnetic resonance imaging (MRI).

A magnetic resonance imaging (MRI) system is used to diagnose maladies in the soft tissues of the body.

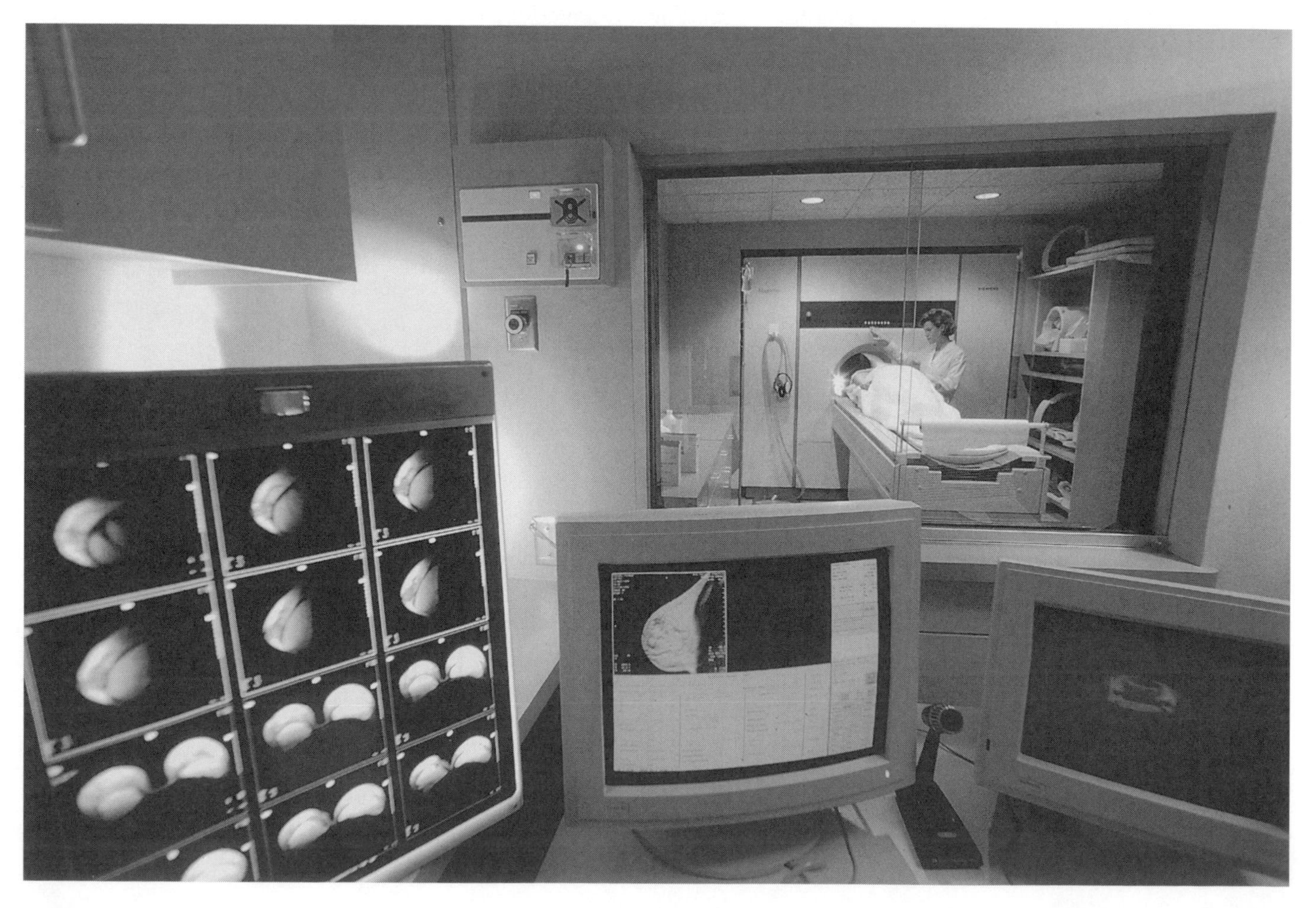

Magnetic Resonance Imaging

Magnetic resonance imaging (MRI) is a spin-off of NMR. The two techniques differ in two important respects. First, in NMR spectrums the individual resonance bands of different frequency are displayed in a spectrum to reveal structural and purity information while in MRI all of the resonance bands are grouped together as a single quantity. Second, in NMR the magnetic field is very homogeneous so that the different frequency bands may be clearly separated; however, with MRI a linear magnetic gradient is superimposed on the main magnetic field so that the frequency of the NMR signal is a function of space instead of structure. The gradient is alternated along different axes so that an image may be constructed. The resulting image is a two-dimensional slice through the sample. Several slices may be accumulated so that a three-dimensional image may then be constructed if needed. Since the 1980s, MRI has grown to be an indispensable tool in the medical diagnosis of many maladies—especially of soft tissue, such as the brain and the spine. In an MRI body scan, the image is constructed primarily from the detected radiation of the hydrogen atoms in water and **lipids** of the various tissues. While the varying concentration of water and lipids in various tissues contribute to the formation of an MRI image, the dominant factor is the variation in the length of time that the hydrogen atoms found in different tissues will radiate. This difference in radiation time is referred to as relaxation time. Contrast, and thus good images, may be achieved by delaying the onset of radio-frequency detection until some tissues have almost quit radiating while other tissues are still radiating strongly. Particularly noteworthy is the difference in relaxation time of normal tissue compared to malignant tissue. MRI may be optimized best for imaging specific tissues using a combination of radio-frequency pulse parameters, delay times, and magnetic gradients. These same parameters may be used to measure the flow of fluids through tissues, such as the flow of blood through muscle tissue. The contrast and thus the quality of the MRI for some specific tissues may be significantly increased by the use of specially engineered compounds called MRI contrast **reagents**. These compounds have two required properties: They must associate specifically with a tissue under investigation and they must be magnetic. The presence of a contrast reagent in a tissue changes the relaxation time of the water in this tissue in such a way that the MRI image is enhanced. SEE ALSO MOLECULAR STRUCTURE; ORGANIC CHEMISTRY; SPECTROSCOPY.

lipids: a nonpolar organic molecule; fatlike; one of a large variety of nonpolar hydrophobic (water-hating) molecules that are insoluable in water

reagent: chemical used to cause a specific chemical reaction

Ben Shoulders

Bibliography

Bloch, F.; Hansen, W. W.; and Packard, M. (1946). "The Nuclear Induction Experiment." *Physics Review* 70:474.

Pound, R. V., and Purcell, E. M. (1946). "Measurement of Magnetic Resonance Absorption by Nuclear Moments in a Solid." *Physics Review* 70:980.

Nuclear Medicine

Nuclear medicine involves the injection of a radiopharmaceutical (radioactive drug) into a patient for either the diagnosis or treatment of disease. The history of nuclear medicine began with the discovery of radioactivity from uranium by the French physicist Antoine-Henri Becquerel in 1896, followed shortly thereafter by the discovery of radium and polonium by the renowned French chemists Marie and Pierre Curie. During the 1920s and 1930s radioactive phosphorus was administered to animals, and for the first time it was determined that a metabolic process could be studied in a living animal. The presence of phosphorus in the bones had been proven using radioactive material. Soon ^{32}P was employed for the first time to treat a patient with leukemia. Using radioactive iodine, thyroid physiology was studied in the late 1930s. Strontium-89, another compound that localizes in the bones and is currently used to treat pain in patients whose cancer has spread to their bones, was first evaluated in 1939.

radioactive decay: process involving emission of subatomic particles from the nucleus, typically accompanied by emission of very short wavelength electromagnetic radiation

isotope: form of an atom that differs by the number of neutrons in the nucleus

A nuclide consists of any configuration of protons and neutrons. There are approximately 1,500 nuclides, most of which are unstable and spontaneously release energy or subatomic particles in an attempt to reach a more stable state. This nuclear instability is the basis for the process of **radioactive decay**, and unstable nuclides are termed *radionuclides.* During the 1940s and 1950s nuclear reactors, accelerators, and cyclotrons began to be widely used for medical radionuclide production. Reactor-produced radionuclides are generally electron-rich and therefore decay by β^--emission. The main application of β^--emitters is for cancer therapy, although some reactor-produced radionuclides are used for nuclear medicine imaging. Cyclotron-produced radionuclides are generally prepared by bombarding a stable target (either a solid, liquid, or gas) with protons and are therefore proton-rich, decaying by β^+-emission. These radionuclides have applications for diagnostic imaging by positron-emission tomography (PET). One of the most convenient methods for producing a radionuclide is by a *generator.* Certain parent–daughter systems involve a long-lived parent radionuclide that decays to a short-lived daughter. Since the parent and daughter nuclides are not **isotopes** of the same element, chemical separation is possible. The long-lived parent produces a continuous supply of the relatively short-lived daughter radionuclide and is therefore called a generator.

photon: a quatum of electromagnetic energy

Currently, the majority of radiopharmaceuticals are used for diagnostic purposes. These involve the determination of a particular tissue's function, shape, or position from an image of the radioactivity distribution within that tissue or at a specific location within the body. The radiopharmaceutical localizes within certain tissues due to its biological or physiological characteristics. The diagnosis of disease states involves two imaging modalities: Gamma (γ) scintigraphy and PET. In the 1950s γ scintigraphy was developed by Hal O. Anger, an electrical engineer at Lawrence Berkeley Laboratory. It requires a radiopharmaceutical containing a radionuclide that emits γ radiation and a γ camera or single **photon** emission computed tomography (SPECT) camera capable of imaging the patient injected with the γ-emitting radiopharmaceutical. The energy of the γ-photons is of great importance, since most cameras are designed for particular windows of energy, generally in the range of 100 to 250 kilo-electron volts (keV). The most widely used radionuclide for imaging by γ scintigraphy is ^{99m}Tc ($T_{\frac{1}{2}}$ = 6 hours), which is produced from the decay of ^{99}Mo ($T_{\frac{1}{2}}$ = 66 hours). In 1959 the Brookhaven National Laboratory (BNL) developed the $^{99}Mo/^{99m}Tc$ generator, and in 1964 the first ^{99m}Tc radiotracers were developed at the University of Chicago. The low cost and convenience of the $^{99}Mo/^{99m}Tc$ generator, as well as the ideal photon energy of ^{99m}Tc (140 keV), are the key reasons for its widespread use. A wide variety of ^{99m}Tc radiopharmaceuticals have been developed during the last forty years, most of them coordination complexes. Many of these are currently used every day in hospitals throughout the United States to aid in the diagnosis of heart disease, cancer, and an assortment of other medical conditions.

PET was developed during the early 1970s by Michel Ter-Pogossian and his team of researchers at Washington University. It requires a radiopharmaceutical labeled with a positron-emitting radionuclide (β^+) and a PET camera for imaging the patient. Positron-decay results in the emission of two 511 keV photons 180° apart. PET scanners contain a circular array of detectors with coincidence circuits designed to specifically detect the 511 keV photons emitted in opposite directions. The positron-emitting radionuclides most frequently used for PET imaging are ^{15}O ($T_{\frac{1}{2}}$ = 2 minutes), ^{13}N ($T_{\frac{1}{2}}$ = 10 minutes), ^{11}C ($T_{\frac{1}{2}}$ = 20 minutes), and ^{18}F ($T_{\frac{1}{2}}$ = 110 minutes). Of these, ^{18}F is most widely used for producing PET radiopharmaceuticals. The most frequently used ^{18}F-labeled radiopharmaceutical is 2-deoxy-2- [^{18}F]fluoro-D-**glucose** (FDG). This agent was approved by the Food and Drug Administration (FDA) in the United States in 1999 and is now routinely used to image various types of cancer as well as heart disease.

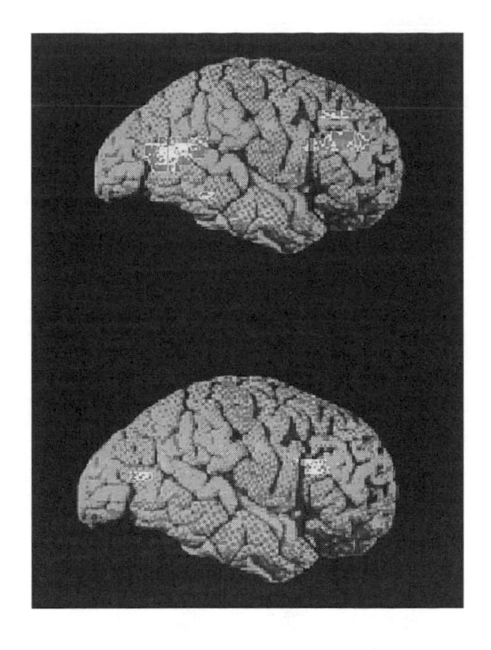

Two positron-emission tomography (PET) scans showing the brain of a depressed person (top) and a healthy person (bottom).

glucose: common hexose monosaccharide; monomer of starch and cellulose; also called grape sugar, blood sugar, or dextrose

The use of radiopharmaceuticals for therapeutic applications (α- or β^--emitters) is increasing. The first FDA-approved radiopharmaceutical in the United States was, in fact, for therapeutic use. Sodium [^{131}I] iodide was approved in 1951 for treating thyroid patients. There are currently FDA-approved radiopharmaceuticals for alleviating pain in patients whose cancer has metastasized to their bones. These include sodium ^{32}P-phosphate, ^{89}Sr-chloride, and ^{153}Sm-EDTMP (where EDTMP stands for ethylenediaminetetramethylphosphate). In February 2002 the first radiolabeled monoclonal **antibody** was approved by the FDA for the radioimmunotherapy treatment of cancer. Yttrium-90-labeled anti-CD20 monoclonal antibody is used to treat patients with non-Hodgkin's lymphoma.

antibody: protein molecule that recognizes external agents in the body and binds to them as part of the immune response of the body

Many branches of chemistry are involved in nuclear medicine. Nuclear chemistry has developed accelerators and reactors for radionuclide production. Inorganic chemistry has provided the expertise for the development of **metal**-based radiopharmaceuticals, in particular, ^{99m}Tc radiopharmaceuticals, whereas organic chemistry has provided the knowledge base for the development of PET radiopharmaceuticals labeled with ^{18}F, ^{13}N, ^{11}C, and ^{15}O. Biochemistry is involved in understanding the biological behavior of radiopharmaceuticals, while medical doctors and pharmacists are involved in clinical studies. Nuclear medicine, which benefits the lives of millions of people every day, is truly a multidisciplinary effort, one in which chemistry plays a significant role. SEE ALSO BECQUEREL, ANTOINE-HENRI; CURIE, MARIE SKLODOWSKA; NUCLEAR CHEMISTRY; NUCLEAR FISSION; RADIATION EXPOSURE; RADIOACTIVITY.

metal: element or other substance the solid phase of which is characterized by high thermal and electrical conductivities

Carolyn J. Anderson

Bibliography

McCarthy, T. J.; Schwarz, S. W.; and Welch, M. J. (1994). "Nuclear Medicine and Positron Emission Tomography: An Overview." *Journal of Chemical Education* 71: 830–836.

Schwarz, S. W.; Anderson, C. J.; and Downer, J. B. (1997). "Radiochemistry and Radiopharmacology." In *Nuclear Medicine Technology and Techniques*, 4th edition, ed. D. R. Bernier, P. Christian, and J. K. Langan. St. Louis, MO: Mosby Year Book.

Internet Resources

"A Brief History of Nuclear Medicine." UNM, Ltd. Available from <http://www.nucmednet.com/history.htm>.

"The History of Nuclear Medicine." Society of Nuclear Medicine. Available from <http://www.snm.org/nuclear/history.html>.

Nucleic Acids

Nucleic acids are a family of macromolecules that includes deoxyribonucleic acid (**DNA**) and multiple forms of ribonucleic acid (**RNA**). DNA, in humans and most organisms, is the genetic material and represents a collection of instructions (genes) for making the organism. This collection of instructions is called the genome of the organism. The primary classes of RNA molecules either provide information that is used to convert the genetic information in DNA into functional proteins, or are important players in the **translational process**, in which the actual process of protein **synthesis** (on **ribosomes**) occurs.

DNA: deoxyribonucleic acid—the natural polymer that stores genetic information in the nucleus of a cell

RNA: ribonucleic acid—a natural polymer used to translate genetic information in the nucleus into a template for the construction of proteins

translational process: transfer of information from codon on m-RNA to anticodon on t-RNA; used in protein synthesis

synthesis: combination of starting materials to form a desired product

ribosome: large complex of proteins used to convert amino acids into proteins

Discovery of and Evidence for DNA as the Genetic Material

DNA was first discovered in 1869 by a Swiss biochemist, Johann Friedrich Miescher. He extracted a gelatinous material that contained organic phosphorus from cells in human pus that was obtained from the bandages of wounded soldiers. He named this material *nuclein.* Ten years later Albrecht Kossel explored the chemistry of nuclein (for which he received the Nobel Prize) and discovered that it contained the organic bases **adenine**, **thymine**, **guanine**, and **cytosine**. In 1889 Richard Altman removed the proteins from the nuclein in yeast cells and named the deproteinized material *nucleic acid.* It was not until about 1910 that it was realized that there were two types of nucleic acid, DNA and RNA. A great deal of chemistry during the early part of the twentieth century focused on characterizing the composition of and the linkages in both DNA and RNA. A chemical test for deoxyribose, developed by Robert Feulgen during the 1920s, was the first test capable of distinguishing DNA from RNA. Because of the simplicity of the composition of DNA, which has only four bases (and early reports indicated erroneously that there were equimolar quantities of each), it was originally thought that DNA molecules functioned in chromosomal stability and maintenance. It was only after Erwin Chargaff, in 1950, showed that the molar amounts of the bases varied widely in different organisms that the notion that DNA might be the genetic material became an attractive idea.

adenine: one of the purine bases found in nucleic acids, $C_5H_5N_5$

thymine: one of the four bases that make up a DNA molecule

guanine: heterocyclic, purine, amine base found in DNA

cytosine: heterocyclic, pyrimidine, amine base found in DNA

The general consensus prior to the mid-1940s was that proteins (which contain twenty different amino acids) were the most logical candidate for the genetic material. Three later, however, findings pointed toward the conclusion that DNA was the genetic material. During the 1920s Frederick Griffith examined the activity of cell extracts in an attempt to identify a "transforming principle" (and a specific molecule related to this principle) in experiments with the bacterium *Streptococcus pneumoniae.* Unfortunately, his experiments failed to identify a specific molecule. In 1944 Oswald Avery, Colin MacLeod, and Maclyn McCarty partially purified cell extracts and presented evidence that the genetic component of these cells was DNA. In 1952 Alfred Hershey and Martha Chase investigated the infection of *Escherichia coli* cells with phage T2 (a virus) and their results were further corroboration that DNA was the genetic material. Since that time, a large body of evidence has confirmed the (nearly) universal truth that DNA is the genetic, heritable material in organisms. (The only exception to this is the case of RNA viruses, such as the AIDS virus, in which RNA is the only nucleic acid present in the virus and the genetic material.)

Modern research took a giant step forward when James Watson and Francis Crick, analyzing the collected findings of a number of laboratories, proposed the double **helix** structure of DNA in 1953. Their announcements motivated scientists to find corroboration for this proposal. During the 1980s detailed x-ray crystallographic analyses of DNA became acknowledged as proof of the structural arrangement that had been described by Watson and Crick, including the Watson–Crick complementary base-pairing arrangements. The elucidation of the structure of DNA led to an enormous and rapid expansion of our understanding of DNA's function in the living cell.

helix: form of a spiral or coil such as a corkscrew

Types of Nucleic Acids: Composition and Structure

All nucleic acids are linear, nonbranching polymers of nucleotides, and are therefore polynucleotides. DNA is double-stranded in virtually all organisms. (It is single-stranded in some viruses.) DNA occurs in many, but not all, small organisms as double-stranded and circular (without any ends). Higher organisms (eukaryotes) have approximately ten million base pairs or more, with the genetic material parceled out into multiple genetic pieces called chromosomes. For example, humans have twenty-three pairs of chromosomes in the nucleus of each **somatic cell**. Within the nucleus, the DNA molecules are found in "looped arrangements" that mimic the circular DNA observed in many prokaryotes.

somatic cell: cells of the body with the exception of germ cells

All RNA molecules are single-stranded molecules. RNA molecules are synthesized from DNA templates in a process known as **transcription**; these molecules have a number of vital roles within cells. It is convenient to divide RNA molecules into the three functional classes, all of which function in the cytoplasm.

transcription: enzyme-catalyzed assembly of an RNA molecule complementary to a strand of DNA

Messenger RNA (mRNA) contains the information (formerly residing in DNA) that is decoded in a way that enables the manufacture of a protein, and migrates from the nucleus to ribosomes in the cytoplasm (where proteins are assembled). A triplet of nucleotides within an RNA molecule (called a codon) specifies the amino acid to be incorporated into a specific site in the protein being assembled. A cell's population of mRNA molecules is very diverse, as these molecules are responsible for the synthesis of the many different proteins found in the cell. However, mRNA makes up only 5 percent of total cellular RNA.

Ribosomal RNA (rRNA) is the most abundant intracellular RNA, making up 80 percent of total RNA. The **eukaryotic** ribonucleoprotein particle (ribosome) is composed of many proteins and four rRNA molecules (which are classified according to size). Ribosomes reside in the cytoplasm and are the "molecular platform" (the actual physical site of) for protein synthesis.

eukaryotic: relating to organized cells of the type found in animals and plants

Transfer RNA (tRNA) molecules contain between seventy-four and ninety-five nucleotides and all tRNAs have similar overall structures. There are twenty individual tRNAs; each one binds to a specific amino acid in the cytoplasm and brings its "activated amino acid" to a ribosome—part of the

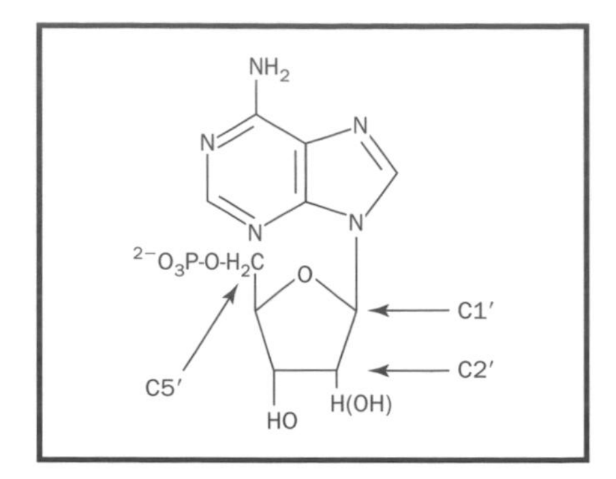

Figure 1.

uracil: heterocyclic, pyrimidine, amine base found in RNA

ester: organic species containing a carbon atom attached to three moieties: an O via a double bond, an O attached to another carbon atom or chain, and a H atom or C chain; the R(C=O)OR functional group

hydrogen bond: interaction between H atoms on one molecule and lone pair electrons on another molecule that constitutes hydrogen bonding

hydrogen bonding: intermolecular force between the H of an N–H, O–H, or F–H bond and a lone pair on O, N, or F of an adjacent molecule

translational machinery that carries out protein synthesis. Transfer RNA makes up the remaining 15 percent of cellular RNA.

All nucleic acids are polynucleotides, with each nucleotide being made up of a base, a sugar unit, and a phosphate. The composition of DNA differs from that of RNA in two major ways (see Figure 1). Whereas DNA contains the bases guanine (G), cytosine (C), adenine (A), and thymine (T), RNA contains G, C, and A, but it contains **uracil** (U) in place of thymine. Both DNA and RNA contain a five-membered cyclic sugar (a pentose). RNA contains a ribose sugar. The sugar in DNA, however, is 2′-deoxyribose.

In DNA, each base is linked by a β-glycosidic bond to the C1′ position of the 2′-deoxyribose, and each phosphate is linked to either the C3′ or C5′ position. The linkages are essentially the same in RNA.

DNA is a right-handed, double-stranded helix, in which the bases essentially occupy the interior of the helix, whereas the phosphodi**ester** backbone (sugar-phosphate backbone) more or less comprises the exterior. The bases on the individual strands form intermolecular **hydrogen bonds** with each other (the complementary Watson–Crick base pairs). An adenine base on one strand interacts specifically with a thymine base on the other, forming two hydrogen bonds and an A–T base pair; while a G–C base pair contains three hydrogen bonds. These interactions possess a specificity that is pivotal to both DNA replication and transcription (see Figure 2).

DNA structure is also described in terms of primary, secondary, and tertiary structures. The primary structure is simply the sequence of nucleotides. The secondary structure refers to the **hydrogen bonding** between A–T and G–C base pairs. The tertiary structure refers to the larger twists and turns of the DNA molecule. Other features of DNA are the major and

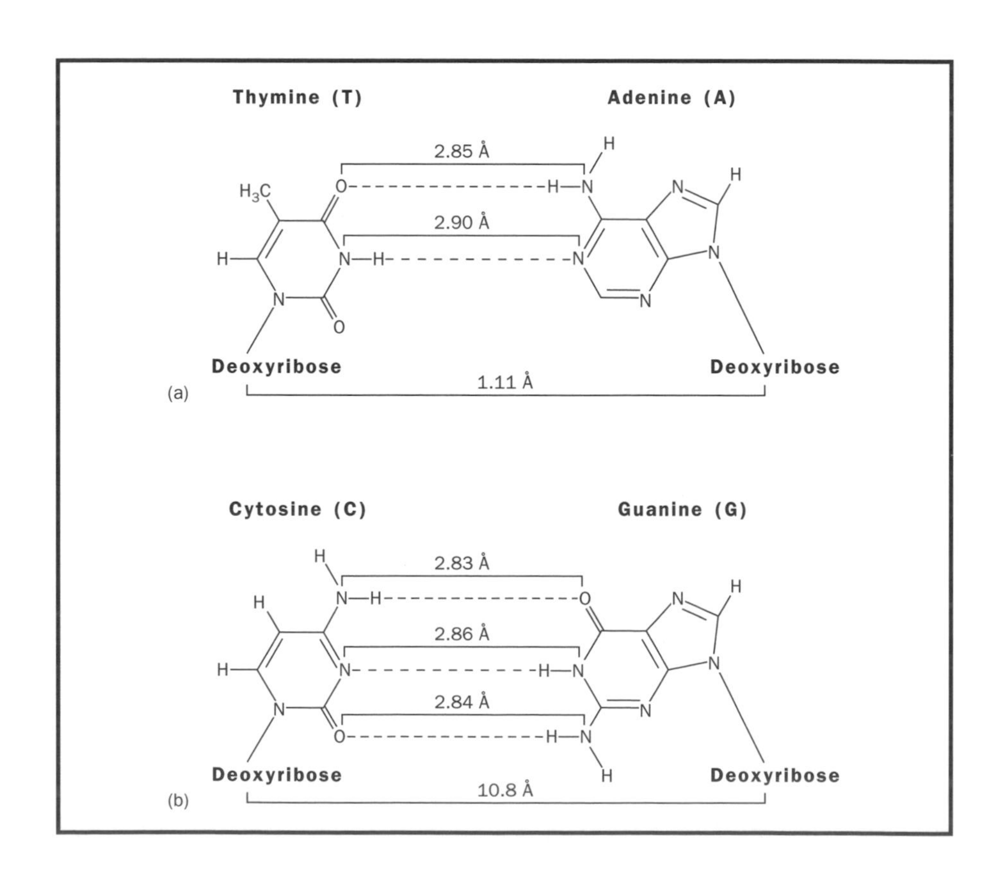

Figure 2.

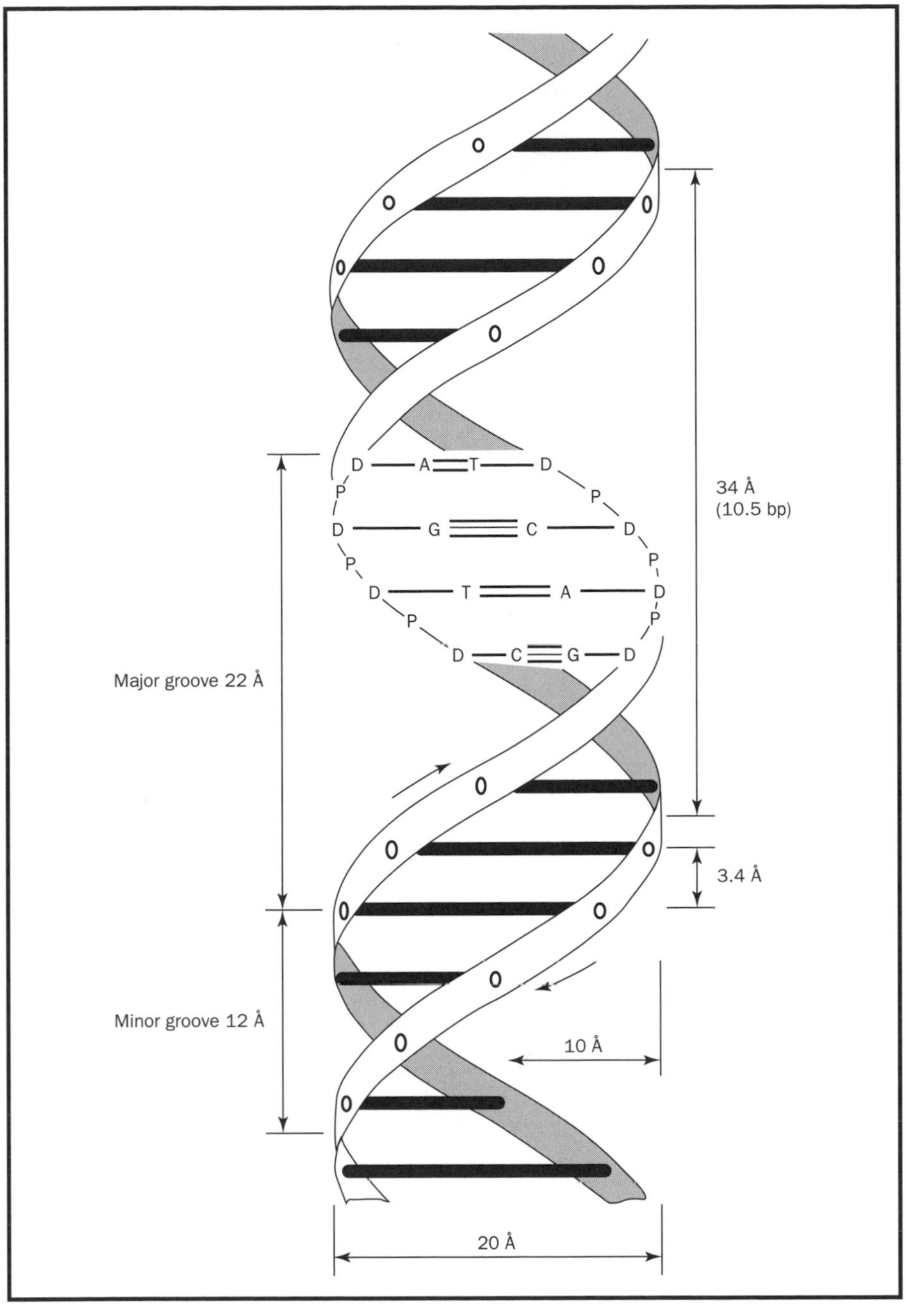

Figure 3.

minor grooves that run along the helix that are the target sites for DNA binding proteins involved in replication and transcription. Although DNA can exist in several alternate structures, the B-form of DNA (see Figure 3) is the biologically relevant form.

As stated previously, DNA is the genetic material in humans and in virtually all organisms, including viruses—with the exception of a few viruses that possess RNA as the genetic material.

In complex multicellular organisms (such as humans), DNA carries within itself the instructions for the synthesis and assembly of virtually all the components of the cell and (therefore) for the structure and function of tissues and organs. Within the approximately 3.2×10^9 base pairs (3.2 Gbps) in human DNA, the Human Genome Project has determined that there are a minimum of about 25,000 individual segments that correspond to individual genes. The genes collectively make up only about 2 to 3 percent of the total DNA, but encode the detailed genetic instructions for the synthesis of proteins. Proteins are the "workhorses" of the cell, and in one way or another are responsible for the functions that permit a cell to communicate with other cells and that define the character of the individual cell. A kidney cell is very different from a heart or eye cell. Although every cell contains the same DNA, different subsets of the 25,000 genes are expressed in the different organs or tissues. The expressed genes determine the type of cell that is produced and a cell's ultimate function in a multicellular organism.

Interestingly, there is only approximately a 0.1 percent difference in DNA among humans. The nucleotide sequences of DNA differs between organisms and is a fundamental difference between individuals and between species. For example, our closest (species) relative, the chimpanzee, has DNA that is 98.5 percent identical to that of humans. SEE ALSO DEOXYRIBONUCLEIC ACID; DOUBLE HELIX; WATSON, FRANCIS DEWEY.

William M. Scovell

Bibliography

Berg, Jeremy M.; Tymoczko, John L.; and Stryer, Lubert (2002). *Biochemistry*, 5th edition. New York: W. H. Freeman.

Kimball, John W. (1994). *Biology*, 6th edition. Dubuque, IA: Wm. C. Brown Publishers. Also available from <http://users.rcn.com/jkimball.ma.ultranet/BiologyPages/>.

Nucleotides

Nucelotides are the repeating building blocks of nucleic acids (which are polynucleotides or polymers of nucleotides). A nucleotide is made up of a heterocyclic base (a purine or pyrimidine), a cyclic sugar unit (ribose or deoxyribose), and a phosphate group. A nucleotide is either a ribonucleotide, the repeating unit in ribonucleic acid (**RNA**), or a deoxyribonucleotide, the repeating unit in deoxyribonucleic acid (**DNA**). Table 1 below lists the names of purine and pyrimidine bases, the nucleosides (base + ribose), the corresponding 5′-nucleotides (base + ribose + phosphate), and the abbreviations.

RNA: ribonucleic acid—a natural polymer used to translate genetic information in the nucleus into a template for the construction of proteins

DNA: deoxyribonucleic acid—the natural polymer that stores genetic information in the nucleus of a cell

Nucleotides are sometimes abbreviated as 5′-NMP, in which N can stand for any of the bases.

Figure 1 illustrates two nucleotides, in which the phosphate is attached to the (ribose) sugar at either the 5′-carbon (5′-AMP) or the 3′-carbon (3′-TMP). The numbering systems for both purine (**guanine** shown here) and pyrimidine (**thymine** shown here) compounds are given, in addition to that for the sugar (ribose or deoxyribose). Note that the linkage between the

guanine: heterocyclic, purine, amine base found in DNA

thymine: one of the four bases that make up a DNA molecule

Base	Nucleoside	Nucleotide	Abbreviation
Adenine	Adenosine	Adenosine-5'-(mono)phosphate	5'-AMP
Guanine	Guanosine	Guanosine-5'-(mono)phosphate	5'-GMP
Cytosine	Cytidine	Cytidine-5'-(mono)phosphate	5'-CMP
Thymine	Thymidine	Thymidine-5'-(mono)phosphate	5'-TMP
Uracil	Uridine	Uridine-5'-(mono)phosphate	5'-UMP

(These nucleotides are generally abbreviated, 5'-NMP, where N can contain any of the bases.)

Table 1. Purine and pyrimidine bases, the nucleosides, nucleotides, and corresponding abbreviations.

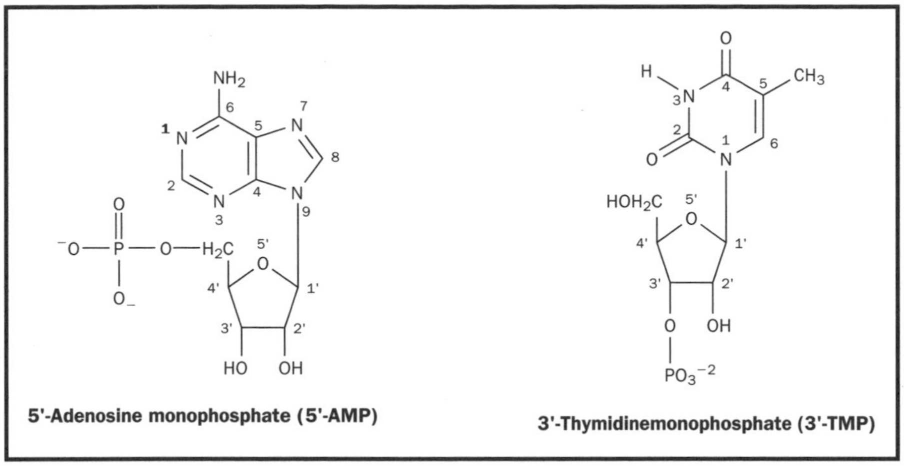

Figure 1.

(deoxy)ribose unit and the base, a β-glycosidic bond, has a different connectivity according to whether the link is to a purine or a pyrimidine. Although the linkage involves a C1′ carbon in both cases, the β-glycosidic bond in the case of a purine nucleotide is a link to the N-9 of the **purine base**, but to the N-1 of the **pyrimidine base**.

purine base: one of two types of nitrogen bases found in nucleic acids

pyrimidine base: one of two types of nitrogen bases found in nucleic acids

Nucleotides are also found in which two or three phosphate groups are linked together, the chain of phosphate groups bonded to the sugar's 5′-position. In these cases, they are nucleoside diphosphates (5′-NDP) and nucleoside triphosphates (5′-NTP).

The bases have very limited solubilities in water, whereas the nucleosides and nucleotides have greater solubilities, due to the presence of polar sugars, or of both sugars and charged phosphate groups, respectively.

The nucleoside triphosphates are of special interest for at least two reasons. First, they are the actual precursor molecules used in the **biosynthesis** of nucleic acids. Second, ATP is a high-energy molecule, produced primarily in mitochondria during oxidative **phosphorylation**. It stores energy, which is released (dG° = −7.3 kcal/mol) during the hydrolysis of ATP to ADP and phosphate (Pi), as shown, and then utilized to power cell reactions.

biosynthesis: formation of a chemical substance by a living organism

phosphorylation: the process of addition of phosphates into biological molecules

$$ATP + H_2O \longrightarrow ADP + Pi$$

SEE ALSO Deoxyribonucleic Acid (DNA); Restriction Enzymes.

William M. Scovell

Bibliography

Berg, Jeremy M.; Tymoczko, John L.; and Stryer, Lubert (2002). *Biochemistry*, 5th edition. New York: W. H. Freeman.

Lehninger, Albert L. (2000). *Lehninger Principles of Biochemistry*, 3rd edition, ed. David L. Nelson and Michael M. Cox. New York: Worth Publishers.

Nutrition ***See Ascorbic Acid; Retinol; Riboflavin; Thiamin.***

Nylon

In 1928 E. I. du Pont de Nemours & Company (Du Pont) launched one of its first basic research programs and hired Wallace Hume Carothers to run it. He was brought to Du Pont in part because his fellow researchers at Harvard University and the University of Illinois had called him the best

Figure 1.

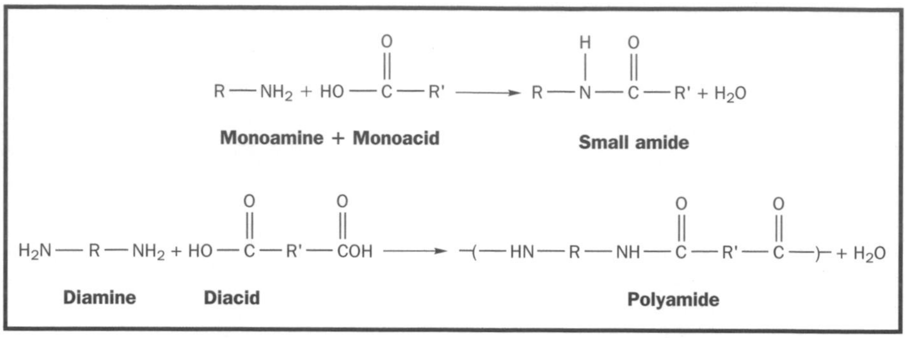

$$R{-}NH_2 + HO{-}\overset{O}{\overset{\|}{C}}{-}R' \longrightarrow R{-}\overset{H}{\overset{|}{N}}{-}\overset{O}{\overset{\|}{C}}{-}R' + H_2O$$

Monoamine + Monoacid → **Small amide**

$$H_2N{-}R{-}NH_2 + HO{-}\overset{O}{\overset{\|}{C}}{-}R'{-}\overset{O}{\overset{\|}{C}}OH \longrightarrow {-}\!\!({-}HN{-}R{-}NH{-}\overset{O}{\overset{\|}{C}}{-}R'{-}\overset{O}{\overset{\|}{C}}{-})\!\!{-} + H_2O$$

Diamine **Diacid** → **Polyamide**

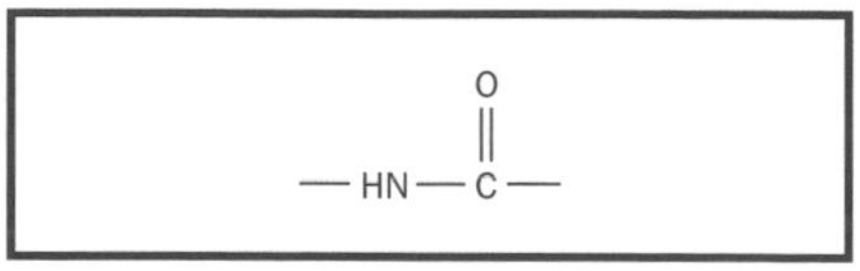

$$-HN-\overset{O}{\overset{\|}{C}}-$$

Figure 2. An amide unit.

synthesis: combination of starting materials to form a desired product

diamine: compound, the molecules of which incorporate two amino groups ($-NH_2$) in their structure, such as 1,2 diamino ethane (sometimes called ethylenediamine) and the three diamine benzene compounds

carboxylic acid: one of the characteristic groups of atoms in organic compounds that undergoes characteristic reactions, generally irrespective of where it occurs in the molecule; the $-CO_2H$ functional group

STEPHANIE KWOLEK (1923–)

Stephanie Kwolek has seventeen patents, the first of which is for Kevlar. After creating a new polymer, she would spin them into fibers for strength and flexibility testing. Material for fibers of Kevlar was cloudy white instead of molasses brown, the first indication that she had uncovered an exceptional polymer.

—Valerie Borek

synthetic chemist they knew. The program he supervised was designed to investigate the composition of natural polymers such as silk, cellulose, and rubber. Many of Carothers's efforts related to condensation polymers were based on his deduction that if a monofunctional reactant reacted in a certain manner in forming a small molecule, then similar reactions that employed a comparable reactant, but with two reactive groups, would form polymers. (See Figure 1.)

The amide unit (found in polyamides) shown in Figure 2 is the same connective grouping that is found in proteins.

Although the Carothers group had worked with both polyesters and polyamides, they initially emphasized their work on the polyesters, as polyesters were more soluble and easier to work with. Julian Hill, a member of the Carothers team, noticed that he could form fibers if he separated a portion of a soft polyester material using a glass stirring rod and pulled it away from the clump. But because the polyesters had softening points that were too low for their use as textiles, the group returned to its work with the polyamides. The researchers found that fibers could also be formed from the polyamides, similar to those formed from the polyesters. The strength of these fibers approached, and in some cases surpassed, the strengths of natural fibers. This new miracle fiber (nylon) was introduced at the 1939 New York World's Fair, in an exhibit that announced the **synthesis** of a wonder fiber from "coal, air, and water"—an exaggeration but nevertheless eye-catching. When the nylon stockings were first offered for sale in New York City, on May 15, 1940, over four million pairs were sold in the first few hours. Nylon stocking sales took a large drop during World War II when it became publicized that nylon was needed to make parachutes.

The polyamides (nylons) were given a special naming system. Nylons made from **diamines** and dicarboxylic acids are designated by two numbers, the first representing the number of carbons in the diamine chain *(a)* and the second the number of carbons in the di**carboxylic acid** *(b)*. (See Figure 3.)

The nylon developed by Carothers at Du Pont was nylon 6,6. Because of the importance of starting out with equal amounts of the two reactants, salts of the diamine and of the diacid are made and then used in the commercial synthesis of nylon 6,6. (See Figure 4.)

Nylon 6,6 (or simply nylon 66) is the largest volume nylon used as fiber, film, and plastic. About 1,134 million kilograms (2,500 million pounds) of nylon 66 were produced for fiber applications in 2000. Nylon 66 is used to make tire cord, rope, clothing, thread, hose, undergarments, rug filament,

$$-(\!-NH-(CH_2-)_a-NH-\overset{\overset{O}{\|}}{C}-(CH_2-)_{b-2}-\overset{\overset{O}{\|}}{C}-)_n-$$

Nylon a,b

Figure 3.

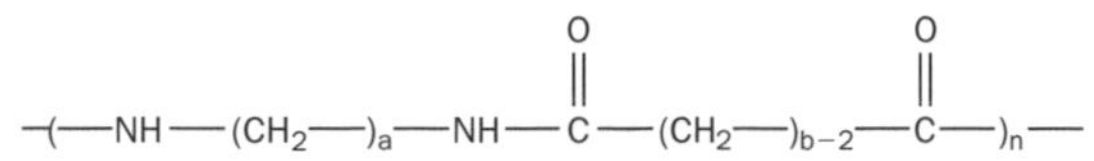

socks, dresses, and more. Because of the presence of polar units in nylons, similar to the presence of polar units in proteins, materials made from nylon have a nice "feel" to them. Nylon materials also attract odors (many everyday odors are polar in nature) and are easily stained. Most textile and fabric products are treated to repel unwanted odors and stainmaking materials.

Nylon 66 was the first engineering thermoplastic, and up until 1953 represented all of engineering thermoplastic sales. The term "thermoplastic" denotes a material that can be melted through heating. The term "engineering thermoplastics" describes a plastic material that can be cut, drilled, or machined. About 680.4 million kilograms (1,500 million pounds) of nylons were produced in the United States in 2000 for thermoplastic use. Nylon 66 plastic is tough and rigid. It has a relatively high use temperature (to about 270°C or 518°F), and is used in the manufacture of products ranging from automotive gears to hairbrush handles. Molded nylon 66 is used to make skate wheels, motorcycle crank cases, bearings, tractor hood extensions, skis for snowmobiles, lawnmower blades, bicycle wheels, and so on.

Most polymers, when heated, progress from a glasslike solid to a softer solid, and then to a viscous "taffylike" material that is most amenable to heat-associated fabrication. In the case of nylon 66, the transition from the solid to the soft stage is abrupt, requiring that fabrication be closely watched.

The presence in nylons of polar groups results in materials that have a relatively high glass transition temperature (T_g, the point at which segmental mobility begins) and high **melting point** (the point at which entire polymer chain mobility begins), so that, unlike many vinyl polymers such as polyethylene and polypropylene (which must be at temperatures above their glass transition temperatures to possess needed flexibility), nylons, and many other condensation polymers, function best in contexts in which strength, and not flexibility, is the desired attribute.

melting point: temperature at which a substance in the solid state undergoes a phase change to the liquid state

Figure 4.

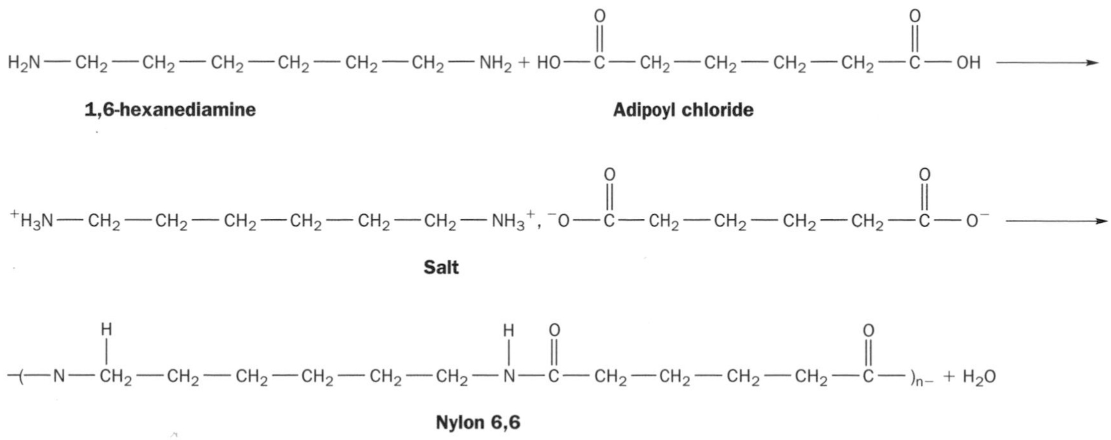

$$H_2N-CH_2-CH_2-CH_2-CH_2-CH_2-CH_2-NH_2 + HO-\overset{\overset{O}{\|}}{C}-CH_2-CH_2-CH_2-CH_2-\overset{\overset{O}{\|}}{C}-OH \longrightarrow$$

1,6-hexanediamine **Adipoyl chloride**

$$^{+}H_3N-CH_2-CH_2-CH_2-CH_2-CH_2-CH_2-NH_3^{+},\ ^{-}O-\overset{\overset{O}{\|}}{C}-CH_2-CH_2-CH_2-CH_2-\overset{\overset{O}{\|}}{C}-O^{-} \longrightarrow$$

Salt

$$-(\!-\overset{\overset{H}{|}}{N}-CH_2-CH_2-CH_2-CH_2-CH_2-CH_2-\overset{\overset{H}{|}}{N}-\overset{\overset{O}{\|}}{C}-CH_2-CH_2-CH_2-CH_2-\overset{\overset{O}{\|}}{C}-)_{n-} + H_2O$$

Nylon 6,6

Figure 5.

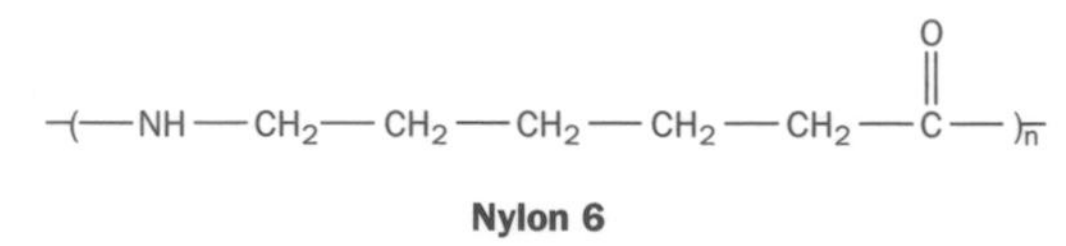

$$-\!\!\left(-NH-CH_2-CH_2-CH_2-CH_2-CH_2-\overset{O}{\overset{\|}{C}}-\right)_n$$

Nylon 6

hydrogen bonding: intermolecular force between the H of an N–H, O–H, or F–H bond and a lone pair on O, N, or F of an adjacent molecule

Because they have these polar groups that also allow for **hydrogen bonding**, nylons and most condensation polymers are also stronger, more rigid and brittle, and tougher in comparison to most vinyl polymers. Nylons are also "lubrication-free," meaning they do not need lubricant for easy mobility; thus they can be used to make mechanical bearings and gears that do not need periodic lubrication.

During the early 1950s George deMestral, after walking in the Swiss countryside, noticed that he had cockleburs caught in his jacket. He examined the cockleburs and noticed that they had tiny "hooks." His cotton jacket had loops that "held" the cockleburs. He began playing with his observations and making combinations of materials—one having rigid hooks and the other having flexible loops or eyes. Today, Velcro, the name given to the nylon-based hook-and-eye combination, uses nylon as both the hook material and the eye material. Polyester is sometimes blended with the nylon to make it stronger. (Polyesters have also been used to make hook-and-eye material.) Velcro is used to fasten shoes, close space suits, and it has many other applications.

Nylon 6, produced via the ring-opening reaction of the compound caprolactam is structurally similar to nylon 66 and has similar properties and uses. It is widely used in Europe in place of nylon 66, but not in the United States. (See Figure 5.)

hydrophobic: part of a molecule that repels water

ductile: property of a substance that permits it to be drawn into wires

Nylon 6,10 and nylon 6,12 are also commercially available. Because of the presence of the additional methylene ($-CH_2-$) groups that are **hydrophobic** (water-hating), these nylons are more resistant to moisture and more **ductile** than nylon 66.

DSM (once called Dutch State Mines) introduced nylon 4,6 (Stanyl) in 1990. It is produced via the condensation reaction between adipic acid and 1,4-diaminobutane, produced from renewable resources. Stanyl can withstand temperatures up to about 300°C (570°F), allowing it to occupy a niche position—between conventional nylons and high-performance materials. (See Figure 6.)

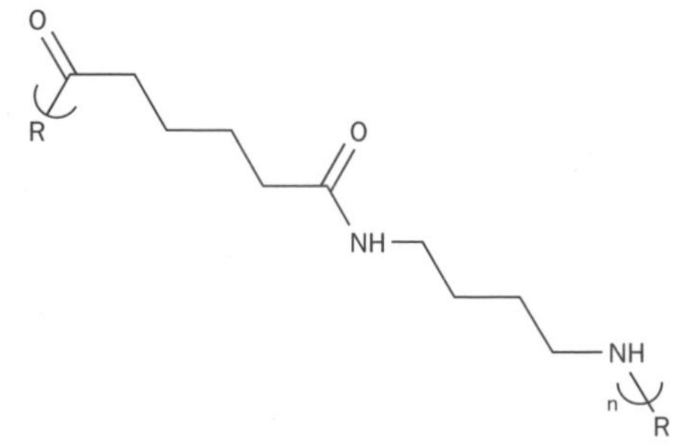

O
R
O
NH
NH
n
R

Nylon 4, 6

Figure 6.

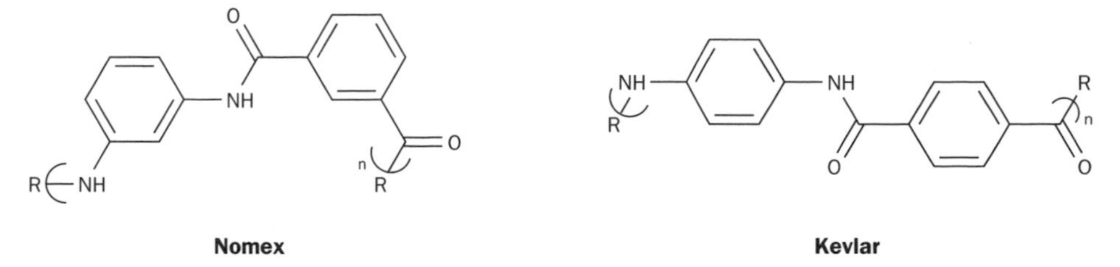

Nomex

Kevlar

Figure 7.

Several new commercial ventures are based on using natural, renewable starting materials (instead of petrochemicals). These products are known as "green" products because they are made from renewable resources and can be composted. The compound 1,4-butanediamine, used to produce nylon 4,6 from natural material, is such a green product.

In general, more crystalline nylons are fibrous whereas less crystalline nylon materials are more plastic in behavior.

Several aromatic polyamides, called aramids, have been produced. These materials are strong, are stable at high temperatures, and have good flame-resistance properties. Nomex (made from *m*-diaminobenzene and isophthalic acid) is used to make flame-resistant clothing and the thin pads used in space shuttles to protect sintered silica-fiber mats from stress and vibration during flight. Kevlar (made from *p*-diaminobenzene and terephthalic acid) is structurally similar to Nomex and by weight is stronger than steel. It is used in the manufacture of so-called bullet-resistant clothing. Because of its outstanding strength to weight ratio, it was used as the skin covering of the human-powered Gossamer Albatross, flown over the English Channel.

Aramids are also widely used as the fibers that are part of space-age composites and in the manufacture of tire cord and tread. (See Figure 7.) SEE ALSO CAROTHERS, WALLACE; MATERIALS SCIENCE; POLYMERS, SYNTHETIC.

Charles E. Carraher Jr.

Bibliography

Allcock, Harry R.; Lampe, Frederick W.; and Mark, James E. (2003). *Contemporary Polymer Chemistry*, 3rd edition. Upper Saddle River, NJ: Pearson/Prentice Hall.

Amato, Ivan (1997). *Stuff: The Materials the World Is Made Of.* New York: Basic Books.

Campbell, Ian M. (2000). *Introduction to Synthetic Polymers*, 2nd edition. New York: Oxford University Press.

Carraher, Charles E., Jr. (2003). *Giant Molecules: Essential Materials for Everyday Living and Problem Solving*, 2nd edition. Hoboken, NJ: Wiley.

Carraher, Charles E., Jr. (2003). *Polymer Chemistry*, 6th edition. New York: Marcel Dekker.

Collier, Billie J., and Tortora, Phyllis G. (2000). *Understanding Textiles.* Upper Saddle River, NJ: Prentice Hall.

Craver, Clara E., and Carraher, Charles E., Jr. (2000). *Applied Polymer Science: 21st Century.* New York: Elsevier.

Elias, Hans-Georg (1997). *An Introduction to Polymer Science.* New York: Wiley.

Morawetz, Herbert (1985). *Polymers: The Origins and Growth of a Science.* New York: Wiley.

Morgan, Paul W., and Kwolek, Stephanie L. (1959). "The Nylon Rope Trick." *Journal of Chemical Education* 36:182–184.

Rodriguez, Ferdinand (1996). *Principles of Polymer Systems*, 4th edition. Washington, DC: Taylor & Francis.

Salamone, Joseph C., ed. (1996). *Polymeric Materials Encyclopedia*. Boca Raton, FL: CRC Press.

Stevens, Malcolm P. (1990). *Polymer Chemistry: An Introduction*, 2nd edition. New York: Oxford University Press.

Thrower, Peter (1996). *Materials in Today's World*, 2nd edition. New York: McGraw-Hill.

Tonelli, Alan E. (2001). *Polymers from the Inside Out: An Introduction to Macromolecules*. New York: Wiley-Interscience.

Oil ***See Fossil Fuels.***

Oppenheimer, Robert

AMERICAN PHYSICIST
1904–1967

The son of a wealthy New York City textile importer (Julius) and a painter (Elle Friedman), Julius Robert Oppenheimer enjoyed an affluent childhood. He graduated from the Ethical Culture School of New York at the top of his class in 1921 and summa cum laude from Harvard in 1925. He then studied at the Cavendish Laboratory in Cambridge, England, and with Max Born at the University of Göttingen, in Germany, where he earned a doctorate in 1927. Although a rising star, he often was plagued by deep self-doubts and dark moods.

In 1929 Oppenheimer moved to California and for many years taught at both the California Institute of Technology (Caltech) and the University of California, Berkeley. He made several contributions to subatomic physics, including (with his former mentor) the Born–Oppenheimer approximation, which posited that the spin and vibration of protons could be ignored in theoretical calculations. Thin, wiry, enigmatic, and charismatic, Oppenheimer was associated with several leftist groups, which he helped fund. In 1940 he married biologist Katherine Harrison, a former member of the Communist Party.

Oppenheimer declared his leftist ties severed soon after he joined a secret group of elite scientists working with Ernest O. Lawrence at Berkeley's radiation laboratory to develop an atomic bomb. In spite of continuing suspicions about his loyalty, the U.S. Army appointed him director of the bomb design unit in October 1942. Oppenheimer's team of hundreds of gifted young scientists was secluded at a facility in Los Alamos, New Mexico. One of the brightest, Edward Teller, became a disaffected rival of Oppenheimer, but most found him a brilliant and inspiring leader.

theoretical physics: branch of physics dealing with the theories and concepts of matter, especially at the atomic and subatomic levels

fissionable: of or pertaining to unstable nuclei that decay to produce smaller nuclei

American physicist Robert Oppenheimer, scientific director of the Manhattan Project in which the atomic bomb was developed.

Pushing the boundaries of **theoretical physics**, Oppenheimer's Los Alamos scientists followed two technological paths simultaneously. One was a "gun assembly" designed to fire two masses of uranium at each other to initiate a chain reaction. But because **fissionable** uranium was exceedingly difficult to refine, other scientists worked on a design using more readily available but less stable plutonium, imploding a hollow sphere of the fuel with high explosives. Both designs worked, and they were used against

EDWARD TELLER (1908–2003)

Edward Teller was a physical chemist and an important, if controversial, voice in the politics of nuclear science. His work contributed to an understanding of both fusion and fission bombs. Regarded as the "father of the H-bomb," he was an ardent anticommunist and cold war warrior and he staunchly advocated the development and stockpiling of nuclear weapons. He opposed treaties limiting nuclear arsenals and testing, and he supported the development of space-based weapons.

—*Valerie Borek*

Japan—a uranium bomb was dropped over Hiroshima, a plutonium one over Nagasaki—in early August 1945 to end World War II.

Initially, Oppenheimer was elated over these technical achievements, but he quickly be came regretful and despondent about a nuclear future. As director of the prestigious Center for Advanced Study at Princeton (1947–1952) and chair of the General Advisory Committee of the Atomic Energy Commission (AEC), he was an outspoken advocate for the international sharing of nuclear technology and for international arms control, and he opposed further development of the hydrogen bomb. After a 1954 hearing, the AEC security board affirmed his loyalty but revoked his security clearance. Although there is little hard evidence that Oppenheimer ever passed atomic secrets to the former Soviet Union, the controversy surrounding this claim continues. Oppenheimer spent his final years sailing in the Virgin Islands and writing about science and Western culture. He died in Princeton, New Jersey, on February 18, 1967. SEE ALSO LAWRENCE, ERNEST; MANHATTAN PROJECT.

David B. Sicilia

Bibliography

Herken, Gregg (2002). *Brotherhood of the Bomb: The Tangled Lives and Loyalties of Robert Oppenheimer, Ernest Lawrence, and Edward Teller.* New York: Henry Holt and Co.

Rhodes, Richard (1986). *The Making of the Atomic Bomb.* New York: Simon & Schuster.

Rhodes, Richard (1995). *Dark Sun: The Making of the Hydrogen Bomb.* New York: Simon & Schuster.

Organic Chemistry

Organic chemistry is the chemistry of carbon compounds. All organic compounds contain carbon; however, there are some compounds of carbon that are not classified as organic. For example, salts such as carbonates (e.g., Na_2CO_3, $CaCO_3$) and cyanides (e.g., NaCN, KCN) are usually designated as inorganic. Perhaps a more useful description might be: Organic compounds are compounds of carbon that usually contain hydrogen and that may also contain other elements such as oxygen, nitrogen, sulfur, phosphorus, or **halogen** (F, Cl, Br, or I). In any case, there are very few carbon compounds that are not organic, while there are millions that are.

halogen: element in the periodic family numbered VIIA (or 17 in the modern nomenclature) that includes fluorine, chlorine, bromine, iodine, and astatine

History

Prehistoric civilizations obtained many useful chemicals from plants and animals. They were familiar with sugar, which they learned to ferment to make wine. Then they found that the wine could turn into vinegar. Ancient Egyptians used blue dye made from the indigo in madder root, and a royal purple dye extracted from a rare kind of mollusk. Soap was made by heating animal fat with base from wood ashes.

During the Middle Ages dry distillation of wood yielded mixtures of methyl alcohol, acetone, and acetic acid. Alchemists isolated cholesterol from gallstones, morphine from opium, and drugs such as quinine, strychnine, and brucine from various plants. Two hundred years ago chemists such as Antoine Lavoisier determined the elemental composition of many of these substances and noted that they all contained carbon and hydrogen, and that many also contained oxygen and nitrogen. It also appeared that there were two classes of materials: the mineral type (generally hard, high-melting, and noncombustible), and the organic type (often soft, liquid or low melting solids, and frequently easily combustible materials). Most organic chemicals could be burned to produce carbon dioxide; and any hydrogen present was converted to water (H_2O). Because organic compounds had for centuries been isolated only from plants and animals, it was commonly believed that some "vital force" in living things was necessary to produce them. This belief persisted until 1828, when Friedrich Wöhler was able to make urea, a chemical found in the urine of animals, from the inorganic salt ammonium cyanate.

$$NH_4^+OCN^- \xrightarrow{\text{heat}} H_2N{-}\overset{\displaystyle O}{\overset{\|}{C}}{-}NH_2$$

(inorganic) **urea (organic)**

Since that time organic chemistry has grown into a vast and ever expanding field that encompasses millions of chemical compounds.

Scope of Organic Chemistry

The field of organic chemistry includes more than twenty million compounds for which properties have been determined and recorded in the literature. Many hundreds of new compounds are added every day. Much more than half of the world's chemists are organic chemists. Some new organic compounds are simply isolated from plants or animals; some are made by modifying naturally occurring chemicals; but most new organic compounds are actually synthesized in the laboratory from other (usually smaller) organic molecules. Over the years organic chemists have developed a broad array of reactions that allow them to make all kinds of complex products from simpler starting materials.

Singular Attributes of Carbon

When one considers the millions of chemical compounds that are known and notes that more than 95 percent of them are compounds of carbon, one realizes that carbon is unique. Why are there so many carbon compounds? It turns out that atoms of carbon are quite remarkable in a number of ways.

Carbon atoms form very strong bonds with other carbon atoms. The bonds are so strong that carbon can form long chains, some containing thousands of carbon atoms. (Carbon is the only element that can do this.)

A carbon atom forms four bonds, therefore carbon not only can form long chains, but it also forms chains that have branches. It is a major reason why carbon compounds exhibit so much isomerism. The simple compound decane ($C_{10}H_{22}$), for example, has 75 different **isomers.**

isomer: molecules with identical compositions but different structural formulas

Carbon atoms can be bonded by double or triple bonds as well as single bonds. This multiple bonding is much more prevalent with carbon than with any other element.

Carbon atoms can form rings of various sizes. The rings may be saturated or unsaturated. The unsaturated 6-membered ring known as the benzene ring is the basis for an entire subfield of "aromatic" organic chemistry.

Carbon atoms form strong bonds not only with other carbon atoms but also with atoms of other elements. In addition to hydrogen, many carbon compounds also contain oxygen. Nitrogen, sulfur, phosphorus, and the halogens also frequently occur in carbon compounds.

Various kinds of **functional groups** occur widely among carbon compounds, and many different kinds of isomers are possible.

functional group: portion of a compound with characteristic atoms acting as a group

Hydrocarbons

Compounds of carbon and hydrogen only are called *hydrocarbons.* These are the simplest compounds of organic chemistry. The most basic group of hydrocarbons are the *alkanes,* which contain only single bonds. The simplest member of the alkane series is methane, CH_4, the main component of natural gas. The names of some alkanes are listed in Table 1. Alkanes sometimes

ALKANES

Formula		Name	
CH_4	CH_4	methane	gases
C_2H_6	CH_3CH_3	ethane	
C_3H_8	$CH_3CH_2CH_3$	propane	
C_4H_{10}	$CH_3CH_2CH_2CH_3$	butane	
C_5H_{12}	$CH_3(CH_2)_3CH_3$	pentane	liquids
C_6H_{14}	$CH_3(CH_2)_4CH_3$	hexane	
C_7H_{16}	$CH_3(CH_2)_5CH_3$	heptane	
C_8H_{18}	$CH_3(CH_2)_6CH_3$	octane	
C_9H_{20}	$CH_3(CH_2)_7CH_3$	nonane	
$C_{10}H_{22}$	$CH_3(CH_2)_8CH_3$	decane	
$C_{11}H_{24}$	$CH_3(CH_2)_9CH_3$	undecane	
$C_{12}H_{26}$	$CH_3(CH_2)_{10}CH_3$	dodecane	
$C_{13}H_{28}$	$CH_3(CH_2)_{11}CH_3$	tridecane	
$C_{14}H_{30}$	$CH_3(CH_2)_{12}CH_3$	tetradecane	
$C_{15}H_{32}$	$CH_3(CH_2)_{13}CH_3$	pentadecane	
$C_{16}H_{34}$	$CH_3(CH_2)_{14}CH_3$	hexadecane	
$C_{17}H_{36}$	$CH_3(CH_2)_{15}CH_3$	heptadecane	
$C_{18}H_{38}$	$CH_3(CH_2)_{16}CH_3$	octadecane	solids
$C_{19}H_{40}$	$CH_3(CH_2)_{17}CH_3$	nonadecane	
$C_{20}H_{42}$	$CH_3(CH_2)_{18}CH_3$	eicosane	

Table 1.

have ring structures. Since a 4-carbon chain of the alkane series is called *butane*, a ring of 4 carbon atoms is called *cyclobutane.*

$CH_3CH_2CH_2CH_3$

butane (C_4H_{10})

$$\begin{array}{ccc} H_2C & - & CH_2 \\ | & & | \\ H_2C & - & CH_2 \end{array}$$

cyclobutane (C_4H_8)

Simple hydrocarbons that contain one or more double bonds are called *alkenes.* They are named like alkanes, but their names end in "−ene." The simplest alkene has two carbon atoms and is called *ethene.* A 3-carbon chain that has a double bond is called *propene.*

$H_2C{=}CH_2$ ethene (C_2H_4) $\quad$ $H_2C{=}CHCH_3$ propene (C_3H_6) $\quad$ $CH_3CH{=}CHCH_2CH_2CH_3$ 2-pentene (C_5H_{10})

A 5-carbon hydrocarbon chain with a double bond is called *pentene,* and if the double bond links the second and third carbons, it is 2-*pentene.* Like cycloalkanes, alkenes have the general formula C_nH_{2n}. Alkenes having ring structures are called *cycloalkenes.* A 5-carbon ring with a double bond is called *cyclopentene.*

Hydrocarbons that contain one or more triple bonds are called *alkynes,* and is the name ending is "−yne." A 2-carbon alkyne is therefore named *ethyne.* (However, the compound is often referred to by its common name, which is *acetylene.*)

Compounds that contain double or triple bonds are said to be "unsaturated"—because they are not "saturated" with hydrogen atoms. Unsaturated compounds are reactive materials that readily add hydrogen when heated over a **catalyst** such as nickel. The reverse reaction also occurs. Heating ethane with steam is an important commercial process for making ethene (or ethylene). This is an important commercial process called "steam cracking."

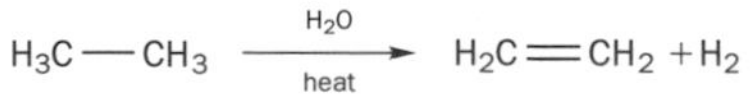

catalyst: substance that aids in a reaction while retaining its own chemical identity

$$H_3C{-}CH_3 \xrightarrow[\text{heat}]{H_2O} H_2C{=}CH_2 + H_2$$

When a 6-carbon ring contains 2 double bonds, it is called *cyclohexadiene,* but when it has 3 double bonds, it is not called cyclohexatriene; this is because a 6-carbon ring with three double bonds takes on a special kind of stability. The double bonds become completely conjugated and no longer behave as double bonds. The ring, known as a "benzene ring," is said to be aromatic.

The removal of a hydrogen atom from a hydrocarbon molecule leaves an alkyl group that readily attaches to a functional group, or forms a branch on a hydrocarbon chain. The groups are named after the corresponding hydrocarbons. For example, CH_3- is named *methyl;* CH_3CH_2-, *ethyl;* $CH_2{=}CH-$, *ethenyl;* $CH_3CH_2CH_2-$, *propyl;* and so on. A benzene ring from which a hydrogen atom has been removed is often referred to as a *phenyl.* The branched molecules shown here would be given names as follows

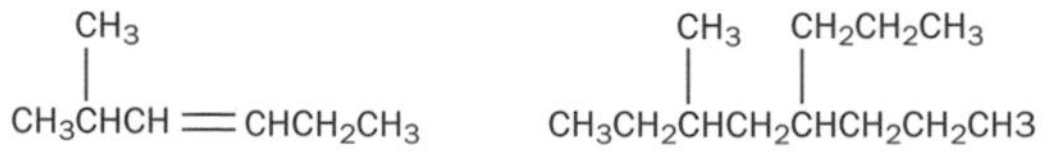

$$\begin{array}{l} CH_3 \\ | \\ CH_3CHCH{=}CHCH_2CH_3 \end{array}$$

2-methyl-3-hexene

$$\begin{array}{l} \quad\quad\quad\; CH_3 \quad CH_2CH_2CH_3 \\ \quad\quad\quad\;\; | \quad\quad\; | \\ CH_3CH_2CHCH_2CHCH_2CH_2CH3 \end{array}$$

3-methyl-5-propyloctane

In a conjugated system, there are alternating double and single bonds, allowing electrons to flow back and forth. Molecules that contain such conjugated systems are said to be stabilized by "resonance." In the benzene ring every other bond is a double bond, all the way around the ring. This results in a special kind of stabilization called "aromaticity," in which the electrons are delocalized and free to travel all around the ring. Certain ring compounds, like benzene, that contain such a conjugated system of double and single bonds are described as "aromatic."

Theoretically there is no limit to the length of hydrocarbon chains. Very large hydrocarbon molecules (polymers) have been made containing as many as 100,000 carbon atoms. However, such molecules are hard to make and very difficult to melt and to shape into useful products.

Hydrocarbons are obtained primarily from fossil fuels—especially petroleum and natural gas. Natural gas is a mixture that is largely methane mixed with varying amounts of ethane and other light hydrocarbons, while petroleum is a complex mixture of many different hydrocarbons. Coal, the other fossil fuel, is a much more complicated material from which many kinds of organic compounds, some of them hydrocarbons, can be obtained.

Functional Groups

Alkane molecules are rather unreactive (except for being very flammable), but alkenes react with many other substances. When a drop of bromine is added to an alkene, for example, the deep orange color of the bromine immediately disappears as the bromine adds across the double bond to form a dibromo derivative. The double bond is called a "functional group" because its presence in a molecule causes reactivity at that particular site. There are a dozen or so functional groups that appear frequently in organic compounds. Some of the most common ones are listed in Table 2. The same molecule may contain several functional groups. Aspirin, for example, is both a **carboxylic acid** and an **ester**, and cholesterol is an alkene as well as an alcohol.

carboxylic acid: one of the characteristic groups of atoms in organic compounds that undergoes characteristic reactions, generally irrespective of where it occurs in the molecule; the $-CO_2H$ functional group

ester: organic species containing a carbon atom attached to three moieties: an O via a double bond, an O attached to another carbon atom or chain, and a H atom or C chain; the R(C=O)OR functional group

Isomerism

Isomers are molecules with the same molecular formula but different structures. There is only one structure for methane, ethane, or propane; but butane, C_4H_{10}, can have either of two different structure:

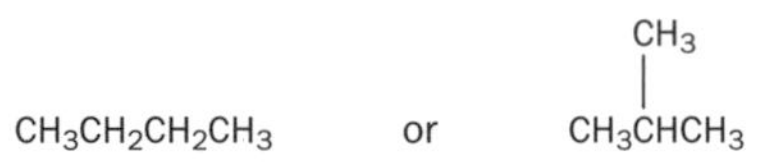

$$\underset{(1)}{CH_3CH_2CH_2CH_3} \quad \text{or} \quad \underset{(2)}{CH_3\overset{\overset{\large CH_3}{|}}{C}HCH_3}$$

The linear molecule (1) is called butane, or *normal* butane (*n*-butane), whereas the branched molecule (2) is methylpropane (rather than 2-methylpropane, as the methyl group has to be in a 2-position). If the methyl group of (2) were attached to a terminal carbon, the resultant molecule would be the same as (1). Methylpropane (2) is also called *iso*butane.

Pentane has 3 isomers: pentane (or *n*-pentane), methylbutane (or *iso*-pentane), and dimethyl propane (or *neo*pentane). Hexane has 5 isomers:

hexane, 2-methylpentane, 3-methylpentane, 2,2-dimethylbutane, and 2,3-dimethylbutane. Heptane has 9 different isomers, octane has 18, nonane has 35, and decane has 75. An increase in the number of carbon atoms greatly increases the possibilities for isomerism. There are more than 4,000 isomers of $C_{15}H_{32}$ and more than 366,000 isomers of $C_{20}H_{42}$. The formula $C_{30}H_{62}$ has more than 4 billion. Of course, most of them have never been isolated as pure compounds (but could be, if there were any point in doing it).

For molecules other than hydrocarbons, still other kinds of isomers are possible. The simple formula C_2H_6O can represent ethyl alcohol or dimethyl ether; and C_3H_6O could stand for an alcohol, an ether, an **aldehyde**, or a **ketone** (among other things). The larger a molecule is, and the greater the variety of atoms and functional groups it contains, the more numerous its isomers.

aldehyde: one of the characteristic groups of atoms in organic compounds that undergoes characteristic reactions, generally irrespective of where it occurs in the molecule; the RC(O)H functional group

ketone: one of the characteristic groups of atoms in organic compounds that undergoes characteristic reactions, generally irrespective of where it occurs in the molecule; the FC(O)R functional group

There is still another kind of isomerism that stems from the existence of "right-" and "left-handed" molecules. It is sometimes referred to as *optical* isomerism because the molecules that make up a pair of these isomers usually differ only in the way they rotate plane polarized light.

Nomenclature

There are so many millions of organic compounds that simply finding names for them all is a major challenge. It was not until the late nineteenth century that chemists developed a logical system for naming organic compounds. Compounds had often been named according to their sources. The 1-carbon carboxylic acid, for example, was first obtained from ants, and so it was called *formic* acid, from the Latin word for ants (*formicae*). The 2-carbon acid was obtained from vinegar (*acetum* in Latin), and was called *acetic* acid.

To bring some order to the naming process an international meeting was held in 1892 at Geneva, Switzerland. The group later became known as the International Union of Pure and Applied Chemistry (IUPAC). Its objective was to establish a naming process that would provide each compound with a unique and systematic name. An initial set of rules was adopted at that first meeting in Geneva, and IUPAC has continued that work. Its systematic naming rules are used by organic chemists all over the world. The names of the alkanes form the basis for the system, with functional groups usually being indicated with appropriate suffixes. Some examples are given in Table 2.

Organic Reactions

Organic chemistry is concerned with the many compounds of carbon, their names, their isomers, and their properties, but it is mostly concerned with their reactions. Organic chemists have developed a huge array of chemical reactions that can convert one organic compound to another. Some reactions involve addition of one molecule to another; some involve decomposition of molecules; some involve substitution of one atom or group by another; and some even involve the rearrangement of molecules, with some atoms moving into new positions. Some reactions require energy in the form of heat or radiation; and some require a special kind of catalyst or some sort

NAMING ORGANIC COMPOUNDS

Functional Group		Type Compound	Example	IUPAC Name	Common Name
C=C	double bond	alkene	$H_2C{=}CH_2$	ethene	ethylene
C≡C	triple bond	alkyne	HC≡CH	ethyne	acetylene
−OH	hydroxyl	alcohol	CH_3OH	methanol	methyl alcohol
−O−	oxy	ether	H_3COCH_3	methoxymethane	methyl ether
H–C(=O)− (−CHO)	carbonyl	aldehyde	$H_2C{=}O$	methanal	formaldehyde
−C(=O)−	carbonyl	ketone	CH_3COCH_3	propanone	acetone
−C(=O)−OH	carboxyl	carboxylic acid	HCOOH	methanoic acid	formic acid
−C(=O)−O−	carboxyl	ester	$HCOOCH_2CH_3$	ethyl methanoate	ethyl formate
$-NH_2$	amino	amine	CH_3NH_2	aminomethane	methylamine
−CN	cyano	nitrile	CH_3CN	ethanenitrile	acetonitrile
−X	halogen	haloalkane	CH_3Cl	chloromethane	methyl chloride

Table 2.

of solvent. Of course, not all organic reactions are highly successful. One reaction might be a very simple one giving essentially 100 percent of the desired product; but another might be a complex multistep process yielding less than 5 percent overall of the wanted product.

Organic reactions can often give remarkable control as to what products should be formed. Adding water to propene for example, produces 2-propanol in the presence of acid, but it yields 1-propanol if treated first with B_2H_6 and then H_2O_2 in the presence of base.

Future Sources of Organic Chemicals

Fossil fuels have been our primary natural source for many organic chemicals for more than a century, but our fossil fuel resources are finite, and they are being rapidly depleted (especially oil and gas). What will be our sources of organic materials in the future? Since fossil fuels are nonrenewable resources, it is believed that the twenty-first century will see a shift toward greater dependence on renewable raw materials. The largest U.S. chemical company has a goal of becoming 25 percent based on renewable resources by 2010. It is already producing 1,3-propanediol from cornstarch using a gene-tailored *E. coli* bacterium. This diol is used in Du Pont's fiber Sorona, which is said to combine the best features of both polyester and nylon fibers. Succinic acid and polyhydroxybutyrate are also obtainable from renewable crops, and the list of such renewable raw materials is destined to grow. For example, ethylene (or ethene), $CH_2{=}CH_2$, which is a highly important commercial chemical used in making many industrial chemicals and polymers, is presently made by steam cracking of ethane obtained from oil or natural gas; but ethylene can also be made by dehydration of ethyl alcohol made by fermentation of sugar. Efforts are even being made to use biowaste materials, such as corn husks, nutshells, and wood chips as industrial raw materials.

Analytical Tools

Organic chemists often need to examine products for identification, purity analysis, or structure determination. There are some marvelous tools available

chromatography: the separation of the components of a mixture in one phase (the mobile phase) by passing it through another phase (the stationary phase) making use of the extent to which the components are absorbed by the stationary phase

spectroscopy: use of electromagnetic radiation to analyze the chemical composition of materials

to help them do these things. **Chromatography**, **spectroscopy**, and crystallography are especially widely used in organic chemistry.

Column chromatography, gas chromatography, and liquid chromatography are all important methods for separating mixtures of organic compounds. Spectroscopic tools include ultraviolet (UV), infrared (IR), nuclear magnetic resonance (NMR), and mass spectroscopy (MS), each capable of providing a different kind of information about an organic compound. Although it is limited to substances that can be prepared as pure crystals, x-ray crystallography is probably the ultimate tool for determining molecular structure.

Careers in Organic Chemistry

Some organic chemists are involved in basic research at government or academic institutions, but most have careers in industry. The industries vary from oil and chemical companies to industries producing food, pharmaceuticals, cosmetics, detergents, paints, plastics, pesticides, textiles, or other kinds of products. Many organic chemists work in laboratories, where they do various kinds of analysis or research, but many others do not. Some are teachers, or writers, or science librarians. Some study law and become patent attorneys; some study medicine and become medical researchers; and some study business and become administrators of companies, colleges, or other institutions. Organic chemistry is an enormous field full of many kinds of career possibilities. SEE ALSO FOSSIL FUELS; LAVOISIER, ANTOINE; ORGANIC HALOGEN COMPOUNDS; WÖHLER, FRIEDRICH.

Kenneth E. Kolb

Bibliography

Atkins, Robert C., and Carey, Francis A. (2002). *Organic Chemistry: A Brief Course*, 3rd edition. Boston: McGraw-Hill.

Bailey, Philip S., Jr., and Bailey, Christina A. (2000). *Organic Chemistry: A Brief Survey of Concepts and Applications*, 6th edition. Upper Saddle River, NJ: Prentice Hall.

Brown, William H., and Foote, Christopher S. (2002). *Organic Chemistry*, 3rd edition. San Diego: Saunders.

Fessenden, Ralph J.; Fessenden, Joan S.; and Logue, Marshall (1998). *Organic Chemistry*, 6th edition. Pacific Grove, CA: Brooks/Cole.

Solomons, T. W. Graham (1997). *Fundamentals of Organic Chemistry*, 5th edition. New York: Wiley.

Organic Halogen Compounds

halogen: element in the periodic family numbered VIIA (or 17 in the modern nomenclature) that includes fluorine, chlorine, bromine, iodine, and astatine

tetrachloride: term that implies a molecule has four chlorine atoms present

Organic **halogen** compounds are a large class of natural and synthetic chemicals that contain one or more halogens (fluorine, chlorine, bromine, or iodine) combined with carbon and other elements. The simplest organochlorine compound is chloromethane, also called methyl chloride (CH_3Cl). Other simple organohalogens include bromomethane (CH_3Br), chloroform ($CHCl_3$), and carbon **tetrachloride** (CCl_4). Some examples of organohalogens are shown in Figure 1.

Synthesis

Organohalogens can be made in various ways. Direct halogenation of hydrocarbons with chlorine gives organochlorines; with bromine, organobromines. Alcohols can be converted into organohalogens by reaction with hydrogen halides.

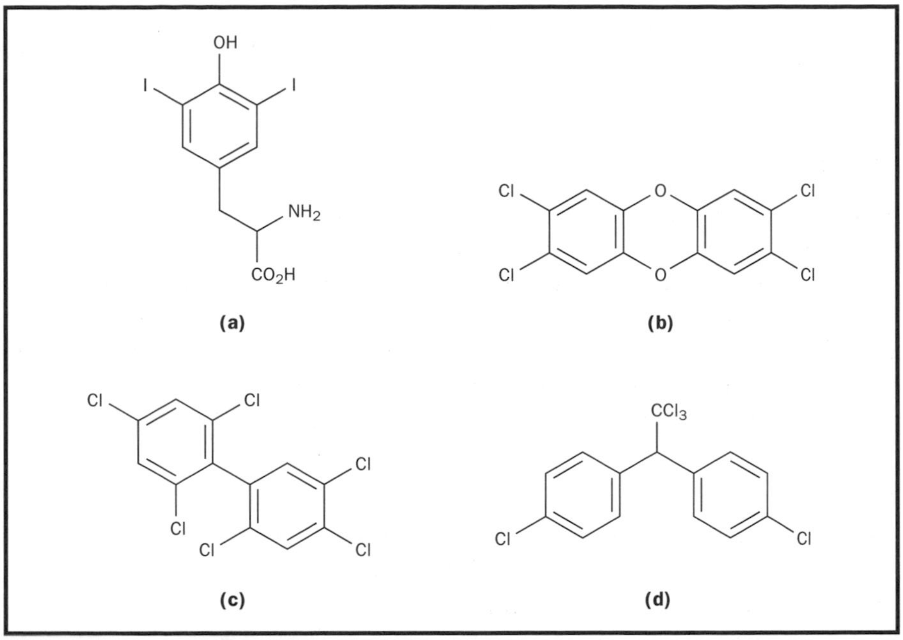

Figure 1. Four representative organohalogens: (a) a thyroid hormone; (b) "dioxin" (2,3,7,8-tetrachlorodibenzo-p-dioxin); (c) a polychlorinated biphenyl (PCB); (d) DDT (dichlorodiphenyltrichloroethane).

Aromatic organohalogens such as chlorobenzene are synthesized by treatment of benzene with halogen and a Lewis acid **catalyst** such as aluminum chloride.

catalyst: substance that aids in a reaction while retaining its own chemical identity

Organohalogen compounds are also produced by adding halogen or hydrogen halide to alkenes and alkynes.

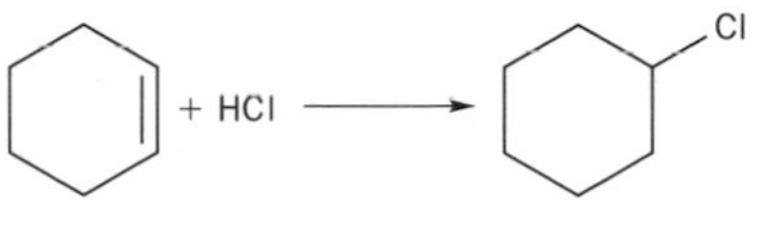

$$\text{cyclohexene} + HCl \longrightarrow \text{chlorocyclohexane} \quad (1)$$

$$CH_2{=}CH_2 + Br_2 \longrightarrow BrCH_2CH_2Br \quad (2)$$

$$HC{\equiv}CH + 2Cl_2 \longrightarrow Cl_2CHCHCl_2 \quad (3)$$

Organoiodines and organofluorines are prepared by displacement reactions.

$$CH_3CH_2Br + KF \longrightarrow CH_3CH_2F + KBr \quad (4)$$

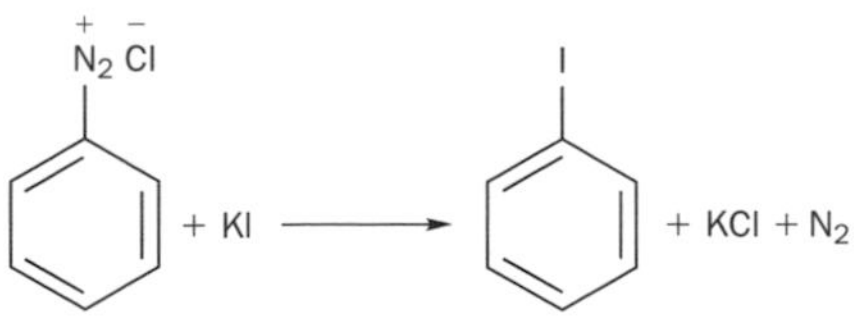

$$C_6H_5\overset{+}{N_2}\overset{-}{Cl} + KI \longrightarrow C_6H_5I + KCl + N_2 \quad (5)$$

Reactivity

The reactivity of organohalogens varies enormously. The war gases phosgene ($ClCOCl$) and mustard ($ClCH_2CH_2SCH_2CH_2Cl$) are very reactive and highly toxic, whereas most other organohalogen compounds are relatively **inert**. Nevertheless, organohalogens undergo many chemical transformations. One common reaction is elimination, induced by the action of a strong base.

inert: incapable of reacting with another substance

$$\text{C}_6\text{H}_{11}\text{Br} \xrightarrow{\text{KOH}} \text{C}_6\text{H}_{10} + \text{KBr} + \text{H}_2\text{O} \quad (6)$$

Toxicity

As with all chemicals, "the dose makes the poison." The chlorine-containing insecticide dichlorodiphenyltrichloroethane (DDT) is highly effective in killing disease-ridden mosquitoes, ticks, and fleas, but it is virtually nontoxic to mammals. The fluorine-containing pesticide "1080," or fluoroacetic acid (FCH_2CO_2H), is highly toxic and often lethal to all mammals. The industrial and **combustion** by-product dioxin is highly toxic to some animals but not to others; in humans, dioxin causes the skin disease chloracne.

combustion: burning, the reaction with oxygen

Use

Organohalogens are widely used in industry and society. Chloromethane is used as a solvent in rubber polymerization. Bromomethane is an important fumigant; the related halons ($CBrClF_2$ and $CBrF_3$) are better fire extinguishants than carbon dioxide.

Eighty-five percent of all pharmaceutical agents and **vitamins** involve chlorine chemistry; many drugs require chlorine, fluorine, or bromine to be effective. Ceclor and Lorabid are used to treat ear infections, Toremifene is a breast-cancer drug, and the natural antibiotic vancomycin is used to fight penicillin-resistant infections. Benzyl chloride is used to synthesize the drugs phenobarbital, benzedrine, and demerol. Inhalation anesthetics include the organofluorines desflurane, sevoflurane, and enflurane ($CHClFCF_2OCHF_2$). Perfluorocarbons, such as perfluorotributylamine ($[CF_3CF_2CF_2CF_2]_3N$), are used as blood substitutes or blood extenders ("artificial blood") and are used for coronary angioplasty. The insecticide DDD (mitotane), related to DDT, is used to treat inoperable adrenal cancer. The chemical advantages to some of these halogenated drugs are shown in Figure 2.

vitamins: organic moleculs needed in small amounts for the normal function of the body; often used as part of an enzyme catalyzed reaction

Vinyl chloride ($CH_2{=}CHCl$), a carcinogenic gas, is polymerized to polyvinyl chloride (PVC), a plastic of great versatility and safety. PVC is an invaluable component of building materials, consumer goods, medical equipment, and many other everyday products. More than 2.2 billion kilograms (5 billion pounds) of PVC are used annually for wire, cable, and other electrical applications. The chlorine in PVC makes this plastic flame retardant and ideal for construction and furnishing applications. Polytetrafluoroethylene (Teflon) is the polymer of tetrafluoroethylene ($CF_2{=}CF_2$). Because of its chemical stability (very strong carbon-fluorine bonds), it has many diverse applications in our society; best known perhaps are the coatings used to make "nonstick" cookware. Trichloroethylene ($CHCl{=}CCl_2$) and tetrachloroethylene ($CCl_2{=}CCl_2$, "Perc") are widely used solvents in the dry cleaning industry.

Organohalogens are essential for crop production and protection as herbicides and insecticides. Ninety percent of grain farms utilize these chemicals in food production. The chemical structures of some of these organohalogens are shown in Figure 3.

Polychlorinated biphenyls (PCBs) were introduced in 1929 as insulators in capacitors and transformers in the electric power industry, as lubricants

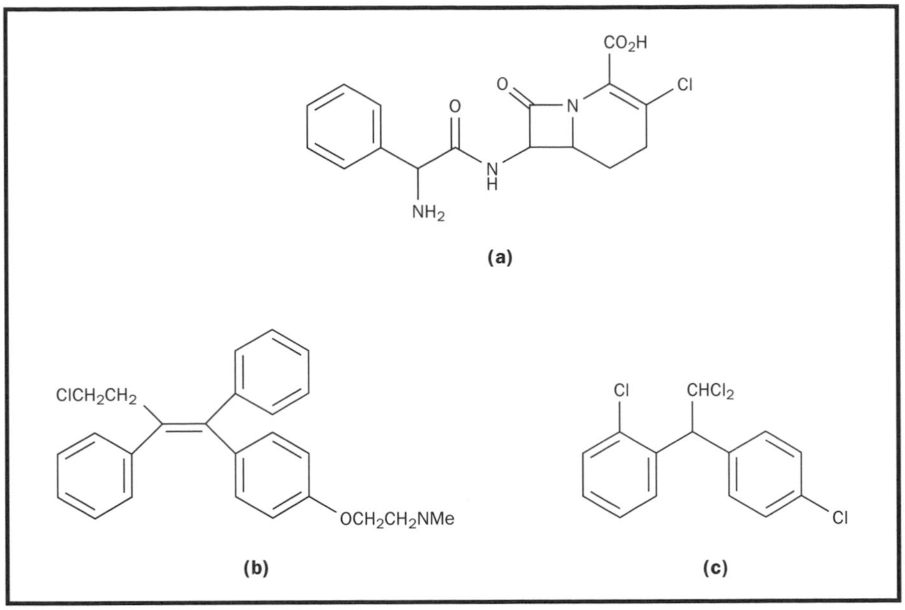

Figure 2. Organohalogens as drugs: (a) Lorabid, antibiotic; (b) toremifene, a breast cancer drug; (c) mitotane, a cancer drug.

and coolants in vacuum pumps, as paint additives, and in food packaging. The manufacture and use of PCBs were discontinued in 1977 because of their adverse effects on the environment. Their effect on humans is still unknown. An example of a PCB is shown in Figure 1.

chlorofluorocarbon (CFC): compound that contains carbon, chlorine, and fluorine atoms, which remove ozone in the upper atmosphere

Chlorofluorocarbons (CFCs or freons) are strongly implicated in causing the ozone hole, and are being phased out of use as refrigerants, dry cleaning solvents, propellants, fire extinguishants, and foam-blowing agents. These chemicals include CFC-11 (CCl_3F), CFC-13 ($CClF_3$), and CFC-112 (CCl_2FCCl_2F). Replacements for CFCs are the hydrochlorofluorocarbons (HCFCs) and the hydrofluorocarbons (HFCs), both of which have no impact on stratospheric ozone and have low global warming potential. Examples include HCFC-21 ($CHCl_2F$) and HFC-152 (FCH_2CH_2F).

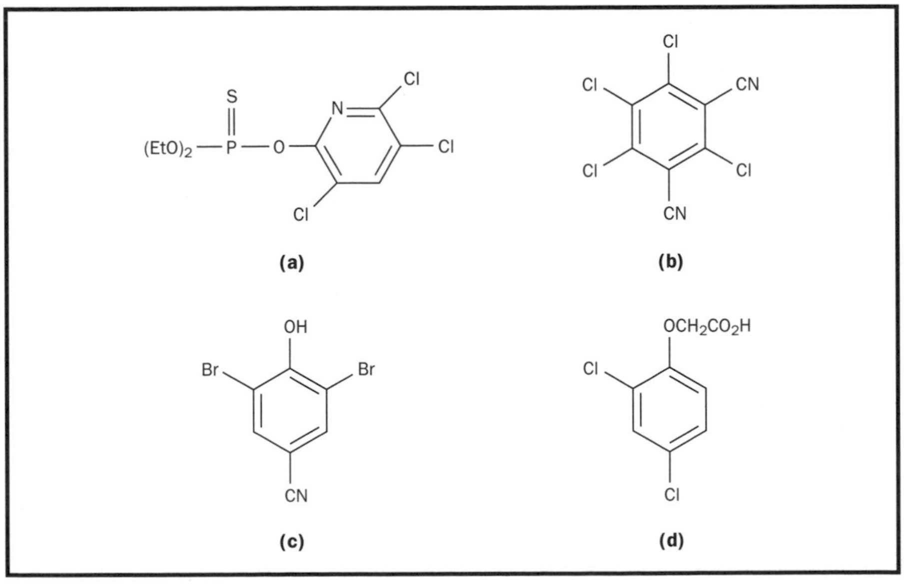

Figure 3. Organohalogens used as pesticides: (a) Dursban, an insecticide; (b) Daconil, a fungicide; (c) Bromoxynil, an herbicide; (d) 2,4-D, an herbicide.

Natural Occurrence

The number of known natural organohalogens has grown from thirty in 1968 to nearly 3,900 during the early 2000s. Many are the same as synthetic chemicals. They are biosynthesized by marine organisms, bacteria, fungi, plants, insects, and some mammals, including humans. Algae, wood-rotting fungi, mushrooms, several trees, phytoplankton, and even potatoes produce chloromethane. Termites are a major producer of chloroform, and several vegetables produce bromomethane. One hundred organohalogens have been found in the favorite edible seaweed of native Hawaiians.

Chloride and bromide salts are normally present in plants, wood, soil, and minerals; as a result, forest fires and volcanoes produce organohalogens. Meteorites contain organochlorines. Global emissions of chloromethane from the **biomass** are 5 million tons per year, whereas synthetic emissions are only 26,000 tons per year. Volcanoes also emit hydrogen chloride (3 million tons/year) and hydrogen fluoride (11 million tons/year), both of which can react with organic compounds to produce organohalogens. Chlorofluorocarbons have been detected in volcanic emissions in Guatemala and Siberia, but a study of volcanoes in Italy and Japan indicates that they may not be a major source of environmental CFCs.

biomass: collection of living matter

Seaweeds contain hundreds of organohalogens (see Figure 4). Telfairine, like the synthetic insecticide lindane, is a powerful insecticide. These organohalogens are used by marine life in chemical defense (natural pesticides). The "smell" of the ocean is likely due to the myriad **volatile** organohalogens produced by seaweeds.

volatile: low boiling, readily vaporized

Organohalogens also serve as hormones. Vegetables such as lentils, beans, and peas synthesize the growth hormone 4-chloro-3-indoleacetic acid. A cockroach produces two chlorine-containing steroids as aggregation pheromones. Female ticks use 2,6-dichlorophenol as a sex attractant. Thyroid hormones (see Figure 1) contain iodine, and an organobromine is involved in the mammalian sleep phenomenon.

Just as iodine is used to treat cuts and chlorine (bleach) to disinfect bathrooms, our white blood cells generate chlorine and bromine to kill germs and fight infection. The sponge **metabolite** spongistatin and the blue-green

metabolites: products of biological activity that are important in metabolism

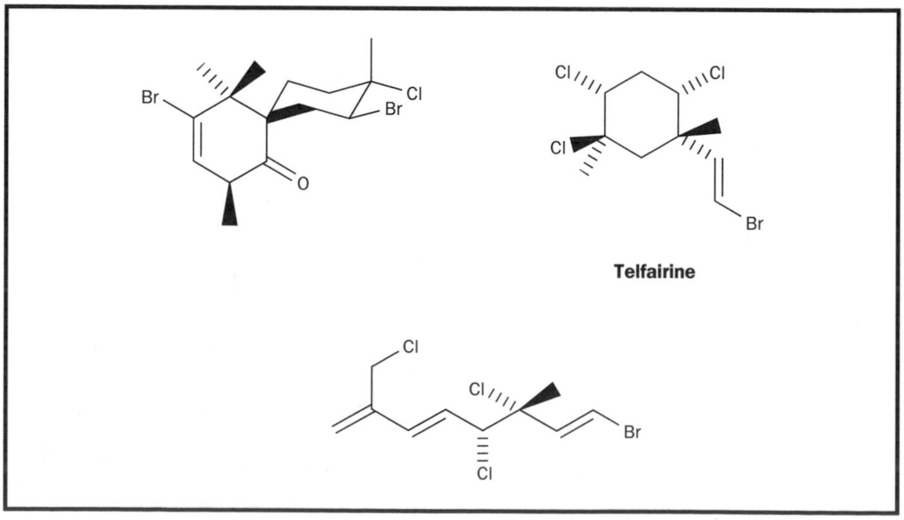

Figure 4. Three of the halogenated terpenes found in red algae, including (at top right) the natural insecticide telfairine.

alga cryptophycin, both of which contain a chlorine atom essential for biological activity, are powerful anticancer drugs. Ambigol, found in terrestrial blue-green alga, is active against human immunodeficiency virus (HIV). An Ecuadorian frog produces an organochlorine that is 500 times more potent than morphine; a synthetic analog is under development as a new anesthetic.

Although some synthetic organohalogens are toxic contaminants that need to be removed from the environment, the vast majority of organohalogens have little or no toxicity. Organic halogen compounds continue to play an essential role in human health and well being as chemists pursue the study of these fascinating chemicals. SEE ALSO ORGANIC CHEMISTRY.

Gordon W. Gribble

Bibliography

Gribble, Gordon W. (1994). "Natural Organohalogens—Many More Than You Think!" *Journal of Chemical Education* 71(11):907–911.

Gribble, Gordon W. (1995). *Chlorine and Health*. New York: American Council on Science and Health.

Gribble, Gordon W. (2003). "The Diversity of Naturally Occurring Organohalogen Compounds." *Chemosphere* 52:289–297.

Howe-Grant, Mary, ed. (1991). *Encyclopedia of Chemical Technology*, 4th edition. New York: Wiley.

Tschirley, Fred H. (1986). "Dioxin." *Scientific American* 254(2):29–35.

Winterton, Neil (2000). "Chlorine: The Only Green Element—Towards a Wider Acceptance of Its Role in Natural Cycles." *Green Chemistry* 2:173–225.

Internet Resources

"Chlorine." Chlorine Chemistry Council. Available from <http://www.c3.org>.

"Chlorine Online: Information Resource." Euro Chlor. Available from <http://www.eurochlor.org>.

Organometallic Compounds

Organometallic compounds have at least one carbon to metal bond, according to most definitions. This bond can be either a direct carbon to metal bond (σ bond or sigma bond) or a metal complex bond (π bond or pi bond). Compounds containing metal to **hydrogen bonds** as well as some compounds containing nonmetallic (**metalloid**) elements bonded to carbon are sometimes included in this class of compounds. Some common properties of organometallic compounds are relatively low melting points, insolubility in water, solubility in ether and related solvents, toxicity, oxidizability, and high reactivity.

hydrogen bond: interaction between H atoms on one molecule and lone pair electrons on another molecule that constitutes hydrogen bonding

metalloid: elements that exhibit properties that are between those of metals and nonmetals; generally considered to include boron, silicon, germanium, arsensic, antimony, tellurium, and polonium

An example of an organometallic compound of importance years ago is tetraethyllead (Et_44Pb) which is an antiknock agent for gasoline. It is presently banned from use in the United States.

The first metal complex identified as an organometallic compound was a salt, $K(C_2H_4)PtCl_3$, obtained from reaction of ethylene with platinum (II) chloride by William Zeise in 1825. It was not until much later (1951–1952) that the correct structure of Zeise's compound (see Figure 1) was reported in connection with the structure of a metallocene compound known as ferrocene (see Figure 2).

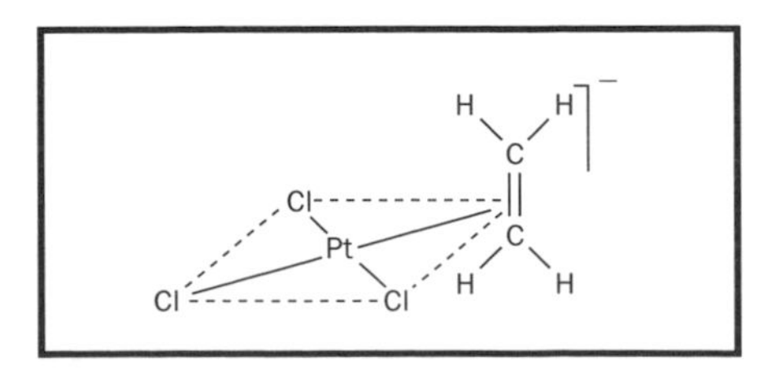

Figure 1. Anion of Zeise's compound

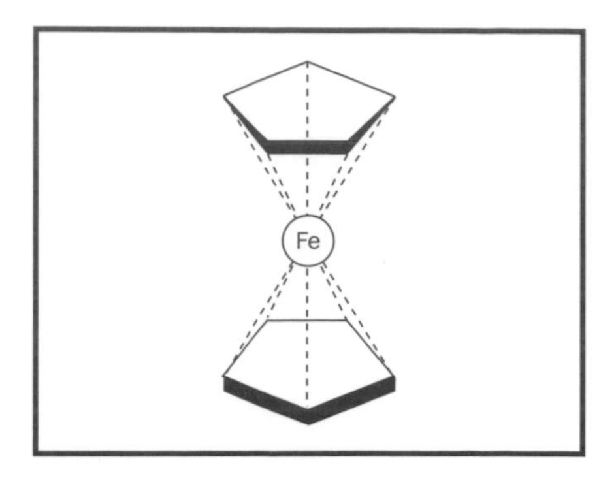

Figure 2. Ferrocene

transition metals: elements with valence electrons in d-sublevels; frequently characterized as metals having the ability to form more than one cation

ligand: molecule or ion capable of donating one or more electron pairs to a Lewis acid

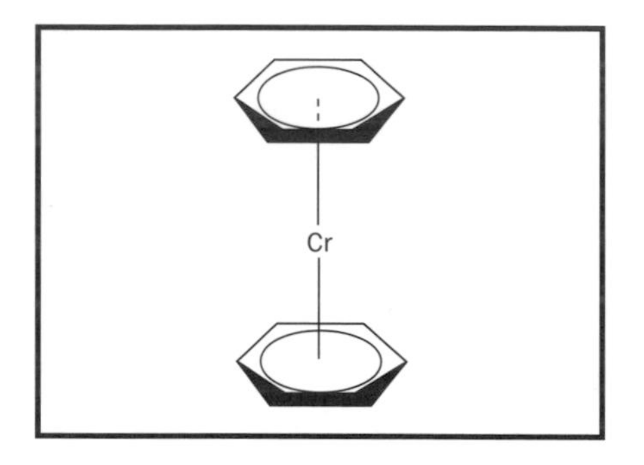

Figure 3. Dibenzenechromium

The name "ferrocene" was coined by one of Harvard University professor R. B. Woodward's postdoctoral students, Mark Whiting. The entire class of transitional metal dicyclopentadienyl compounds quickly became known as "metallocenes" and this has since been expanded for compounds $[(H_5\text{-}C_5H_5)\ 2M]$ in general. G. Wilkinson and Woodward published their results on ferrocene in 1952.

Preparation of ferrocene was reported at about the same time by two research groups, and a sandwich structure was proposed, based on ferrocene's physical properties (Kauffman, pp. 185–186). The sandwich structure was confirmed by x-ray diffraction studies. Since then, other metallocenes composed of other metals and other carbon ring molecules, such as dibenzenechromium (see Figure 3) and uranocene (see Figure 4), have been prepared.

Possibly the first scientist to synthesize an organometallic compound was Edward Frankland, who prepared diethylzinc by reaction of ethyl iodide with zinc metal in 1849 (Thayer 1969b, pp. 764–765).

$$2\ CH_3CH_2I + 2\ Zn \rightarrow CH_3CH_2ZnCH_2CH_3 + ZnI_2$$

In organometallic compounds, most p-electrons of **transition metals** conform to an empirical rule called the 18-electron rule. This rule assumes that the metal atom accepts from its **ligands** the number of electrons needed in order for it to attain the electronic configuration of the next **noble gas**. It assumes that the **valence** shells of the metal atom will contain 18 electrons. Thus, the sum of the number of d electrons plus the number of electrons supplied by the ligands will be 18. Ferrocene, for example, has 6 d electrons from Fe(II), plus 2×6 electrons from the two 5-membered rings, for a total of 18. (There are exceptions to this rule, however.)

Possibly the earliest biomedical application of an organometallic compound was the discovery, by Paul Ehrlich, of the organoarsenical Salvarsan, the first antisyphilitic agent. Salvarsan and other organoarsenicals are sometimes listed as organometallics even though **arsenic** is not a true metal. **Vitamin** B_{12} is an organocobalt complex essential to the diet of human beings. Absence of or deficiency of B_{12} in the diet (or a body's inability to absorb it) is the cause of pernicious anemia.

Use as Reagents or Catalysts

Organometallic compounds are very useful as catalysts or reagents in the **synthesis** of organic compounds, such as pharmaceutical products. One of the major advantages of organometallic compounds, as compared with organic or inorganic compounds, is their high reactivity. Reactions that cannot be carried out with the usual types of organic reagents can sometimes be easily carried out using one of a wide variety of available organometallics. A second advantage is the high reaction selectivity that is often achieved via the use of organometallic catalysts. For example, ordinary free-radical polymerization of ethylene yields a waxy low-density polyethylene, but use of a special organometallic **catalyst** produces a more ordered linear polyethylene with a higher density, a higher **melting point**, and a greater strength. A third advantage is that many in this wide range of compounds are stable, and many of these have found uses as medicinals and pesticides. A fourth advantage is the case of recovery of pure metals. Isolation of a pure sample of an organometallic compound containing a desired metal can be readily accomplished, and the pure metal can then be easily obtained from the compound. (This is generally done via preparation of a pure metal carbonyl, such as $Fe[CO]_5$ or $Ni[CO]_4$, followed by thermal decomposition.) Other commonly used organometallic compounds are organolithium, organozinc, and organocuprates (sometimes called Gilman reagents).

Grignard Reagents

One of the most commonly used classes of organometallic compounds is the organomagnesium halides, or Grignard reagents (generally RMgX or ArMgX, where R and Ar are alkyl and aryl groups, respectively, and X is a **halogen** atom), used extensively in synthetic organic chemistry. Organomagnesium halides were discovered by Philippe Barbier in 1899 and subsequently developed by Victor Grignard. They are usually prepared by reaction of magnesium metal with alkyl or aryl halides. Other commonly used organometallic compounds are the organolithium and organozinc compounds.

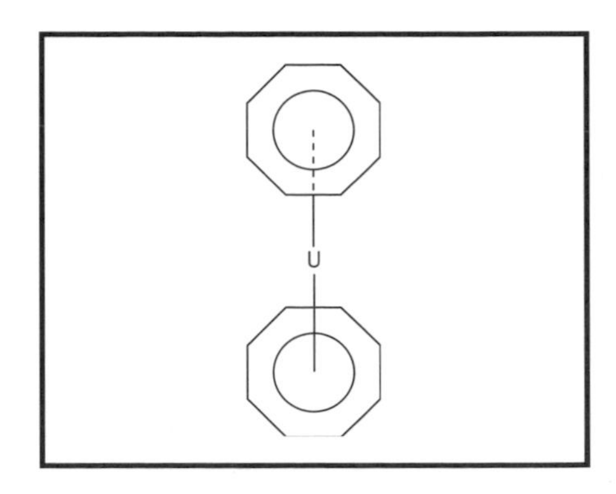

Figure 4. Uranocene

noble gas: element characterized by inert nature; located in the rightmost column in the Periodic Table

valence: combining capacity

arsenic: toxic element of the phosphorus group

vitamins: organic molecules needed in small amounts for the normal function of the body; often used as part of an enzyme catalyzed reaction

synthesis: combination of starting materials to form a desired product

catalyst: substance that aids in a reaction while retaining its own chemical identity

melting point: temperature at which a substance in the solid state undergoes a phase change to the liquid state

halogen: element in the periodic family numbered VIIA (or 17 in the modern nomenclature) that includes fluorine, chlorine, bromine, iodine, and astatine

Carbenes

Carbenes are the electrons of free carbenes that have two spin states, singlet and triplet. The electrons are paired as a sp^2 lone pair in the singlet ($:CH_2$); there is one electron in each of the sp^2 and p orbitals in the triplet ($\cdot CH_2$). Carbenes are generally unstable in the free state, but are stable when bonded to metal atoms. Metal-carbene complexes have the general structure $L_nM=CXY$, where L_nM is the metal fragment with *n* ligands, and X and Y are alkyl groups, aryl groups, hydrogen atoms, or heteroatoms (O, N, S, or halogens). The first carbene complex $[(CO)_5W=CPh(OMe)]$ was reported by E. O. Fischer and A. Maasbol in 1964 (Dunitz, Orgel, and Rich, pp. 373–375). In 1974 Richard R. Schrock prepared compounds in which the substituents attached to carbon were hydrogen atoms or alkyl groups; these complexes are known as Schrock-type carbene complexes. The two types of carbene complexes differ in their reactivities. Fischer-type complexes tend to undergo attack at carbon atoms by nucleophiles (negatively charged species) and are electrophilic (electron-attracting). Schrock-type complexes undergo attack at carbon atoms by electrophiles and are considered to be nucleophilic species. SEE ALSO CATALYSIS AND CATALYSTS; CHIRALITY; EHRLICH, PAUL.

A. G. Pinkus

Bibliography

Abel, Edward F.; Stone, F. Gordon A.; and Wilkinson, Geoffrey, eds. (1995). *Comprehensive Organometallic Chemistry II: A Review of the Literature 1982–1994, Cumulative Indexes*, Vol. 14. New York: Pergamon.

Cotton, F. Albert (2002). "A Half-Century of Nonclassical Organometallic Chemistry: A Personal Perspective." *Inorganic Chemistry* 41:643–658.

Crabtree, Robert H. (2001). *The Organometallic Chemistry of the Transition Metals.* New York: Wiley.

Dunitz, J. D.; Orgel, L. E.; and Rich, R. A. (1956). *Acta Crystallographica* 9:373–375.

Jones, Paul R., and Southwick, Everett (1970). "V. Grignard: Concerning Organomagnesium Compounds in Solution and Their Application to the Synthesis of Acids, Alcohols, and Hydrocarbons." *Journal of Chemical Education* 47:290–299.

Kauffman, George B. (1983). "The Discovery of Ferrocene, the First Sandwich Compound." *Journal of Chemical Education* 60:185–186.

Long, Nicholas J. (1998). *Metallocenes.* London: Blackwell Science.

Miller, Samuel A; Tebboth, John A.; and Themaine, John F. (1952) "Dicyclopentadienyl Iron." *Journal of the Chemical Society* 632–635.

Omae, Iwao (1998). *Applications of Organometallic Compounds.* New York: Wiley.

Schlosser, Manfred, ed. (2002). *Organometallics in Synthesis: A Manual.* New York: Wiley.

Spessard, Gary O., and Miessler, Gary L. (1997). *Organometallic Chemistry.* Upper Saddle River, NJ: Prentice-Hall.

Thayer, John S. (1969a). "Historical Origins of Organometallic Chemistry Part 1: Zeise's Salt." *Journal of Chemical Education* 46:442–443.

Thayer, John S. (1969b). "Historical Origins of Organometallic Chemistry Part 2: Edward Frankland." *Journal of Chemical Education* 46:764–765.

Opiate Peptides *See Endorphins.*

Osmium

MELTING POINT: 3,127°C
BOILING POINT: 5,303°C
DENSITY: 22.590 g/cm^3
MOST COMMON IONS: $OsCl_6^{3-}$, $OsCl_6^{2-}$

The element osmium was discovered in 1804 by English chemist Smithson Tennant (1761–1815) in the black residue that remained after crude platinum was dissolved in aqua regia. The average abundance in Earth's crust is very low, about 0.005 grams (0.00018 ounces) per metric ton, and only four osmium-containing minerals, all extremely rare, are known: erlichmanite, OsS_2; omeiite, $(Os,Ru)As_2$; and osarsite and anduoite, (Os,Ru)AsS. Osmium also occurs in natural **alloys** with iridium and/or ruthenium (e.g., iridosmium). Osmium is obtained as a by-product of refining nickel and the more common platinum group **metals**. Worldwide production is very small, approximately 500 kilograms (1,102 pounds) per year (versus 2,500,000 kilograms, or 5,512,000 pounds, per year for gold). Despite its rarity, osmium is only 30 percent more expensive than gold because it has few commercial uses. Osmium metal is lustrous, bluish-white, hard, and brittle; it melts at 3,127°C (5,661°F) and boils at 5,303°C (9,577°F). It is the densest element known: Its density is 22.59 grams (0.8 ounces) per cubic centimeter (twice that of lead). Osmium is combined with other platinum group elements to yield extremely hard alloys, which find limited use as electrical contacts, wear-resistant instrument pivots and bearings, and tips for high-priced ink pens. Osmium forms compounds in all of its **oxidation** states, from +8 to −2. Its chemistry closely resembles that of ruthenium. The most important compound is osmium tetroxide, OsO_4, a pale yellow solid used as a stain in microscopy, in fingerprint detection, and as a **catalyst** in the production of some pharmaceuticals. Osmium tetroxide has an unpleasant chlorinelike odor, which prompted Tennant to name the element using the Greek word *osme*, "a smell."

alloy: mixture of two or more elements, at least one of which is a metal

metal: element or other substance the solid phase of which is characterized by high thermal and electrical conductivities

oxidation: process that involves the loss of electrons (or the addition of an oxygen atom)

catalyst: substance that aids in a reaction while retaining its own chemical identity

Gregory S. Girolami

Bibliography

Emsley, John (2001). *Nature's Building Blocks: An A–Z Guide to the Elements.* New York: Oxford University Press.

Internet Resources

Winter, Mark. "WebElements." Available from <http://www.webelements.com>.

Ostwald, Friedrich Wilhelm

COFOUNDER OF MODERN PHYSICAL CHEMISTRY
1853–1932

Friedrich Wilhelm Ostwald, born in Riga, Latvia, Russia, almost single-handedly established physical chemistry as an acknowledged academic discipline. In 1909 he was awarded the Nobel Prize in chemistry for his work on **catalysis**, chemical equilibria, and reaction velocities.

German chemist Friedrich Ostwald, recipient of the 1909 Nobel Prize in chemistry, "in recognition of his work on catalysis and for his investigations into the fundamental principles governing chemical equilibria and rates of reaction."

catalysis: the action or effect of a substance in increasing the rate of a reaction without itself being converted

aqueous solution: homogenous mixture in which water is the solvent (primary component)

Ostwald graduated with a degree in chemistry from the University of Dorpat (now Tartu, Estonia) and was appointed professor of chemistry at Riga in 1881, before he moved from Russia to Germany to become chair of the physical chemistry department at the University of Leipzig in 1887. For about twenty years he made Leipzig an international center of physical chemistry: by establishing an instruction and research laboratory that attracted virtually the entire next generation of physical chemists; by editing the first journal in the field (*Zeitschrift für physikalische Chemie*); and by writing numerous textbooks. In 1906 he retired from the university and devoted the rest of his life to various topics, including the history and philosophy of science, color theory, painting, the writing of textbooks and popular books about science, the international organization of science, and the formation of an artificial language for the international exchange of ideas.

Throughout his career as a chemist Ostwald followed the general approach of applying physical measurements and mathematical reasoning to chemical issues. One of his major research topics was the chemical affinities of acids and bases. To that end, he studied the points of equilibria in reaction systems where two acids in an **aqueous solution** compete with each other for a reaction with one base and vice versa. Because chemical analysis would have changed the equilibria, he skillfully adapted the measurement of physical properties, such as volume, refractive index, and electrical conductivity, to that problem. From his extensive data he derived for each acid and base a characteristic affinity coefficient independent of the particular acid–base reactions.

To understand different chemical affinities, Ostwald drew on a new, but then hardly accepted and not yet fully developed, theory advanced by the Swedish physical chemist Svante Arrhenius. According to this theory of electrolytic dissociation, electrolytes such as acids, bases, and salts dissociated in solution into oppositely charged ions to a certain degree, such that at infinite dilution dissociation was complete. Ostwald recognized that if all acids contained the same active ion, their specific chemical affinities must correspond to the number of these active ions in solution, which depended on their specific degree of dissociation at each concentration, and which could be measured through electric conductivity studies. By applying the law of mass action to the dissociation reaction, a simple mathematical relation was derived between the degree of dissociation α, the concentration of the acid c, and an **equilibrium** constant specific for each acid K:

equilibrium: condition in which two opposite reactions are occurring at the same speed, so that concentrations of products and reactants do not change

$$K = \frac{\alpha^2}{(1 - \alpha)c}$$

This is Ostwald's famous dilution law from 1888, which he proved by measuring the electric conductivities of more than 200 organic acids at various concentrations. He substantiated the dissociation theory not only to explain the different activity of acids, but also as a general theory of electrolytes in solution. The theory gained further support from the Dutch physical chemist Jacobus Hendricus van't Hoff who, at the same time, advanced it on a general thermodynamic basis to explain his law of osmotic pressure of solutions as well as Raoult's laws of vapor pressure lowering and freezing point depression. Thus, the new physical chemistry grew to a comprehensive theory of solutions, based on both thermodynamics and dissociation theory.

propagating: reproducing; disseminating; increasing; extending

Ostwald was particularly successful in systematizing and **propagating** these new ideas, applying them to other fields, and organizing a school of physical chemistry. Many chemists rejected the dissociation theory because it predicted wrong values at high concentrations and for strong electrolytes. Despite his concessions about its restricted validity, Ostwald provided numerous proofs of its broad usefulness in his textbooks on general, inorganic, and, particularly, analytical chemistry.

Originally, and incorrectly, Ostwald studied reaction velocities as a measure of chemical affinity. Later, he broadly investigated the time (or kinetic) aspects of chemical reactions and provided a system for the study of chemical kinetics. He first recognized catalysis as the change of reaction velocity by a foreign compound, which allowed him to measure catalytic activities. He distinguished catalysis from triggering and from autocatalysis, which he considered essential to biological systems. His most famous contribution to applied chemistry was on the catalytic **oxidation** of ammonia to nitric acid, which became widely used in the industrial production of fertilizers. SEE ALSO ACID-BASE CHEMISTRY; ARRHENIUS, SVANTE; EQUILIBRIUM; PHYSICAL CHEMISTRY; VAN'T HOFF, JACOBUS.

oxidation: process that involves the loss of electrons (or the addition of an oxygen atom)

Joachim Schummer

Bibliography

Hiebert, Erwin N., and Körber, Hans-Günther (1978). "Ostwald, Friedrich Wilhelm." In *Dictionary of Scientific Biography*, Vol. XV, Supplement I, ed. Charles C. Gillispie. New York: Scribners.

Ostwald, Wilhelm (1926–1927). *Lebenslinien: eine Selbstbiographie.* 3 vols. Berlin: Klasing.

Rodnyj, N. I., and Solowjew, Ju. I. (1977). *Wilhelm Ostwald.* Leipzig: Teubner. (Russian original: *Vilgelm Ostvald.* Moscow: Nauka, 1969.)

Servos, John W. (1990). *Physical Chemistry from Ostwald to Pauling: The Making of a Science in America.* Princeton, NJ: Princeton University Press.

Oxygen

MELTING POINT: −218.4°C
BOILING POINT: −182.96°C
DENSITY: 1.429 g/L
MOST COMMON IONS: OH^-, OH_2^-, O^{2-}

Joseph Priestley and Carl Scheele (each working independently) are credited with the isolation and "discovery" in 1774 of the element oxygen. A few years later Antoine Lavoisier showed that oxygen is a component of the atmosphere. Oxygen is the most abundant element on Earth, constituting about half of the total material of its surface (47 percent by weight of the lithosphere and 89 percent by weight of the ocean) and about 21 percent by volume of the air. Under ordinary conditions (STP) on Earth, oxygen is a colorless, odorless, tasteless gas that is only slightly soluble in water. Oxygen has a pale blue color in the liquid and the solid phases. Ordinary oxygen gas (O_2) exists as diatomic molecules. It also exists in another allotropic form, the triatomic molecule ozone (O_3). Although eight **isotopes** of oxygen are known, atmospheric oxygen is a mixture of only three: those having mass numbers 16, 17, and 18.

isotope: form of an atom that differs by the number of neutrons in the nucleus

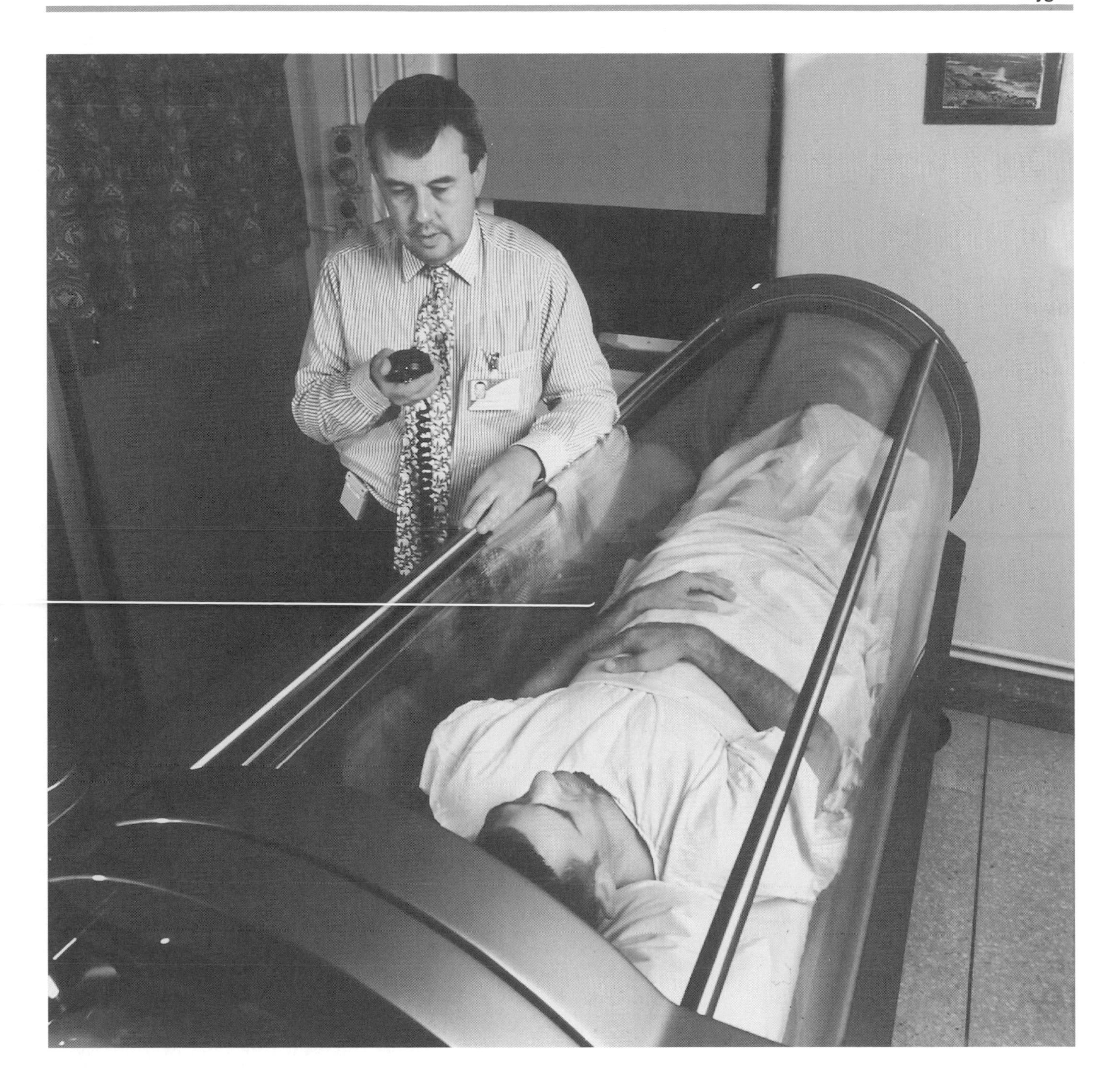

A patient is undergoing hyperbaric oxygen therapy.

Oxygen is very reactive. Its reaction with another substance to form an oxide is called **oxidation**. It is a constituent of a number of compound groups, such as acids, hydroxides, carbonates, chlorates, nitrates and nitrites, and phosphates and phosphites—as well as carbohydrates, proteins, fats, and oils. The respiration of animals and plants is actually a form of oxidation, essential to the production of energy within these organisms. The burning of substances in air is a rapid form of oxidation called **combustion**. In the eighteenth century the idea of combustion replaced the idea (phlogiston theory) that a colorless, odorless, tasteless, and weightless substance named phlogiston was given off during the burning of a substance. SEE ALSO LAVOISIER, ANTOINE; PRIESTLEY, JOSEPH; SCHEELE, CARL.

Ágúst Kvaran

oxidation: process that involves the loss of electrons (or the addition of an oxygen atom)

combustion: burning, the reaction with oxygen

Bibliography

Lane, Nick (2002). *Oxygen: The Molecule that Made the World.* Oxford: Oxford University Press.

Ozone

Earth's ozone layer plays a critical role in protecting Earth's surface from the Sun's harmful ultraviolet (UV) radiation. Every ozone molecule, which consists of three oxygen atoms, has the ability to absorb a certain amount of UV radiation. Under normal circumstances, the ozone layer, which is located in the **stratosphere** between 15 and 50 kilometers (9 and 31 miles) above Earth, remains in a continuous balance between natural processes that both produce and destroy ozone.

stratosphere: layer of the atmosphere where ozone is found; starts about 10 km (6.2 mi) above ground

Ozone is produced in the upper atmosphere through a two-step chemical process that involves oxygen and UV radiation.

$$O + \text{UV radiation} \rightarrow O + O$$

$$O + O_2 \rightarrow O_3$$

The process begins with UV radiation breaking apart molecular oxygen (O_2), thus producing two oxygen (O) atoms. In the second step, an oxygen atom (O) recombines with an oxygen molecule (O_2) to form an ozone (O_3) molecule.

Ozone can also be naturally destroyed through reactions with chlorine, nitrogen, and hydrogen. For example, chlorine can be a very effective destroyer of ozone via the following set of reactions.

$$Cl + O_3 \rightarrow ClO + O_2$$

$$ClO + O \rightarrow Cl + O_2$$

In this process, a chlorine atom (Cl) reacts with ozone (O_3) to produce chlorine monoxide (ClO) and an oxygen molecule (O_2). ClO can then combine with an oxygen atom (O) to reform Cl and O_2. In this reaction set, because chlorine is reformed after destroying ozone, the cycle can repeat itself very quickly.

In recent years global chlorine levels have increased due to the use of **chlorofluorocarbons (CFCs)**, a large class of chemicals useful in a variety of industries. Under certain circumstances, even a single chlorine atom released from a CFC's molecule can destroy many thousands of ozone molecules through a chemical chain reaction. Current declines in global ozone levels and the development of the Antarctic ozone hole have both been linked to CFC use.

chlorofluorocarbon (CFC): compound that contains carbon, chlorine, and fluorine atoms, which remove ozone in the upper atmosphere

Although ozone concentrations in the upper atmosphere play an important role in protecting Earth's surface from harmful UV radiation, ozone at its surface is a pollutant harmful to human health. Enhanced levels of surface ozone are often the result of automobile exhaust and pose a serious health risk. Fortunately, current levels of surface ozone (also known as smog) over most major cities have declined to healthier levels due in part to domestic and international governmental regulations. SEE ALSO ATMOSPHERIC CHEMISTRY.

Eugene C. Cordero

Bibliography

Graedel, T. E., and Crutzen, Paul J. (1993). *Atmospheric Change: An Earth System Perspective.* New York: W. H. Freeman.

Internet Resources

Stratospheric Ozone: An Electronic Textbook. Available from <http://www.ccpo.odu.edu/SEES/ozone/oz_class.htm>.

Palladium

MELTING POINT: 1,552°C
BOILING POINT: 3,760°C
DENSITY: 12.0 g/cm^3
MOST COMMON IONS: Pd^{2+}

The element palladium was isolated and identified by William Wollaston in 1803. Its name comes from the asteroid Pallas. (Pallas was another name for Athena, the Greek goddess of wisdom.) Palladium in pure form is not found in nature. The preparation of the element is via a series of reactions. Platinum **metal** ore concentrates (65% of which come from the Merensky Reef in South Africa) are treated with aqua regia (giving copper and nickel as by-products). The solutions, containing H_2PdCl_4 with platinum and gold complexes, are treated with $FeCl_2$ (which precipitates gold) and then with excess of NH_4OH followed by HCl to precipitate the impure $[Pd(NH_3)_2Cl_2]$. This compound is purified by dissolution in NH_4OH and precipitation with HCl. The pure $[Pd(NH_3)_2Cl_2]$ is ignited to palladium metal.

metal: element or other substance the solid phase of which is characterized by high thermal and electrical conductivities

Palladium metal, like platinum metal, is silvery-white and lustrous and has malleable and **ductile** properties. It has the face-centered cubic crystal structure. It forms a fluoride, PdF_4 (brick-red), and other halides: PdF_2 (pale violet), α-$PdCl_2$ (dark red), $PbBr_2$ (red black), and PdI_2 (black). Pd metal can absorb up to 935 times its own volume of hydrogen molecules. When the composition reaches about $PdH_{0.5}$, the substance becomes a semiconductor.

Palladium can form complexes in a variety of **oxidation** states. Table 1 contains some examples.

Palladium has extensive use as a **catalyst** in hydrogenation and dehydrogenation reactions, due to its capacity of combination with hydrogen. Palladium films are used as electrical contacts in connectors. Palladium-silver and palladium-nickel **alloys** are used to substitute for gold in jewelry.

Lea B. Zinner

ductile: property of a substance that permits it to be drawn into wires

oxidation: process that involves the loss of electrons (or the addition of an oxygen atom)

catalyst: substance that aids in a reaction while retaining its own chemical identity

alloy: mixture of two or more elements, at least one of which is a metal

Oxidation states	Complexes
(0)	$K_4[Pd(CN)_4]$ (yellow)
(I)	$[PdCl(CO)]_x$ (reddish-violet)
(II)	$Na_2[PdCl_4]$
(IV)	$K_2[PdF_6]$ (bright yellow)

Table 1. Some palladium complexes.

Bibliography

Allred, A. L. (1961). *Journal of Inorganic Nuclear Chemistry* 17:215.

Greenwood, Norman N., and Earnshaw, A. (1984). *Chemistry of the Elements.* New York: Pergamon Press.

Livingstone, Stanley E. (1973). "The Platinum Metals." In *Comprehensive Inorganic Chemistry*, Vol. 3, ed. J. C. Bailar Jr.; H. J. Emeléus; Sir Ronald Nyholm; and A. F. Trotman-Dickenson. Oxford, U.K.: Pergamon Press.

Paracelsus

GERMAN PHYSICIAN, ALCHEMIST, AND SCIENTIST
1493–1541

Paracelsus was born Theophrastus Bombastus von Hohenheim. He was a contemporary of Martin Luther and Nicolaus Copernicus. He adopted his

A woodcut from *Works* of the German alchemist Paracelsus.

pseudonym based on his assertion that he was a better physician than Celsus, the first century C.E. Roman author on medicine acclaimed in Renaissance Europe (he was "Para-Celsus," or beyond Celsus). His self-promotion as "The Most Highly Experienced and Illustrious Physician . . . " has given us the word "bombastic," derived from his birth name.

Paracelsus gained his early medical knowledge from his father, who was a physician. He followed this education with formal medical training at the University of Ferrara in Italy. Finding his formal training disappointing, Paracelsus embarked on a life of travel and study combined with medical practice. According to Paracelsus, he collected medical knowledge anywhere he could find it without regard to academic authority. He acknowledged his consultations with peasants, barbers, chemists, old women, quacks, and magicians. Paracelsus developed his notions of disease and treatment away from any established medical faculty and promoted the idea that academic medical training had reached a state deeply in need of reform.

Paracelsus believed in the four "Aristotelian" elements of earth, air, fire, and water. His medical theory was based on the notion that earth is the fundamental element of existence for humans and other living things. Paracelsus believed that earth generated all living things under the rule of three "principles": salt, sulfur, and mercury. He therefore believed these substances to be very potent as chemical reactants, as poisons, and as medical treatments. (Indeed, salt and sulfur can yield strong mineral acids, for example, hydrochloric acid and sulfuric acid, and mercury is a strong poison.) Finally, Paracelsus believed in the "Philosopher's Stone." The Philosopher's Stone (which he sometimes claimed to possess) was supposed to cure all ills and to enable the transformation of any **metal** into gold. Such a stone, it was believed, would be the strongest chemical reactant and the strongest medicine possible.

metal: element or other substance the solid phase of which is characterized by high thermal and electrical conductivities

Paracelsus advocated the direct observation of a patient's medical condition and the assessment of his or her surroundings. He was one of the first physicians to describe occupational diseases. He described several lung diseases of miners and recommended improved ventilation as a means of their prevention. He emphasized that the legitimacy of a treatment was whether or not it worked, not its recommendation by an ancient authority in an ancient text. Paracelsus promoted the use of mineral treatments. Because small amounts of mercury salts were effective against some illnesses, these medicines were judged to be very strong.

Paracelsus's exalted claims for himself and his abrasive personality often brought him into conflict with civil authorities. His methods of trial and error and observation led him to reject the use of sacred relics as medical treatment. It brought him into conflict with religious authorities. His calls for reformation of the medical profession offended medical authorities. As a consequence he was on the move often. Paracelsus held an academic post only once, and it lasted only a year. Although he wrote a great deal, only one of his manuscripts was published in his lifetime. Most of his manuscripts were left in a variety of cities and were published several years after his death. Within these manuscripts are inconsistencies and contradictions. Paracelsus never established any one strong school of thought or medical practice. He did, however, influence future generations of iatrochemists (physician-chemists, *iatro* being Greek for "physician"), who continued to apply chemistry to questions of medical practice. SEE ALSO ALCHEMY.

David A. Bassett

Bibliography

Jacobi, Jolande, ed. (1942, reprint 1988). *Paracelsus, Selected Writings*, tr. Norbert Gutman. Princeton, NJ: Princeton University Press.

Partington, J. R. (1961). *A History of Chemistry*, Vol. 2. New York: Martino Publishing.

Sigerist, Henry E., ed. (1941, reprint 1996). *Paracelsus: Four Treatises.* Baltimore: Johns Hopkins University Press.

Particles, Fundamental

Fundamental particles are the elementary entities from which all matter is made. They have no known smaller parts. As recently as 1900 most people believed that atoms were the tiniest particles in the universe.

By the 1930s, however, it was clear that atoms were made up of even smaller particles—protons, neutrons, and electrons, then considered to be the fundamental particles of matter. (A proton is a positively charged particle that weighs about one atomic mass unit [1.0073 AMU]; a neutron has about the same mass [1.0087 AMU] but no charge; and an electron has a much smaller mass [0.0005 AMU] and a negative charge.) Protons and neutrons make up the tiny nucleus of an atom, while electrons exist outside the atomic nucleus in discrete energy levels within an electron "cloud."

By 1970 it began to appear that matter might contain even smaller particles, an idea suggested in 1963 by American physicist Murray Gell-Mann (who called the particles *quarks*) and independently by American physicist George Zweig (who called them *aces*). There are in actuality hundreds of subatomic particles that have been observed, but many of them are unstable.

Fermions

At the start of the twenty-first century, scientists believe that all matter is made up of tiny particles called fermions (named after American physicist Enrico Fermi). Fermions include quarks and leptons. Leptons include electrons (along with muons and neutrinos); they have no measurable size, and they are not affected by the strong nuclear force. Quarks, on the other hand, are influenced by the strong nuclear force. They are the fundamental particles that make up protons and neutrons (as well as mesons and some other particles). Both protons and neutrons are classified as baryons, composite particles each made up of three quarks.

Quarks come in six different types, or "flavors": up and down, top and bottom, and charm and strange. Protons and neutrons are made of up (u) quarks (which have a charge of $+\frac{2}{3}$) and down (d) quarks (which have a charge of $-\frac{1}{3}$). A proton is made from two u quarks $(+\frac{2}{3})(+\frac{2}{3})$ and one d quark $(-\frac{1}{3})$, giving a total charge of +1. A neutron contains one u quark $(+\frac{2}{3})$ and two d quarks $(-\frac{1}{3})(-\frac{1}{3})$ for a total charge of zero.

Fundamental Forces

There are also fundamental forces acting on matter; these have their own sets of fundamental particles. The forces are the strong nuclear force (or strong interaction), the weak nuclear force (or weak interaction), and electromagnetism (which includes light, x rays, and all the other electromagnetic forces). All these forces are transmitted by particles called *fundamental bosons* (named after Indian physicist S. N. Bose).

Fundamental bosons differ from fermions in spin and the number of quarks they contain. Fermions have spins measured in half numbers, and they contain an odd number of quarks. Bosons have whole integer spins, and they contain an even number of quarks. The bosons that transmit the strong nuclear force are called *gluons,* those that transmit electromagnetic forces are *photons*, and those transmitting the weak force are known as *weak bosons*. A fourth force, the gravitational force, is believed to be transmitted by particles called gravitons; however, the particles have not yet been observed. Still another kind of boson, called a *Higgs boson*, is thought to be the source of mass in other particles, but this particle also has not actually been observed.

Particle Accelerators

The study of fundamental particles often involves speeding up charged particles, such as protons or electrons, and then letting them collide with targets so as to produce other particles for further study. The particle accelerators used to do this are devices that force the charged particles to jump over longer and longer space gaps per unit of time, until the particles are moving at speeds approaching the speed of light.

The earliest of such devices were the linear Cockcroft-Walton accelerator (1929), the circular cyclotron (1930), and the Van de Graaff generator (1931). Modern *synchrotrons* are large machines that have both linear and curved sections. The most powerful synchrotron is the *Tevatron* proton accelerator at the Fermilab located near Batavia, Illinois (just outside of Chicago); it lies inside an underground circular tunnel that measures almost 6.4 kilometers (4.0 miles) around. The longest accelerator is the collider at the CERN research center in Geneva, Switzerland—it has a circumference of about 27.3 kilometers (17.0 miles).

Particle Detectors

Detection of fundamental particles is difficult because the particles are so extremely tiny. The earliest detector was just photographic film, since particles passing through would expose the film and become evident when it was developed. The first device designed for the purpose of detecting tiny particles was the "cloud chamber" (invented by Scottish physicist Charles Wilson in 1911). It was a glass container filled with air saturated with water (or alcohol) vapor. Charged particles passing through the chamber formed ions leaving fog tracks—the heavier the particles, the wider their tracks.

The "bubble chamber" (invented by American physicist Donald Glaser in 1952) was similar to a cloud chamber, except that it was filled with a liquid (usually liquefied helium or hydrogen) held at a temperature just below its boiling point. Moving particles would disturb the liquid, causing bubbles to form along their paths. There was also a "spark chamber" (invented in Japan in 1959) that contained a series of parallel metal plates and produced an electrical discharge along the ion trail left by a charged particle. Although all of these devices were once important for detecting subatomic particles, they have largely been replaced by more modern detectors.

In the twenty-first century fundamental particles are studied using detectors such as tracking chambers (which trace the path of a particle with electrical signals), sampling calorimeters (which track the particle's path by its energy of motion), scintillators (which give off light when particles strike them), or magnetic detectors (which cause charged particles to move in curved paths). Many instruments use combinations of these various kinds of detectors.

C. T. R. WILSON (1869–1959)

The inspiration for C. T. R. Wilson's expansion, or cloud, chamber came from his interest in meteorological sciences. His initial intention was to recreate cloud formations. This led to an interest in studying atmospheric electric fields and the vapor trail of ions. For his work he shared the Nobel Prize in 1927.

—Valerie Borek

As high-energy particles pass through a bubble chamber, bubbles are formed, which leave tracks as they move. This is important in the study of subatomic collisions, which normally are not visible to the naked eye.

Antimatter

To further complicate the subject of subatomic particles, each kind of particle has an antiparticle. For example, for each kind of quark there is an antiquark of the same mass and spin, but of opposite charge. The first antiparticle to be observed was the positron, an electron with a positive charge. An antiproton is like a proton, but it has a negative charge. Antiparticles can be observed, and molecules of antimatter can even be generated. A positron orbiting an antiproton, for example, is an antihydrogen atom.

Many scientists believe that there must be some areas of the universe that are completely made up of antimatter, the exact opposite of the kind of matter found on Earth. If that is true, such areas would not be very compatible with areas made of matter—when a particle and its antiparticle make contact, they destroy each other and are converted into energy. According to Einstein's special theory, $E = mc^2$, which means that energy is equivalent to mass times the speed of light, squared. In other words, a tiny speck of matter can be converted to a considerable amount of energy.

The conversion can also go the other way. Large releases of energy that occur when high-energy particles collide can produce new particles and antiparticles of matter. Much modern research in particle physics involves high-energy collisions between beams of particles, such as protons, so as to generate other kinds of particles. Some collisions involve interactions of particles with antiparticles (e.g., electrons with positrons). Particle accelerators have been turned into giant colliders in which beams of particles moving at speeds approaching the speed of light collide with each other, producing other kinds of particles.

By the early 2000s several hundred subatomic particles were known, almost all of them being made of quarks. The few remaining ones are the leptons, the electron being the best known. The other leptons are still rather mysterious. The muons and tau particles are negatively charged like the electron, while the neutrinos (which have no detectable mass but often pair up with the heavier leptons) have no charge. The electron was discovered in 1897, but there is still much to learn about other fundamental particles. SEE ALSO ATOMIC STRUCTURE; FERMI, ENRICO.

Doris K. Kolb

Bibliography

Boslough, John (1985). "Worlds within the Atom." *National Geographic* 167(5): 634–663.

Brehm, John J. (1989). *Introduction to the Structure of Matter.* West Sussex, U.K.: John Wiley & Sons.

Halzen, Francis, and Martin, Alan D. (1984). *Quarks and Leptons.* Weinheim, Germany: Wiley-VCH.

Martin, B. R., and Shaw, G. (1997). *Particle Physics*, 2nd edition. New York: Wiley.

Quigg, C. (1985) "Elementary Particles and Forces." *Scientific American* April:84–95.

Serway, Raymond K., and Faughn, Jerry S. (2003). *College Physics*, 6th edition. Pacific Grove, CA: Brooks/Cole.

Pasteur, Louis

FRENCH CHEMIST AND MICROBIOLOGIST
1822–1895

Louis Pasteur was born in 1882 in Dole, France. Many people are unaware of the fact that he was a chemist. Pasteur received his schooling at the École Normale Supérieure in Paris—a school specifically designed to foster the development of students in the sciences and letters. He was, perhaps, the most accomplished of these students.

Pasteur's first major contribution to chemistry occurred when he was only 26 years old, working with French Chemist Antione Balard (1802–1876) in the new field of crystallography. Organic molecules—at the time thought to be made exclusively by living beings—were a particularly important area of study and Pasteur was both fortunate and perceptive when working with a compound called tartaric acid—a chemical found in the sediments of fermenting wine.

Pasteur, as well as other scientists of his time, used the rotation of plane-polarized light as one means for studying crystals. Polarized light can be thought of as occupying a single plane in space. If such light is passed through a solution with dissolved tartaric acid, the angle of the plane of light is rotated. Many organic acids display this feature. What made Pastuer's work with tartaric acid and polarized light so important was his careful observation of crystals.

In addition to tartaric acid another compound named paratartaric acid was found in wine sediments. Chemical analysis showed this compound to have the same composition as tartaric acid, so most scientists assumed the two compounds were identical. Strangely enough, however, paratartaric acid did not rotate plane-polarized light. Pasteur would not accept the idea that such an experimental result could be an accident or unimportant. He guessed that even though the two compounds had the same chemical composition, they must somehow have different structures—and he set out to find evidence to prove his hypothesis.

French chemist Louis Pasteur.

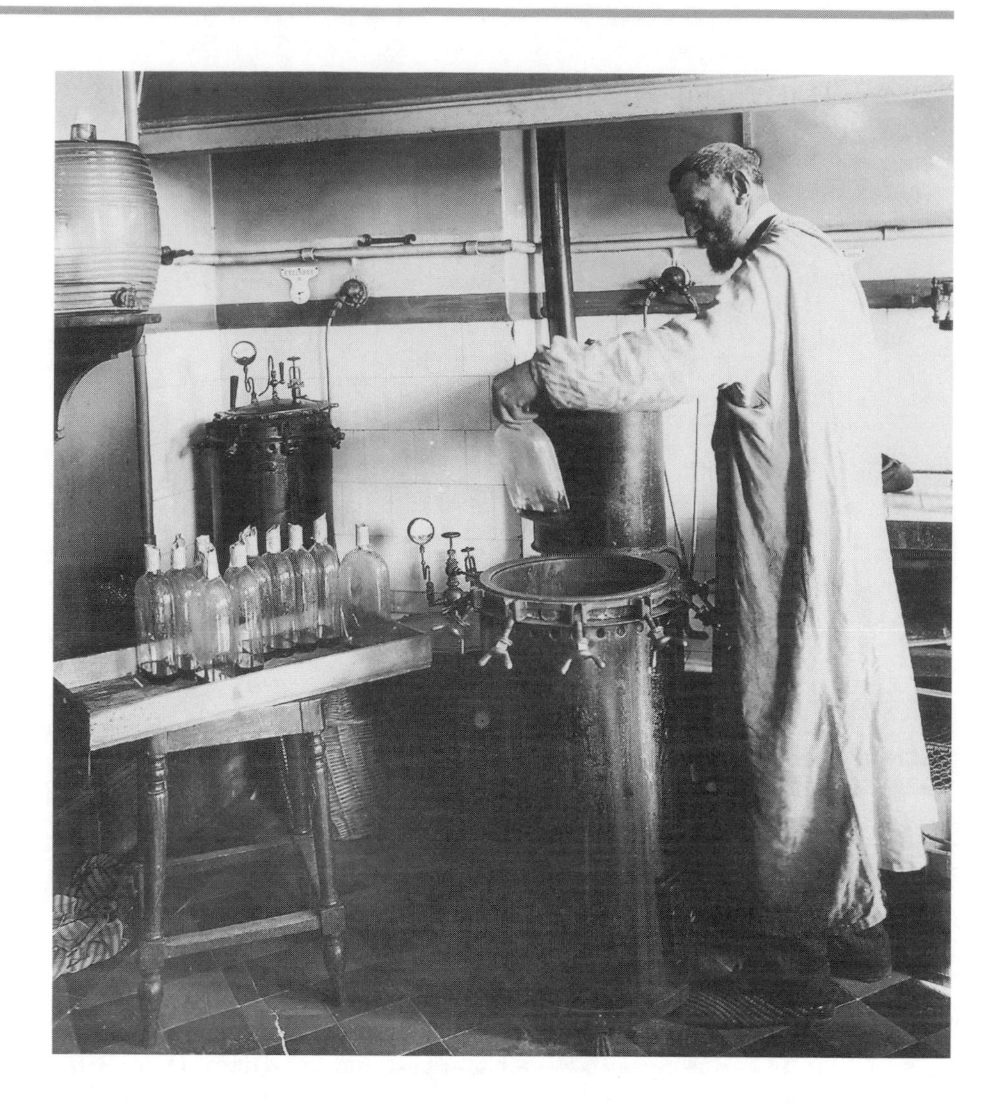

First, Pasteur carefully observed the paratartaric acid under a microscope. Looking at the tiny crystals, he noticed two different types. While almost identical, they were actually mirror images of each other as depicted in Figure 1. Pasteur's next step required incredibly meticulous work. Again, working with the microscope, he separated the two types of crystals into two piles. After separating the crystals, Pasteur made two solutions—one with each of the piles—and tested how they interacted with polarized light. He found that both solutions rotated the light—*but in opposite directions.* When the two types of crystals were together in the solution of paratartaric acid the effect of rotation of the light was canceled.

stereochemistry: the study of the three dimensional shape of molecules and the effects of shape upon the properties of molecules

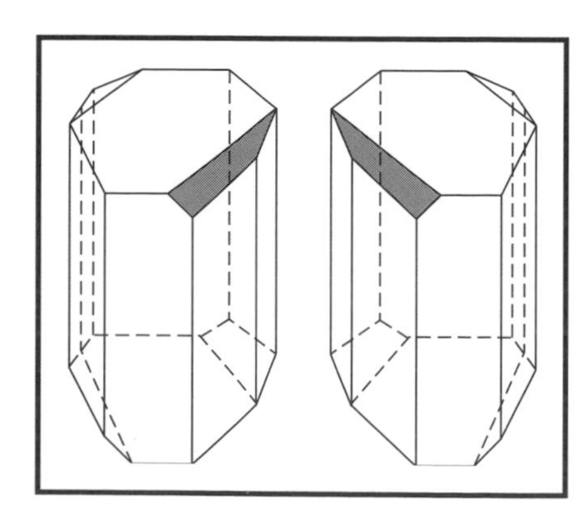

Figure 1. Structure of paratartaric crystals.

Most importantly for the development of chemistry, these experiments by Pasteur established that composition alone does not provide all the information needed to understand how a chemical behaves. His work allowed chemists to start thinking about the structure of molecules in terms of their **stereochemistry**, a field that remains important in chemistry research.

The discovery of stereochemistry was not the last chemical work carried out by Pasteur. Seven years after he first started working in crystallography (in 1854) he was became a professor of chemistry in Lille, France. Among the main commercial interests in Lille was the production of alcohol in distilleries. One of Pasteur's students was the son of a distillery owner who was encountering troubles with his factory. Too often the product of their efforts was lactic acid rather than alcohol. Once again, Pasteur would need to contradict current scientific beliefs to answer a chemical question.

At the time of his work in Lille, the scientific community knew that the alcohol produced by fermentation came from the breakdown of sugars (found in grapes for wine-making). However, they believed that the breakdown was caused by something in the sugar itself that they called unstabilizing vibrations. These unstabilizing vibrations could be transferred from one vat to a new batch of freshly squeezed grapes to make more wine. What this notion did not explain, however, was why some batches of grapes produced lactic acid rather than alcohol.

Pasteur approached this problem much like the earlier crystallography dilemma—by using his microscope to make careful observations. He observed microbes in the wines and noticed that different shaped microbes were present when lactic acid was formed versus when alcohol was formed. He also observed that some of the compounds rotated plane-polarized light, so Pasteur concluded that the microbes were living (because it was thought that stereochemistry was related to living systems only.) Ultimately he was able to help isolate the yeast that was responsible for good fermentation and he solved the chemical problem of lactic acid formation and at the same time invented the field of microbiology.

Pasteur went on to make many more advances in microbiology. He also realized the importance of making science an international endeavor and advocated for a scientific approach to the betterment of the human condition. He once remarked, "Do not put forward anything that you cannot prove by experimentation." Pasteur died in 1893, two years after the first international Pasteur Institute was established in Saigon in what was then French Indochina (now Ho Chi Min City in Vietnam). SEE ALSO CHIRALITY; COORDINATION CHEMISTRY; ISOMERISM; ORGANIC CHEMISTRY.

Thomas A. Holme

Bibliography

Debré P. (1998). *Louis Pasteur.* Baltimore, MD: Johns Hopkins University Press.

Geison, Gerald L. (1995). *The Private Science of Louis Pasteur.* Princeton, NJ: Princeton University Press.

Pauli, Wolfgang

AMERICAN THEORETICAL PHYSICIST
1900–1958

Wolfgang Ernst Pauli was born in Vienna, Austria, where his father, regarded as one of the founders of colloid chemistry, was employed at the University of Vienna. His godfather was Ernst Mach, a famous physicist, philosopher, and one of the founders of logical positivism; he had a significant influence on Pauli's thinking. In high school Pauli was an outstanding student with a special talent for mathematics and physics. His parents fostered Pauli's appetite for science by hiring a private tutor. The tutor was so successful that within twelve months of beginning his studies at the University of Munich in 1918, Pauli had submitted three original papers on the

American physicist Wolfgang Pauli, recipient of the 1945 Nobel Prize in physics, "for the discovery of the Exclusion Principle, also called the Pauli Principle."

exclusion principle: principle that states that no two electrons can have the same four quantum numbers

theoretical physics: branch of physics dealing with the theories and concepts of matter, especially at the atomic and subatomic levels

theory of relativity to a leading physics periodical; all were published before his twentieth birthday.

Pauli received his doctorate in 1921 for theoretical work on the hydrogen molecule ion. He then became an assistant to Max Born at Göttingen. While at Göttingen, Pauli met Niels Bohr, who invited him to work for a year with his group in Copenhagen, Denmark. Once there, Pauli began work on the problem of the anomalous Zeeman effect (how the energy levels of a multielectron atom are split in a magnetic field), work that he continued when in 1923 he moved to a new position at the University of Hamburg. By 1924 he had decided that the current model of atomic structure used by Bohr, which assumed only two numbers and which allowed many electrons to have identical quantum numbers, needed to be modified. He also found that the currently accepted idea that it was the magnetic moment of the core of the atom that was responsible for the splitting of the electron energy levels of the outer electrons, was incorrect. Instead, Pauli proposed a new model that had as its consequence his famous **exclusion principle**.

The new model had its origins in a new classification of electron levels published in 1924 by Edmond C. Stoner, an English physicist at the University of Leeds who was an expert on the magnetic properties of matter. This classification divides the electrons of an atom into electronic shells using three quantum numbers (n, k1, k2). The first two number are the same as those used by Bohr, and the third one, the inner quantum number k2, was chosen so that twice the sum of the individual k2 numbers became the number of electrons in a subgroup. It was Pauli's genius that allowed him to extend this classification by adding a fourth quantum number (m1), which could have only two values (+1/2 and −1/2). As a result, Pauli was able in 1925 to arrive at the first statement of his exclusion principle, that stated that there cannot be two or more equivalent electrons in an atom for which in strong fields the values of all quantum numbers n, k1, k2, and m1 are the same. Initially, Pauli rejected the notion that the two-valuedness of m1 was due to spin, but after discussing the matter of electron spin with fellow physicists Samuel Goudsmit and George Uhlenbeck, he accepted the idea. The term "exclusion principle" had its origin in Pauli's insistence on each electron having a unique set of quantum numbers. This requirement immediately solved many problems in the interpretation of observed atomic spectra, because it prevented many lines that, according to prior theories, should be seen but never were, to become forbidden.

In 1928 Pauli became professor of **theoretical physics** at the Federal Institute of Technology, Zurich; largely through his efforts it became a leading center for research in theoretical physics. In 1931 he observed that when an electron was emitted from a nucleus, a loss of energy occurred that could not be explained by then-current theories. He proposed that it was due to the existence of another particle which carried no charge and had very low mass. The American physicist Enrico Fermi named this particle the "neutrino"; it was eventually discovered some twenty-five years later.

During World War II Pauli worked at the Institute for Advanced Studies at Princeton in New Jersey; he then returned to Zurich, where he died in 1958. SEE ALSO BOHR, NIELS; FERMI, ENRICO.

John E. Bloor

Bibliography

McMurray, Emily J., ed. (1995). *Notable Twentieth-Century Scientists.* Detroit: Gale Research.

Peierls, Rudolf E. (1959). "Wolfgang Ernst Pauli." In *Biographical Memoirs of Fellows of the Royal Society*, Vol. 5, pp. 175–192. London: Royal Society.

Internet Resources

"Wolfgang Pauli and Modern Physics." Available from <http://www.ethbib.ethz.ch/exhibit/pauli>.

Pauling, Linus

AMERICAN CHEMIST
1901–1994

Linus Carl Pauling was born in Portland, Oregon, on February 28, 1901, the first of three children of pharmacist Herman W. Pauling and Lucy Isabelle Pauling (née Darling). An internationally acclaimed scientist, educator, humanitarian, and political activist, the only person to have received two unshared Nobel Prizes (for chemistry in 1954; for peace in 1962), Pauling was once characterized by *New Scientist* as one of the twenty greatest scientists of all time, on a par with Isaac Newton, Charles Darwin, and Albert Einstein. His magnum opus, *The Nature of the Chemical Bond* (1939), was one of the most influential and frequently cited scientific books of the twentieth century. His advocacy of megadoses of **vitamin** C for the common cold, cancer, and AIDS is still controversial, and the work for which he is best known. His life and career were characterized by controversy, and almost everything about him was larger than life.

vitamins: organic molecules needed in small amounts for the normal function of the body; often used as part of an enzyme catalyzed reaction

Pauling majored in chemical engineering at Oregon Agricultural College (now Oregon State University), where he developed the belief that would guide his lifetime of research: Atomic arrangements must be responsible for the chemical and physical properties of material substances. He received his B.S. degree in 1922 and entered the California Institute of Technology (Caltech) at Pasadena, where he worked with Roscoe G. Dickinson and adopted the relatively new technique of x-ray crystallography to explore the structure of crystals. In 1925 Pauling received his Ph.D. and was awarded a Guggenheim fellowship to pursue postgraduate research in Europe with the seminal atomic theorists Arnold Sommerfeld, Niels Bohr, and Erwin Schrödinger. The first to realize the ramifications of the new quantum mechanics within chemistry, he used this body of ideas to explain and predict the properties of atoms and ions, and thus to revolutionize chemistry. In 1927 Pauling returned to Pasadena to join the faculty of Caltech, where he stayed until 1963. There he used x-ray diffraction to measure the lengths and angles of atomic bonds in the three-dimensional structures of, first, inorganic crystals and, later, organic compounds.

American chemist Linus Carl Pauling, recipient of the 1954 Nobel Prize in chemistry, "for his research into the nature of the chemical bond and its application to the elucidation of the structure of complex substances," and the 1962 Nobel Peace Prize.

One of the key concepts of Pauling's quantum theory of chemical bonding, introduced in 1931, was resonance: In many cases an ion or molecule could not be represented, conceptually or on paper, as one classical structure, but required what he called a "hybridization" of two or more of these structures. The single classical structure simply did not describe the chemical bond(s). In less than a decade he had transformed the earlier, somewhat simplistic theory of the chemical bond into a powerful, highly sophisticated theory and research tool.

During the mid-1930s Pauling turned his attention to molecules present in living things. His interest in the binding of oxygen to hemoglobin (the protein molecule that carries oxygen via the bloodstream to cells throughout the body) provoked a more general interest in proteins, the nitrogen-containing organic compounds required in all of animal **metabolism**. In 1948, while in bed with influenza, Pauling occupied himself with making a paper model of linked amino acids, the basic building blocks of proteins. In this way he received the inspiration that led to his discovery of the α-**helix**—a crucial concept that helped James Watson and Francis Crick to determine the structure of **DNA**, one of the discoveries of the century. And this landmark discovery of Watson and Crick led, ultimately, to the Human Genome Project and the current revolution in genetic engineering.

metabolism: the complete range of biochemical processes that take place within living organisms; comprises processes that produce complex substances from simpler components, with a consequent use of energy (anabolism), and those that break down complex food molecules, thus liberating energy (catabolism)

helix: form of a spiral or coil such as a corkscrew

DNA: deoxyribonucleic acid—the natural polymer that stores genetic information in the nucleus of a cell

After World War II Pauling studied sickle cell anemia, and theorized that it was the result of a genetically based defect in the patient's hemoglobin molecules. In 1949 he and Harvey Itano confirmed this theory; they had identified what they called a "molecular disease," one that could be defined by a molecular abnormality. In 1954 Pauling received the Nobel Prize in chemistry "for his research on the chemical bond and its application to the elucidation of the structure of complex substances."

Less well-known is the record of Pauling's evolution from ivory tower scientist to ardent and articulate advocate of nuclear disarmament and of the social responsibility of scientists. His eventual clashes with political and ideological adversaries, including the U.S. government, which denied him research grants and a passport, consumed much of his time and energy. His being chosen for the 1962 Nobel Peace Prize was criticized by many, and the American Chemical Society, which he had served as president in 1949, at around this time chose to slight him.

In 1963 Pauling left Caltech to become research professor at the Center for the Study of Democratic Institutions at Santa Barbara, California, at which time he began to divide his time between chemistry and world peace. In Santa Barbara he became greatly interested in what he called "orthomolecular medicine"—a biochemical approach to human health that included the central idea that large amounts of some chemical compounds normally present in the body could be used to treat or prevent disease. In 1973, following professorships at the University of California, San Diego (1967–1969) and Stanford University (1969–1974), he founded the Institute of Orthomolecular Medicine (later named the Linus Pauling Institute of Science and Medicine), an organization of which he was director of research at the time of his death. He died of cancer at his Deer Flat Ranch near Big Sur, California, on August 19, 1994, at the age of ninety-three.

Pauling has been called one of the two greatest scientists of the twentieth century (the other being Einstein) and the greatest chemist since Antoine-Laurent Lavoisier, the eighteenth-century founder of modern chemistry. Pauling's multifaceted life and activities, scientific and personal, spanned almost the entire twentieth century. SEE ALSO BOHR, NIELS; EINSTEIN, ALBERT; HEMOGLOBIN; LAVOISIER, ANTOINE; NEWTON, ISAAC; PROTEINS; SCHRÖDINGER, ERWIN; WATSON, JAMES DEWEY.

George B. Kauffman

Bibliography

Goertzel, Ted, and Goertzel, Ben (1995). *Linus Pauling: A Life in Science and Politics.* New York: Basic Books.

Hager, Thomas (1995). *Force of Nature: The Life of Linus Pauling.* New York: Simon & Schuster.

Hager, Tom (1998). *Linus Pauling and the Chemistry of Life.* New York: Oxford University Press.

Kauffman, George B., and Kauffman, Laurie M. (1996). "An Interview with Linus Pauling." *Journal of Chemical Education* 73:29–31.

Marinacci, Barbara, ed. (1995). *Linus Pauling: In His Own Words: Selected Writings, Speeches, and Interviews.* New York: Simon & Schuster.

Marinacci, Barbara, and Krishnamurthy, Ramesh, eds. (1998). *Linus Pauling on Peace: A Scientist Speaks Out on Humanism and World Survival; Writings and Talks by Linus Pauling.* Los Altos, CA: Rising Sun Press.

Mead, Clifford, and Hager, Thomas (2001). *Linus Pauling: Scientist and Peacemaker.* Corvallis: Oregon State University Press.

Newton, David E. (1994). *Linus Pauling: Scientist and Advocate.* New York: Facts on File.

Pauling, Linus (1958, 1983). *No More War!* New York: Dodd, Mead & Co.

Pauling, Linus (1964). "Modern Structural Chemistry." In *Nobel Lectures Including Presentation Speeches and Laureates' Biographies, Chemistry 1942–1962.* New York: Elsevier. Also available from <http://www.nobel.se/chemistry/laureates/>.

Serafini, Anthony (1989). *Linus Pauling: A Man and His Science.* New York: Paragon House.

Internet Resources

Pauling, Linus. "Science and Peace." Available from <http://www.nobel.se/peace/laureates>.

Penicillin

Penicillin was discovered accidentally in 1929 when Sir Alexander Fleming observed bacterial **cultures** contaminated with a mold that inhibited bacterial growth. The antibiotic penicillin was subsequently isolated from cultures of the *Penicillium* mold. In 1938 two other British scientists, Howard

culture: living material developed in prepared nutrient media

Penicillin is a mold used to treat bacterial infections.

GRAM-POSITIVE AND GRAM-NEGATIVE BACTERIA

Bacteria can be broadly classified into two groups; the gram-positive bacteria, which are stained purple after the gram staining procedure, and the gram-negative bacteria, which are stained red. The difference in staining reflects differences in the structure of the cell walls between gram-positive and gram-negative bacteria. Pathogenic gram-positive bacteria include *Staphylococcus aureus* and *Bacillus anthracis*, and pathogenic gram-negative bacteria include *Escherichia coli* and *Neisseria gonorrhoeae.*

Florey and Ernst Chain, first used purified preparations of penicillin to treat bacterial infections. Penicillin may have been present in folk remedies used as early as 600 B.C.E., at around which time molded soybean curd was used by the Chinese to treat boils and carbuncles, and moldy cheese was used by Chinese and Ukrainian peasants to treat infected wounds.

Initially it was thought that penicillin was a pure substance, but further studies revealed that a number of closely related compounds were present in *Penicillium* cultures. Naturally occurring penicillins, such as penicillin G, are most effective against gram-positive bacteria, but much less effective against gram-negative bacteria. A further limitation to the use of Penicillin G is that it is not well absorbed when administered orally. Research programs to produce chemically modified penicillins with improved properties have resulted in a large number of clinically useful penicillin derivatives. Examples of such penicillin derivatives include ampicillin and amoxicillin, which have much greater efficacy against gram-negative bacteria than penicillin, retain good activity against gram-positive bacteria, and are well absorbed when administered orally. The principal adverse reaction associated with the penicillins is the occurrence of allergic response.

biosynthesis: formation of a chemical substance by a living organism

lysis: breakdown of cells; also the favorable termination of a disease

The molecular targets for the antibacterial activity of the penicillin and related β-lactam antibiotics such as the cephalosporins are a group of bacterial enzymes known as *penicillin-binding proteins* (PBPs). The PBPs are essential to the final stages of bacterial cell wall **biosynthesis**. Penicillin and other β-lactam antibiotics inhibit PBPs, thereby inhibiting bacterial cell wall biosynthesis, which eventually results in bacterial cell **lysis**. (Vancomycin and cycloserine are nonpenicillin antibiotics that also inhibit bacterial cell wall biosynthesis through other mechanisms.)

cleave: split

The penicillins and related antibiotics have been among the most widely used therapeutic agents since their introduction into clinical practice in the 1940s. However, the widespread use of these antibiotics has resulted in the emergence and spread of bacteria that are resistant to these agents. A major mechanism of resistance to the penicillin and other β-lactam antibiotics is the bacterial production of β-lactamases, enzymes that **cleave** the β-lactam antibiotics and render them inactive before they can inhibit their PBP targets. Significant efforts have been made to develop β-lactam antibiotics resistant to the β-lactamases, and toward finding inhibitors of the β-lactamases to allow β-lactam antibiotics to be useful antibacterial agents against β-lactamase producing bacteria. SEE ALSO ANTIBIOTICS; ENZYMES; FLEMING, ALEXANDER; INHIBITORS.

William G. Gutheil

Bibliography

Mandell, Gerald L., and Petri, William A. (1996). "Antimicrobial Agents: Penicillins, Cephalosporins, and Other β-Lactam Antibiotics." In *Goodman & Gilman's Pharmacological Basis of Therapeutics*, 9th edition, ed. Joel G. Hardman and Lee E. Limbird. New York: McGraw-Hill.

Nicolaou, K. C., and Boddy, Christopher N. C. (2001). "Behind Enemy Lines." *Scientific American* 284(5):54–61.

Peptide Bond

A peptide bond is a linkage between the building blocks of proteins called amino acids (shorter strings of linked amino acids are known as peptides). A peptide bond forms when the **carboxylic acid** group (R-C[O]OH) of one amino acid reacts with the amine group (R-NH_2) of another. The resulting molecule is an amide with a C–N bond (R-C(O)-NH-R).

While drawn as a single bond, the peptide bond has partial double bond character that enforces a well-defined flat structure. The O atom of the amide has a partial negative charge and is a good **hydrogen bond** acceptor, while the NH is partially positive and a good hydrogen bond donor. Hydrogen bonds between amides are critical to protein folding, as well as to the structure of deoxyribonucleic acid (**DNA**).

The **synthesis** of proteins involves the formation of many peptide bonds. Cleavage of peptide bonds, involved in digestion of proteins and in many regulatory processes, is carried out by enzymes known as proteases. One such protease is subtilisin, the enzyme frequently added to laundry detergent to **cleave** many protein contaminants. **Angiotensin**-converting enzyme (ACE) is an enzyme that targets a specific peptide bond, forming a chemical signal that increases blood pressure. Some blood pressure medications act by blocking ACE. SEE ALSO PRIMARY STRUCTURE; PROTEINS.

Alan Schwabacher

carboxylic acid: one of the characteristic groups of atoms in organic compounds that undergoes characteristic reactions, generally irrespective of where it occurs in the molecule; the –CO_2H functional group

hydrogen bond: interaction between H atoms on one molecule and lone pair electrons on another molecule that constitutes hydrogen bonding

DNA: deoxyribonucleic acid—the natural polymer that stores genetic information in the nucleus of a cell

synthesis: combination of starting materials to form a desired product

cleave: split

angiotensin: chemical that causes a narrowing of blood vessels

Periodicity *See Meyer, Lothar; Periodic Table.*

Periodic Table

The Periodic Table places the symbols of chemical elements, sequenced by **atomic number**, in rows and columns that align similar properties.

atomic number: the number assigned to an atom of an element that indicates the number of protons in the nucleus of that atom

Antiquity through the Renaissance

A few thousand years ago, primitive chemistry focused mostly on converting one substance into another. The word "chemistry" itself is arguably traced to the name of a region of ancient Egypt where such transformation attempts were practiced. Over the centuries, philosophers tried to come to terms with the growing variety of known substances. They postulated the role of fundamental entities that could not be broken down further but formed simple materials when combined. By the time of ancient Greece, Democritus, Leucippus, and Empedocles expounded the nature of matter in terms of constituent elements, the simple substances—earth, air, fire, and water—of which all materials were compounded. The term "atom" first appears in this context.

A millennium or so later, Arab civilizations made great strides in laboratory techniques. Subsequently, during the Renaissance period, these techniques were adopted in trying to transform one element into another, most notably into gold from less costly substances like lead. This gave the Arabic term **"alchemy"** its modern mystical connotation.

alchemy: medieval chemical philosophy having among its asserted aims the transmutation of base metals into gold

Post-Renaissance

By the mid-1700s about twenty elements were known. Science was beginning to get more sophisticated as measurements and instrumentation improved rapidly and theories based on observation grew more advanced and more compelling. Chemists, however, continued to anguish over the inability to easily categorize the elements.

What was likely the first attempt at sorting the elements was a table of simple substances, prepared in 1772 by French chemist Louis-Bernard Guyton de Morveau. French chemist Antoine-Laurent Lavoisier was most influential in developing an experimental approach, which is acknowledged to have laid the foundation for modern chemistry. In 1789, Lavoisier published a list of pure substances that included the known elements but also some compounds and light and heat. By the early 1800s, following the introduction of English chemist John Dalton's **atomic theory** and the concept of atomic weights, the number of known elements had grown. Although properties were carefully measured, confusion held sway when it came to agreeing on the composition of compound substances and the related assignment of atomic weights.

atomic theory: physical concept that asserts that matter is composed of microscopically tiny particles called atoms and that various elements differ from each other by having different atoms

In 1829 German chemist Johann Döbereiner noted that there were triplets of elements in which the central species' properties were almost exactly midway between the outer two. The first example of such a triplet included chlorine, bromine, and iodine. Properties such as atomic weights, color, and reactivity followed this "law of triads" for several such groupings, but not for the entire collection of known elements.

In 1860 Italian chemist Stanislao Cannizzaro presented analyses at an international chemistry meeting that, when merged with previously ridiculed hypotheses by fellow Italian Amedeo Avogadro, yielded unambiguous atomic weights. These eliminated most of the disharmony among property determinations. In attendance were German chemist Lothar Meyer and Russian chemist Dimitri Mendeleev, both of whom were inspired to give the presentation further thought.

Industrial Age

In 1862 French geologist Alexandre-Émile Beguyer de Chancourtois arranged the elements in order of increasing atomic weights, wrapping the series around a cylinder in a helical display. He noted that elements with similar properties lined up, one over the other. His idea was obscured by its publication in a nonchemistry journal, the inclusion of compounds and **alloys** in the discussion, and the publisher's decision not to include an essential diagram.

alloy: mixture of two or more elements, at least one of which is a metal

JOHN NEWLANDS (1837–1898)

John Newlands compared elements to musical notes with his law of octaves. As on a scale, every eighth element would share similar properties when arranged by increasing atomic weight. Newlands did not account for exceptions, however, and it was only upon establishment of the Periodic Table that his theory gained credibility.

—Valerie Borek

но въ ней, мнѣ кажется, уже ясно выражается примѣнимость выставляемаго мною начала ко всей совокупности элементовъ, пай которыхъ извѣстенъ съ достовѣрностію. На этотъ разъ я и желалъ преимущественно найдти общую систему элементовъ. Вотъ этотъ опытъ:

			Ti=50	Zr=90	?=180.
			V=51	Nb=94	Ta=182.
			Cr=52	Mo=96	W=186.
			Mn=55	Rh=104,4	Pt=197,4
			Fe=56	Ru=104,4	Ir=198.
			Ni=Co=59	Pl=106,6,	Os=199.
H=1			Cu=63,4	Ag=108	Hg=200.
	Be=9,4	Mg=24	Zn=65,2	Cd=112	
	B=11	Al=27,4	?=68	Ur=116	Au=197?
	C=12	Si=28	?=70	Sn=118	
	N=14	P=31	As=75	Sb=122	Bi=210
	O=16	S=32	Se=79,4	Te=128?	
	F=19	Cl=35,5	Br=80	I=127	
Li=7	Na=23	K=39	Rb=85,4	Cs=133	Tl=204
		Ca=40	Sr=87,6	Ba=137	Pb=207.
		?=45	Ce=92		
		?Er=56	La=94		
		?Yt=60	Di=95		
		?In=75,6	Th=118?		

а потому приходится въ разныхъ рядахъ имѣть различное измѣненіе разностей, чего нѣтъ въ главныхъ числахъ предлагаемой таблицы. Или же придется предполагать при составленіи системы очень много недостающихъ членовъ. То и другое мало выгодно. Мнѣ кажется притомъ, наиболѣе естественнымъ составить кубическую систему (предлагаемая есть плоскостная), но и попытки для ея образованія не повели къ надлежащимъ результатамъ. Слѣдующія двѣ попытки могутъ показать то разнообразіе сопоставленій, какое возможно при допущеніи основнаго начала, высказаннаго въ этой статьѣ.

Li	Na	K	Cu	Rb	Ag	Cs	—	Tl
7	23	39	63,4	85,4	108	133		204
Be	Mg	Ca	Zn	Sr	Cd	Ba	—	Pb
B	Al	—	—	—	Ur	—	—	Bi?
C	Si	Ti	—	Zr	Sn	—	—	—
N	P	V	As	Nb	Sb	—	Ta	—
O	S	—	Se	—	Te	—	W	—
F	Cl	—	Br	—	J	—	—	—
19	35,5	58	80	190	127	160	190	220.

The first Mendeleev periodic table, 1869.

A few years later, British chemist John Newlands also arranged the elements in order of increasing atomic weights. He was apparently the first to assign hydrogen the weight of "1." Newlands noted that properties repeated when the sequence was broken into periods of seven and referred to his system as the "law of octaves."

During the mid-1860s Meyer took the newly established atomic weights of many elements and arranged them into families that bore similarities in properties, including the ability of an atom to combine with other atoms (valency).

atomic weight: weight of a single atom of an element in atom ic mass units (AMU)

In 1869 Mendeleev presented his table of the elements (sixty-three by now) arrayed in periods of seven for the lighter elements and opening up to seventeen for the heavier elements. Furthermore, Mendeleev had the foresight and confidence to break the **atomic weight** sequence by occasionally forcing elements out of order so as to fall in an appropriate location as determined by their properties. He left gaps in the arrangement at several places and used implied trends to predict characteristics of undiscovered elements needed to complete the table.

Mendeleev's Periodic Table was not well received at first, but was shortly helped by the discovery of the element gallium, which filled such a gap and had nearly exactly the atomic weight, density, and valency predicted. Other affirmations followed. It is for these reasons that Mendeleev is given most of the credit for the invention of the Periodic Table.

In 1892 Scottish chemist William Ramsay discovered two more elements, argon and helium. These unreactive gases did not fit into the Periodic Table. In short order, Ramsay also discovered three more unreactive gases. These gases represented a new family of elements that had to be inserted as an eighth main column in Mendeleev's table.

The Modern Periodic Table

As the twentieth century approached, elements of similar properties were arranged in eight main vertical columns referred to as chemical families. The first such family, or Group I, is collectively termed the "alkali metals," commencing with lithium. The next column, Group II, is designated the "alkaline earths," commencing with beryllium. Groups III through V are commonly referred to as the boron, carbon, and nitrogen families, respectively. The next group, the oxygen family, is technically called the "chalcogens." Group VII, the **"halogens,"** begins with fluorine. Finally, the elements of Group VIII, starting with helium, are called the **"noble gases."** Because of their relative unreactivity, they had once also been termed **inert** gases, a label no longer acceptable.

halogen: element in the periodic family numbered VIIA (or 17 in the modern nomenclature) that includes fluorine, chlorine, bromine, iodine, and astatine

noble gas: element characterized by inert nature; located in the rightmost column in the Periodic Table

inert: incapable of reacting with another substance

In the absence of any understanding as to why the periodic arrangement appeared as it did, or whether or not there were yet more surprises, the science of chemistry remained incomplete, although very important and practical. At almost exactly this time, just before the start of the twentieth century, three findings were announced that changed the course of science: x rays were discovered by German physicist Wilhelm Röntgen in 1895, radioactivity by French physicist Antoine-Henri Becquerel in 1896, and the electron by British physicist Joseph John Thomson in 1897.

First Model

What soon emerged was a nuclear model of the atom, first proposed by New Zealand-born physicist Ernest Rutherford. In this view, an element's identity was determined by its atomic number, the amount of positive charge in the very small core nucleus that also contained almost all of the atom's mass. The light electrons were held in orbits by electrostatic **attraction** to the positive core.

attraction: force that brings two bodies together, such as two oppositely charged bodies

Rutherford's view was extended by Danish physicist Niels Bohr in 1913. Bohr modeled that electrons moved in fixed orbits around the nucleus, much as planets orbit the Sun. Furthermore, not only were the locations of these orbits fixed, but so were the speeds of the electrons in each orbit and the number of electrons that could be accommodated in each orbit, a description called the electron configuration. By explaining the quota of electrons allowed in each fixed orbit, Bohr resorted to a new physical idea called quantization. As a consequence, Bohr was able to reproduce the Periodic Table, adding one electron at a time as one stepped to the next higher element. Bohr argued that orbits of increasing radius could accommodate up to a maximum number of electron numbers that, when reached, corresponded to observed horizontal periods of two, eight, eight, eighteen, eighteen, and thirty-two. Bohr acknowledged the unattractiveness of this approach in that it was merely mimicking an observed pattern rather than addressing the underlying science.

Modern Theory

The mid-1920s witnessed a necessary breakthrough. The revolutionary wave concept of matter was incorporated into a mathematical framework, a new quantum theory, that explained all the properties of a bound electron: its energy, the description of where it could be found, and configuration restraints.

An electron could have only certain energies determined by the value of an integer (a whole number), traditionally symbolized by n with values 1, 2, 3, and so on. Electron energy with $n = 1$ is the lowest possible, $n = 2$ being the next lowest, and so on. The region of space where the electron might be found—called an orbital because it replaced Bohr's planetary fixed orbit idea—could be characterized by its size, shape, and orientation (how the shape might be tilted). For each n, there was a determined set of shapes and orientations with letters used to indicate the shapes. For $n = 1$, only a spherical shape is allowed, symbolized by *s*; since a sphere has no orientation, that is the only $n = 1$ orbital. It is abbreviated as *1s.* For $n = 2$, there are larger orbitals: another *s*, the *2s*, and also dumbbell-shaped orbitals with opposing lobes. These are symbolized as *p* orbitals and have three possible orientations for the *2p* and all other *p* orbitals. By the time $n = 3$ is considered, there is a third shape, *d*, with five orientations. For $n = 4$ there is a fourth shape, *f*, with seven orientations in addition to the *4s*, *4p* and *4d.* The sequence of filling follows a relatively simple pattern shown by arrows in Figure 1.

Very early in the development of modern quantum mechanics, German physicist Wolfgang Pauli realized that each of the substates characterized by n, shape, and orientation was permitted to have no more than two electrons, a feature sometimes pictured as if the electron were spinning and where only two spin orientations were allowed: clockwise and counterclockwise.

The predicted sequence of electron filling might be best illustrated by looking at some examples. Hydrogen is $1s^1$, the superscript referring to the number of electrons in the *1s* substate. Lithium (three electrons) and sodium (eleven electrons) are $1s^2 2s^1$ and $1s^2 2s^2 2p^6 3s^1$, respectively. The latter configuration, for example, corresponds to one pair of *1s*-electrons, one pair of

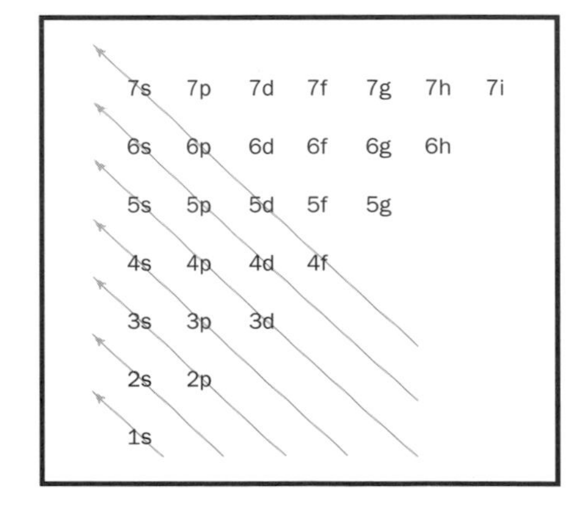

Figure 1. The sequence of substates, indicated by arrows, normally followed in determining electron configurations of the elements. Example: *1s* then *2s* then *2p* then *3s*, etc.

$2s$-electrons, three pairs of $2p$-electrons (six total), and a final $3s$-electron. Neon and argon are $1s^2 2s^2 2p^6$ and $1s^2 2s^2 2p^6 3s^2 3p^6$, respectively. They complete the horizontal periods of length eight.

transition metals: elements with valence electrons in d-sublevels; frequently characterized as metals having the ability to form more than one cation

The periods in which the *d* substates are filling are known as the *d*-block elements or **transition metals**. These ten elements increase the period length to eighteen elements. Some new Periodic Tables have adopted the convention of numbering the columns one through eighteen as a result.

rare earth elements: older name for the lanthanide series of elements, from lanthanum to lutetium

lanthanides: a family of elements (atomic number 57 through 70) from lanthanum to lutetium having from 1 to 14 4f electrons

The *f*-block, whose existence was recognized by American chemist Glenn Seaborg, has two rows containing nearly one-quarter of all the elements. The first row is known as the **rare earth elements** or **lanthanides**. The second *f*-block row is referred to as the actinides. The most common form of the Periodic Table, the Mendeleev-Seaborg form, has the *f*-elements at the bottom. Fourteen *f*-block elements increase the period length to thirty-two.

For nearly three centuries, a new element has been discovered every two-and-one-half years, on average. Undoubtedly, more will be found. Although their names and their discoveries will likely involve controversies, their place at the table is already set. SEE ALSO ALCHEMY; AVOGADRO, AMEDEO; BECQUEREL, ANTOINE-HENRI; BOHR, NIELS; CANNIZZARO, STANISLAO; DALTON, JOHN; LAVOISIER, ANTOINE; MENDELEEV, DIMITRI; MEYER, LOTHAR; PAULI, WOLFGANG; RAMSAY, WILLIAM; RÖNTGEN, WILHELM; RUTHERFORD, ERNEST; SEABORG, GLENN THEODORE; THOMSON, JOSEPH JOHN.

Paul J. Karol

Bibliography

Marshall, James L. (2000). "A Living Periodic Table." *Journal of Chemical Education* 77:979–983.

Mazurs, Edward G. (1974). *Graphic Representations of the Periodic System during One Hundred Years*, revised 2nd edition. University: University of Alabama Press.

van Spronsen, J. W. (1969). *The Periodic System of Chemical Elements: A History of the First Hundred Years*. New York: Elsevier.

Internet Resources

Winter, Mark. "WebElements Period Table." The University of Sheffield and WebElements Ltd., U.K. Available from <http://www.webelements.com>.

Perkin, William Henry

ENGLISH CHEMIST AND CHEMICALS MANUFACTURER
1838–1907

synthesis: combination of starting materials to form a desired product

William Henry Perkin was an entrepreneur and a self-made millionaire at an early age, long before the era of personal computers and dot-coms. His serendipitous **synthesis** of the purple dye mauve (also known as mauveine or aniline purple) in 1856 brought brightly colored clothing to the masses and laid the foundation for today's chemical and pharmaceutical industries.

Perkin was born on March 12, 1838, in London, England. He was a curious boy who liked to play with instruments, tools, and paint. Perkin saw something wonderful in chemistry and dropped his other pursuits after a friend performed for him chemical experiments that yielded crystalline products. A few years later he enrolled at the City of London School and attended chemistry lectures given by Thomas Hall, an instructor at the school. Hall recognized Perkin's ability and arranged for him to enroll at the Royal College of Science, where the German chemist August von Hofmann was a teacher.

Hofmann appointed the seventeen-year-old Perkin as his personal assistant and guided him to work on the synthesis of the antimalarial drug quinine. Perkin had his own ideas for the synthesis of quinine and pursued them in his lab at his parents' home. During Easter break 1856 Perkin ran a reaction with aniline (a compound derived from coal tar) and potassium dichromate that produced a black sludge. Dissolving the sludge in ethyl alcohol, Perkin found that the solution took on an intense purple color. Instead of synthesizing quinine, Perkin had made the first synthetic dye derived from coal tar: mauve.

Perkin undoubtedly appreciated the significance of his discovery, as the worldwide dye and textile industry was the largest chemical industry at that time. Most dyes were derived from natural sources (plants or insects), and chemists were only just beginning to investigate synthetic dyes. Purple was an especially desired color, as expensive natural purple dyes made purple-dyed cloth too expensive for most people. Perkin's discovery was also especially timely, as mauve mania had hit the world a year earlier. Demand for the natural purple dye derived from lichen hit manic proportions (and a cheap, synthetic substitute would be worth vast sums of money).

Perkin left school after patenting his discovery, but promised himself that he would return to research one day. He and members of his family soon formed a company to mass-produce mauve from coal tar, and in 1859 the Perkin and Sons factory commenced production. Mauve mania, however, was short-lived, and within a few years the red dyes fuchsia (or magenta) and alizarin were the craze. Perkin was quick to capitalize on these manias, and made an immense fortune in the process.

But Perkin was not alone: Dye companies quickly sprang up in Austria, England, France, Germany, and Switzerland, and competition became intense. Companies created research subsidiaries that employed hundreds of chemists and found new uses for the flood of compounds being synthesized in their labs. Some of the subsidiaries eventually manufactured pharmaceuticals and explosives.

In 1874 Perkin retired from manufacturing and returned to chemical research. He discovered a reaction (the Perkin reaction) for producing unsaturated carboxylic acids. He also synthesized coumarin, an accomplishment that laid the foundation for the synthetic perfume industry. Perkin died on July 14, 1907, at the age of sixty-nine. SEE ALSO DYES.

Thomas M. Zydowsky

Bibliography

McGrayne, Sharon B. (2001). *Prometheans in the Lab: Chemistry and the Making of the Modern World.* New York: McGraw-Hill.

Internet Resources

"Sir William Henry Perkin." Chemical Heritage Foundation. Available from <http://www.chemheritage.org/perkin>.

Pesticides

Pesticides are chemicals that kill pests, and are categorized by the types of pests they kill. For example, insecticides kill insects, herbicides kill weeds, bactericides kill bacteria, fungicides kill fungi, and algicides kill algae.

A worker is spraying pesticides on ferns in order to eradicate insects and other pests.

Approximately 90 percent of all pesticides used worldwide are used in agriculture, food storage, or shipping. Because of a growing world population, there is pressure to increase and preserve the food supply by using pesticides and other agricultural chemicals.

History of Pesticides

Throughout history, various types of pests, such as insects, weeds, bacteria, rodents, and other biological organisms, have bothered humans or threatened human health. People have been using pesticides for thousands of years to try to control these pests. The Sumerians used sulfur to control insects and mites 5,000 years ago. The Chinese used mercury and arsenic compounds to control body lice and other pests. The Greeks and Romans used oil, ash, sulfur, and other materials to protect themselves, their livestock, and their crops from various pests. And people in various cultures have used smoke, salt, spices, and insect-repelling plants to preserve food and keep pests away.

Classes of Pesticides

Although the use of pesticides is not new, the types of substances people have used as pesticides have changed over time. The earliest pesticides were inorganic substances such as sulfur, mercury, lead, arsenic, and ash. Some of these inorganic pesticides are still used today. For example, sulfur is still used as a fungicide, copper is used as an algicide, lead and arsenic were used as insecticides until World War II, and chromium, copper, and arsenic have been used as wood preservatives to prevent microorganisms from causing wood decay. Even though many of these substances are effective pesticides, the use of some of these materials has been banned or restricted because of health and environmental concerns. Lead and arsenic are no longer used as insecticides, the use of mercury as a fungicide has been restricted, and the U.S. Environmental Protection Agency (EPA) is phasing out the use of arsenic as a wood preservative.

The modern era of chemical pest control began around the time of World War II, when the synthetic organic chemical industry began to develop. The first synthetic organic pesticides were organochlorine compounds, such as dichlorodiphenyltrichloroethane (DDT). Commercial production of DDT began in 1943. At that time, DDT was considered to be a wonderful invention. It was cheap to produce, very toxic to insects, and much less toxic to mammals. DDT and other organochlorine insecticides were used for many years to control mosquitoes and as a broad-spectrum insecticide against insect pests that damaged food and crops. Unfortunately, scientists learned later that many organochlorine insecticides were persistent in the environment (they did not **degrade** readily) and were bioaccumulating in birds, humans, and other animals. In 1962 Rachel Carson wrote the book *Silent Spring,* in which she reported that DDT was causing eggshell thinning in bird eggs and thus was leading to the near extinction of bird species such as peregrine falcons and bald eagles. Today most of the organochlorine pesticides have been banned in the United States by the EPA because of the tendency of these compounds to persist in the environment and bioaccumulate in animals.

degrade: to decompose or reduce to complexity of a chemical

Other classes of insecticides include the organophosphates, carbamates, pyrethroids, and biopesticides. These other classes of pesticides are not as persistent in the environment as the organochlorine pesticides. The organophosphate and carbamate pesticides affect the nervous system by disrupting the enzyme that regulates **acetylcholine**, a neurotransmitter. However, carbamate pesticides are less toxic to humans because their interactions with important enzymes are reversible. As a group, the organophosphate and carbamate pesticides are probably the most widely used insecticides, although many are being restricted by the EPA because of their toxicity.

acetylcholine: neurotransmitter with the chemical forumula $C_7H_{17}NO_3$; it assists in communication between nerve cells in the brain and central nervous system

Pyrethroid pesticides were developed as synthetic versions of the naturally occurring pesticide pyrethrin, which is found in chrysanthemums. Most pyrethroids are safer than the organochlorines, organophosphates, and carbamates, although some synthetic pyrethroids are toxic to the nervous system. Pyrethroids have been modified to increase their stability in the environment, and many different pyrethroids are being used today.

Biopesticides are substances that are derived from such natural materials as animals, plants, bacteria, and certain minerals. For example, canola oil and baking soda have pesticidal applications and are considered biopesticides. Biopesticides fall into three major classes, including microbial pesticides, plant-incorporated protectants, and biochemical pesticides. Microbial pesticides contain microorganisms, such as bacteria, fungi, and viruses, as their active ingredient. The most widely used microbial pesticides are strains of *Bacillus thuringiensis,* or Bt. Plant-incorporated protectants are pesticidal substances that plants produce from genetic material that has been added

RACHEL CARSON (1907–1964)

The role of chemistry became irreversibly intertwined with the environment in 1962 when the term "ecosystem" was introduced in Rachel Carson's *Silent Spring.* One of four works written by Carson, it targeted the now banned pesticide dichlorodiphenyltrichloroethane (DDT), spawning a movement that resulted in the formation of the U.S. Environmental Protection Agency (EPA).

—Valerie Borek

to the plant. Biochemical pesticides are naturally occurring substances that control pests by nontoxic mechanisms. Conventional pesticides, by contrast, are generally synthetic materials that directly kill or inactivate the pest. Biochemical pesticides include substances, such as insect sex pheromones, that interfere with mating, as well as various scented plant extracts that attract insect pests to traps. Because it is sometimes difficult to determine whether a substance meets the criteria for classification as a biochemical pesticide, the EPA has established a special committee to make such decisions.

Pesticide Residues

Pesticide residues are the materials that remain on plants and food when a crop is treated with a pesticide. The U.S. government establishes safe residue levels, called tolerances or maximum residue levels, for each food commodity. However, the presence of pesticide residues in food has been a public concern. There has also been a concern about pesticide residues in water, air, and soil. In response to this concern, the U.S. Congress passed the Food Quality Protection Act in 1996, which has had an impact on safety standards for pesticides.

Approaches to pest management have changed significantly since the 1950s and will continue to change as scientists learn more about the toxicity and environmental behavior of pesticides. Scientists will continue to develop newer approaches to insect pest management that are considered to be safer than the use of broad-spectrum pesticides. The most effective strategy for controlling pests may be to combine methods in an approach known as integrated pest management (IPM), which emphasizes preventing pest damage. In IPM information about pests and available pest control methods is used to manage pest damage by the most economical means and causing the least possible hazard to people, property, and the environment. Methods for pest management will continue to evolve as scientists conduct research and develop new information. SEE ALSO HERBICIDES; INSECTICIDES.

Cynthia Atterholt

Bibliography

Cardé, R. (1990). "Principles of Mating Disruption." In *Behavior-Modifying Chemicals for Insect Management: Applications of Pheromones and Other Attractants*, ed. Richard L. Ridgway, Robert M. Silverstein, and May N. Inscoe. New York: Marcel Dekker.

Copping, L., and Hewitt, H. G. (1998). *Chemistry and Mode of Action of Crop Protection Agents.* Cambridge, U.K.: The Royal Society of Chemistry.

Cunningham, W., and Saigo, B. (2001). "Pest Control." In *Environmental Science: A Global Concern*, 6th edition. New York: McGraw-Hill.

Ecobichon, D. (1991). "Toxic Effects of Pesticides." In *Casarett and Doull's Toxicology: The Basic Science of Poisons*, 4th edition, ed. M. Amdur, J. Doull, and C. Klaassen. Elmsford, NY: Pergamon Press.

Farrell, K.; Flint, M.; Lyons, J.; Madden, J.; Schroth, M.; Weinhold, A.; White, J.; Zalom, F.; and Jaley, M. (1992). *Beyond Pesticides: Biological Approaches to Pest Management in California: Executive Summary.* Oakland: University of California.

Kydonieus, A., and Beroza, M. (1982). "Pheromones and Their Use." In *Insect Suppression with Controlled Release Pheromone Systems*, Vol. 1, ed. A. Kydonieus and M. Beroza. Boca Raton, FL: CRC Press.

Leonhardt, Barbara, and Beroza, Morton, eds. (1982). *Insect Pheromone Technology: Chemistry and Applications.* Washington, DC: American Chemical Society.

Lewis, D., and Cowsar, D. (1977). "Principles of Controlled Release Pesticides." In *Controlled Release Pesticides: A Symposium*, ed. Herbert Scher. Washington, DC: American Chemical Society.

Rice, R., and Kirsch, P. (1990). "Mating Disruption of Oriental Fruit Moth in the United States." In *Behavior-Modifying Chemicals for Insect Management: Applications of Pheromones and Other Attractants*, ed. Richard L. Ridgway, Robert M. Silverstein, and May N. Inscoe. New York: Marcel Dekker.

Scher, H., ed. (1977). *Controlled Release Pesticides.* Washington, DC: American Chemical Society.

Weatherston, I. (1990). "Principles of Design of Controlled-Release Formulations." In *Behavior-Modifying Chemicals for Insect Management: Applications of Pheromones and Other Attractants*, ed. Richard L. Ridgway, Robert M. Silverstein, and May N. Inscoe. New York: Marcel Dekker.

Wheeler, W. (2002). "Role of Research and Regulation in 50 Years of Pest Management in Agriculture." *Journal of Agricultural and Food Chemistry* 50:4151–4155.

Internet Resources

U.S. Environmental Protection Agency. "Pesticides." Available from <http://www.epa.gov/pesticides>.

Petroleum

Petroleum is a naturally occurring complex mixture made up predominantly of carbon and hydrogen compounds, but also frequently containing significant amounts of nitrogen, sulfur, and oxygen together with smaller amounts of nickel, vanadium, and other elements. Solid petroleum is called asphalt; liquid, crude oil; and gas, natural gas. Its source is biological. Organic matter buried in an oxygen-deficient environment and subject to elevated temperature and pressure for millions of years generates petroleum as an **intermediate** in the transformation that ultimately leads to methane and graphite. The first successful drilled oil well came in 1859 in Pennsylvania. This is considered to be the beginning of the modern oil industry. Continuous distillation of crude oil began in Russia in 1875.

intermediate: molecule, often short-lived, that occurs while a chemical reaction progresses but is not present when the reaction is complete

Occurrence

Oil is the largest segment of our energy raw materials use, being 40 percent, while coal use accounts for 27 percent, gas 21 percent, and hydroelectric/nuclear 12 percent. Although there are 20,000 petroleum fields known worldwide, more than half of the known reserves are contained in the 51 largest fields. The Middle East has 66 percent of the known world reserves. The United States has only 2 percent of the known world reserves. Hence the need for imports. The Organization of Petroleum Exporting Countries (OPEC) is important to the international trade and distribution of this crude oil. There is a growing dependence of the United States on imports. Although U.S. domestic production has not grown since the 1950s, imports have grown dramatically, from 0.3 billion barrels of oil in 1955 to 3.0 billion barrels in 1997. The United States has increased its percentage of imports, from approximately 13 percent in 1970 to 55 percent in 2000. It uses approximately 18 million barrels of oil per day. Worldwide production is about 56 million barrels per day. With known reserves, this level of worldwide production could remain constant for only 43 years. But there are large volumes of unconventional petroleum reserves, such as heavy oil, tar sands, and oil shale. These are located in the Western Hemisphere. Improvements in recovery methods must be made, and the cost of production must decrease, for these sources to become more important providers of energy.

The world's first oil well, near Titusville, Pennsylvania, 1863.

Composition

Crude oils vary dramatically in color, odor, and flow properties. There are light and heavy crude oils; they are sweet or sour (i.e., have high or low sulfur content, with an average of 0.65%). Several thousand compounds are present in petroleum. The number of carbon atoms in these compounds can vary from one to over a hundred. Few are separated as pure substances. Many of the demands for petroleum can be served by certain fractions obtained from the distillation of crude oil. Typical distillation fractions and their uses are given in Table 1. The complexity of the molecules, their molecular weights, and their carbon numbers increase with the boiling point. The higher-boiling fractions are usually distilled in vacuo at temperatures lower than their atmospheric boiling points to avoid excessive decomposition to tars.

Each fraction of distilled petroleum is a complex mixture of chemicals, but these mixtures can be somewhat categorized. A certain sample of straight-run gasoline (light naphtha) might contain nearly 30 aliphatic (containing no benzene ring), noncyclic hydrocarbons; nearly 20 cycloaliphatic hydrocarbons (mainly cyclopentanes and cyclohexanes), sometimes called

Table 1.

FRACTIONS OF PETROLEUM

Approximate bp (°C)	Name	Uses
<20°C	Gases	Similar to natural gas and useful for fuel and chemicals.
20–150°C	Light naphtha (C_5–C_6)	Fuel and chemicals, especially gasoline.
150–200°C	Heavy naphtha (C_7–C_9)	Fuel and chemicals.
175–275°C	Kerosene (C_9–C_{16})	Jet, tractor, and heating fuel.
200–400°C	Gas oil (C_{15}–C_{25})	Diesel and heating fuel. Catalytically cracked to naphtha and steam-cracked to alkenes.
>350°C	Lubricating oil	Lubrication. May be catalytically cracked to lighter fractions.
>350°C	Heavy fuel oil	Boiler fuel. May be catalytically cracked to lighter fractions.
	Asphalt	Paving, coating, and structural uses.

SOURCE: Wittcoff, Harold A., and Reuben, Bryan G. (1996). *Industrial Organic Chemicals.* New York: John Wiley.

naphthenes; and 20 aromatic compounds (such as benzene, toluene, and xylene). Examples of compounds found or used in petroleum and mentioned in this article are given in Figure 1.

When any fraction of petroleum is used as a source of energy and burned to CO_2 and H_2O, the sulfur is converted into SO_2 in the air. The SO_2 is a major air contaminant, especially in larger cities. With air moisture it can form H_2SO_4 and H_2SO_3. Much of the sulfur-containing material must be taken out of petroleum before it can be used as fuel. The current maximum percentage allowable in gasoline is 0.10 percent S.

Octane Number

One cannot talk about the chemistry of gasoline without understanding octane numbers. When gasoline is burned in an internal **combustion** engine to CO_2 and H_2O, there is a tendency for many gasoline mixtures to burn unevenly. Such nonconstant and unsmooth combustion creates a "knocking" noise in the engine. Knocking signifies that the engine is not running as efficiently as it could. It has been found that certain hydrocarbons burn more smoothly than others in a gasoline mixture. In 1927 a scale that attempted to define the "antiknock" properties of gasolines was created. At that time, 2,2,4-trimethylpentane (commonly called "isooctane") was the hydrocarbon that, when burned pure in an engine, gave the best antiknock properties (caused the least knocking). This compound was assigned the number 100, meaning it was the best hydrocarbon to use. The worst hydrocarbon researchers could find in gasoline (which when burned pure gave the most knocking) was *n*-heptane, assigned the number 0. When isooctane and heptane were mixed, they gave different amounts of knocking depending on their ratio: The higher the percentage of isooctane in the mixture, the lower was the amount of knocking. Gasoline mixtures obtained from petroleum were burned for comparison. If a certain gasoline has the same amount of knocking as a 90 percent isooctane, 10 percent heptane (by volume) mixture, we now say that its "octane number" is 90. Hence, the octane number of a gasoline is the percent isooctane in an isooctane-heptane

combustion: burning, the reaction with oxygen

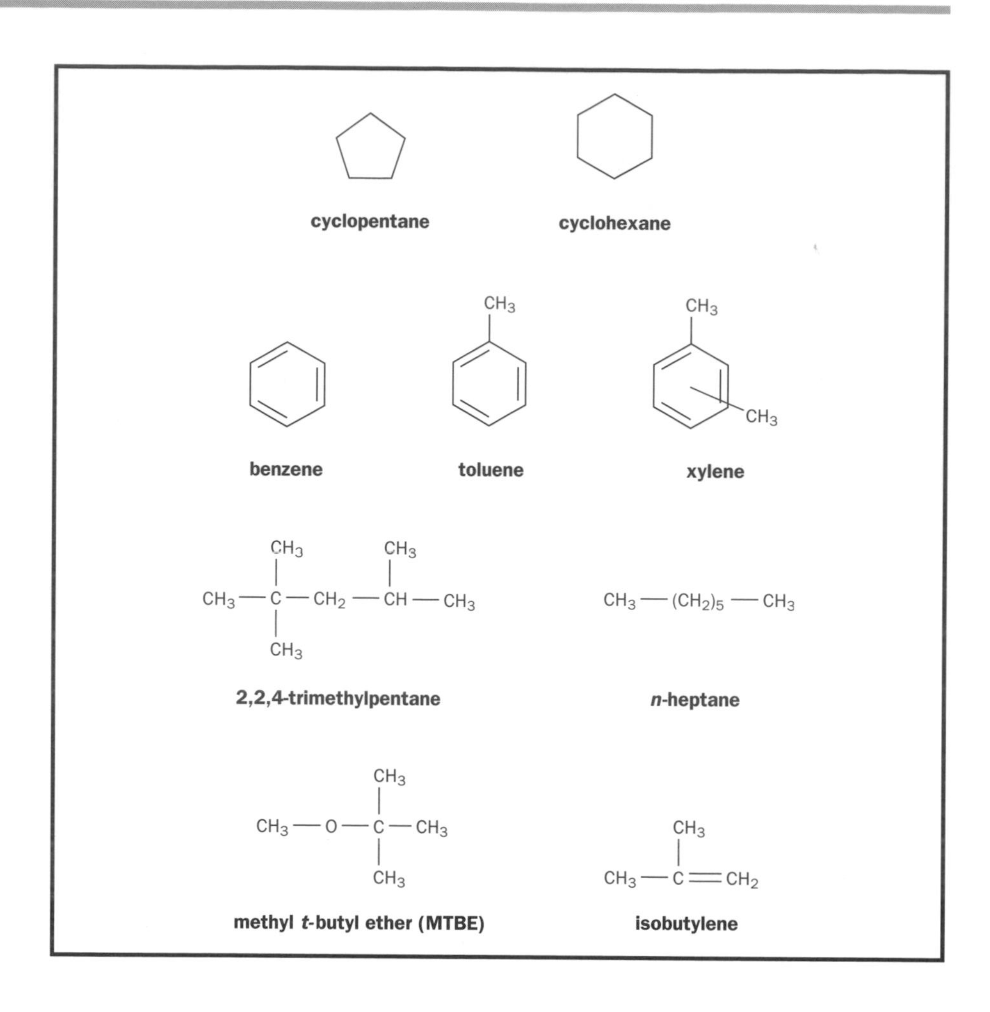

Figure 1. Some compounds found or used in petroleum.

mixture that gives the same amount of knocking as the gasoline being measured. Thus, a high octane number means a low amount of knocking.

Presently there are two octane scales, a research octane number (RON) and a motor octane number (MON). RON values reflect performance at 600 rpm, 148.8°C (125°F), and low speed. MON is a performance index of driving with 900 rpm, 51°C (300°F), and high speed. The station pumps now give the (R + M)/2 value. Regular is usually 87 to 89 and premium about 92 on this scale.

Certain rules have been developed for predicting the octane number of different types of gasoline, depending on the ratio of different types of hydrocarbons in the mixtures:

1. The octane number increases as the amount of branching or the number of rings increases.
2. The octane number increases as the number of double and triple bonds increases.

Additives

In 1922 two chemists working at General Motors, Midgley and Boyd, were looking at different substances that would aid the combustion of gasoline and help the knocking problems of engines. In other words, they were seeking methods of increasing the octane rating of gasoline without altering the

An oil refinery at Cap Bon, Tunisia.

hydrocarbon makeup. They were also interested in cleaning up the exhaust of automobiles by eliminating pollutants such as unburned hydrocarbons and carbon monoxide through more complete combustion. By far the best substance that they found was tetraethyllead. Lead in this form aids in breaking carbon-carbon and carbon-**hydrogen bonds**. But the lead oxide formed in the combustion is not **volatile** and would accumulate in the engine if dibromoethane and dichloroethane were not added. In the environment the lead dihalide formed undergoes reaction by sunlight to elemental lead and **halogen**, both of which are serious pollutants.

hydrogen bond: interaction between H atoms on one molecule and lone pair electrons on another molecule that constitutes hydrogen bonding

volatile: low boiling, readily vaporized

halogen: element in the periodic family numbered VIIA (or 17 in the modern nomenclature) that includes fluorine, chlorine, bromine, iodine, and astatine

For the past several years other additives have been tried. Ethyl alcohol has become popular. When 10 percent ethyl alcohol is mixed with gasoline it is called gasohol and it is popular in states with good corn crops, as the alcohol can be made from corn fermentation. An attractive alternative to tetraethyllead is now methyl *t*-butyl ether (MTBE). MTBE has been approved at the 7 percent level since 1979. From 1984 to 1995 its production grew by 25 percent per year, the largest increase of any of the top chemicals. The Clean Air Act of 1991 specifies that the gasoline must be at the 2.0 percent oxygen level. Thus, MTBE, ethyl *t*-butyl ether (ETBE), ethanol, methanol, and other ethers and alcohols had to be added to gasoline at higher levels. The product is called reformulated gasoline (RFG), and it may cut carbon monoxide levels and may help to alleviate ozone depletion. But improved

Figure 2. Petroleum refining processes.

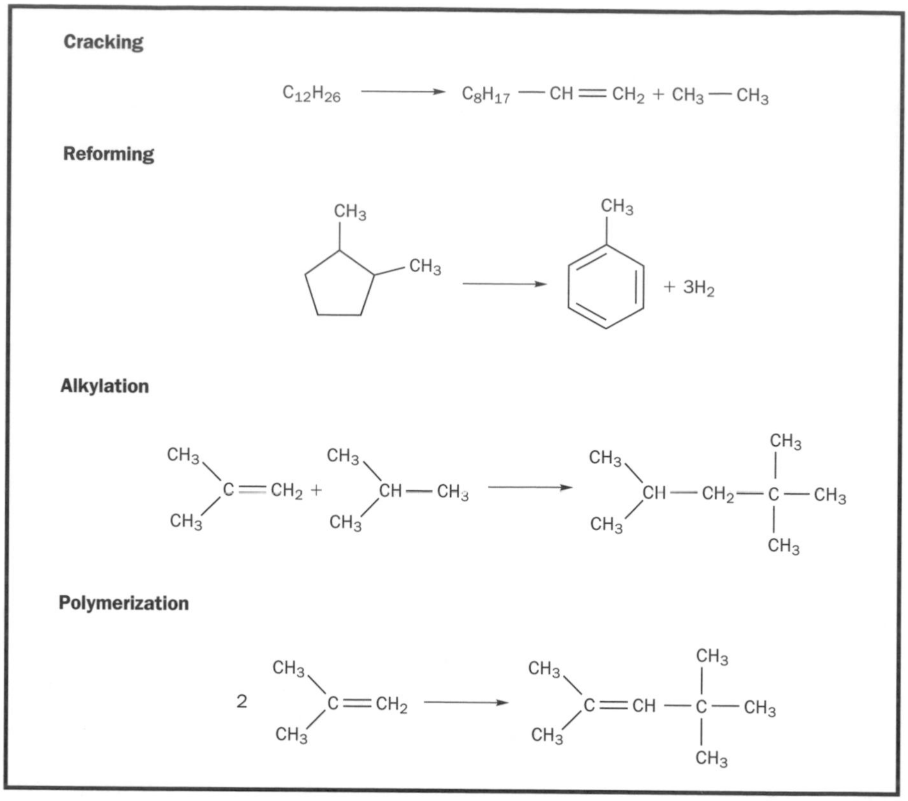

Cracking

$$C_{12}H_{26} \longrightarrow C_8H_{17}-CH=CH_2 + CH_3-CH_3$$

Reforming

CH_3, CH_3 (cyclopentane ring) $\longrightarrow$ CH_3 (benzene ring) $+ 3H_2$

Alkylation

$$(CH_3)_2C=CH_2 + (CH_3)_2CH-CH_3 \longrightarrow (CH_3)_2CH-CH_2-C(CH_3)_2-CH_3$$

Polymerization

$$2\ (CH_3)_2C=CH_2 \longrightarrow (CH_3)_2C=CH-C(CH_3)_2-CH_3$$

emission control systems may make this high-level input unnecessary. Currently MTBE accounts for 85 percent of the additive market, with 7 percent being ethanol and the remaining 8 percent split by other chemicals. In 1999 California took steps toward banning MTBE. In 2000 some factions called for a U.S. ban on MTBE and for increased use of ethanol to meet the oxygenate requirement. MTBE has been found in drinking water. But ethanol cannot be blended into gasoline at the refinery because it is hygroscopic and picks up traces of water in pipelines and storage tanks. Also, ethanol shipped away from the Midwest, where it is made by corn fermentation, would add to the cost of gasoline. Gasohol may increase air pollution because gasoline containing ethanol evaporates more quickly. Studies and debate continue.

Refinery Processes

There are processes that are used to refine petroleum into useful products. These are important processes for the gasoline fraction because they increase the octane rating. Some of these processes are used to increase the percentage of crude oil that can be used for gasoline. They were developed in the 1930s when the need for gasoline became great with the growing automobile industry. These processes are also keys in the production of organic chemicals. An example of each of these processes is given in Figure 2. One process is cracking. In **catalytic** cracking, as the name implies, petroleum fractions of higher molecular weight than gasoline can be heated with a **catalyst** and cracked into smaller molecules. This material can then be blended into the refinery gasoline feed.

catalysis: the action or effect of a substance in increasing the rate of a reaction without itself being converted

catalyst: substance that aids in a reaction while retaining its own chemical identity

feedstock: mixture of raw materials necessary to carry out chemical reactions

Catalytic reforming leaves the number of carbon atoms in the **feedstock** molecules usually unchanged, but the resultant mixture contains a higher number of double bonds and aromatic rings. Reforming has become the principal process for upgrading gasoline. High temperatures with typical catalysts of platinum and/or rhenium on alumina and short contact times are used. A typical example is the reforming of dimethylcyclopentane to toluene. Straight-run gasoline can be reformed to as high as 40 to 50 percent aromatic hydrocarbons, of which 15 to 20 percent is toluene.

Although cracking and reforming are by far the most important refinery processes, especially for the production of petrochemicals, two other processes deserve mention. In alkylation, alkanes (hydrocarbons with no double or triple bonds) react with alkenes (hydrocarbons with double bonds) in the presence of an acid catalyst to give highly branched alkanes. In polymerization an alkene can react with another alkene to generate dimers, trimers, and tetramers of the alkene. As an example, isobutylene (C_4) reacts to give a highly branched C_8 alkene dimer.

Natural Gas

Natural gas can be as high as 97 percent methane, the remainder being hydrogen, ethane, propane, butane, nitrogen, hydrogen sulfide, and heavier hydrocarbons. A typical mixture contains 85 percent methane, 9 percent ethane, 3 percent propane, 1 percent butanes, and 1 percent nitrogen. Uses of natural gas by all industry include fuel (72%) and the manufacture of: inorganic chemicals including ammonia (15%), organic chemicals (12%), and carbon black (1%). The ethane and propane are converted to ethylene and propylene. The methane is purified and used to make a number of other chemicals. SEE ALSO Energy Sources and Production; Fire, Fuels, Power Plants; Fossil Fuels; Gasoline; Industrial Chemistry, Organic.

Philip J. Chenier

Bibliography

Chenier, Philip J. (2002). *Survey of Industrial Chemistry*, 3rd edition. New York: Kluwer Academic/Plenum Publishers.

Wittcoff, Harold A., and Reuben, Bryan G. (1996). *Industrial Organic Chemicals*. New York: Wiley.

Pharmaceutical Chemistry

Pharmaceutical chemists are involved in the development and assessment of therapeutic compounds. Pharmaceutical chemistry encompasses drug design, drug **synthesis**, and the evaluation of drug efficacy (how effective it is in treating a condition) and drug safety. Prior to the nineteenth century, schools of pharmacy trained pharmacists and physicians how to prepare medicinal remedies from natural organic products or inorganic materials. Herbal medications and folk remedies dating back to ancient Egyptian, Greek, Roman, and Asian societies were administered without any knowledge of their biological mechanism of action. It was not until the early 1800s that scientists began extracting chemicals from plants with purported therapeutic properties to isolate the active components and identify them. By discovering and structurally characterizing compounds with medicinal activity, chemists are able to design new drugs with enhanced potency and decreased adverse side effects.

synthesis: combination of starting materials to form a desired product

Drug discovery is the core of pharmaceutical chemistry. The drug discovery process includes all the stages of drug development, from targeting a disease or medical condition to toxicity studies in animals, or even, by some definitions, testing the drug on human subjects. Typically, conditions that affect a larger percentage of the population receive more attention and more research funding. Antiulcer drugs and cholesterol-**reducing agents** are currently the therapeutic areas of greatest emphasis. To develop a drug to target a specific disease, researchers try to understand the biological mechanism responsible for that condition. If the biochemical pathways leading up to the disease are understood, scientists attempt to design drugs that will block one or several of the steps of the disease's progress. Alternatively, drugs that boost the body's own defense mechanism may be appropriate.

reducing agent: substance that causes reduction, a process during which electrons are lost (or hydrogen atoms gained)

How do chemists "discover" drugs? Often there is an existing remedy for a condition, and scientists will evaluate how that drug exerts its actions. Once the drug's structure is known, the drug can serve as a prototype or "lead compound" for designing more effective therapeutic agents of similar chemical structure. Lead compounds are molecules that have some biological activity with respect to the condition under investigation. However, the lead compound may not be effective in combating the disease, or it may produce undesirable side effects. Lead optimization involves chemical modifications to the lead compound to produce a more potent drug, or one with fewer or decreased adverse effects.

Computers have transformed the drug discovery process. Rational drug design involves computer-assisted approaches to designing molecules with desired chemical properties. Rational drug design is based on a molecular understanding of the interactions between the drug and its target in biological systems. Molecular modeling software depicts three-dimensional images of a chemical. Mathematical operations adjust the positions of the atoms in the molecule in an attempt to accurately portray the size and shape of the drug, and the location of any charged groups. Chemists can vary the atoms or groups within the model and predict the effect the transformation has on the molecular properties of the drug. In this way, new compounds can be designed.

Advances in technology have made it possible for medicinal chemists to synthesize a vast number of compounds in a relatively short time, a process referred to as combinatorial chemistry. In this technique, one part of a molecule is maintained, as different chemical groups are attached to its molecular framework to produce a series of similar molecules with distinct structural variations. Combinatorial libraries of these molecular variants are thus created.

Every chemical that is synthesized must be tested for biological activity. **In vitro** testing involves biological assays outside a living system. For example, if the desired effect of a drug is to inhibit a particular enzyme, the enzyme can be isolated from an organ and studied in a test tube. New technologies have made it possible to assay large numbers of compounds in a short period. High-throughput drug screening allows pharmaceutical chemists to test between 1,000 and 100,000 chemicals in a single day! A compound that demonstrates some biological activity will undergo further tests, or it may be chemically modified to enhance its activity. As a consequence, chemical libraries consisting of potentially therapeutic compounds are developed. Each of these compounds can then serve as leads for the development of new drugs to be screened.

in vitro: Latin, meaning "in glass" and applied to experiments done under artificial conditions, that is, in a test tube

Once a drug shows promise in vitro as a therapeutic agent, it must also be screened for toxic properties. Adverse drug side effects are often due to the interaction of the drug with biological molecules other than the desired target. It is very rare that a drug interacts with only one type of molecule in a living system. Drug selectivity refers to the ability of the compound to interact with its target, not with other proteins or enzymes in the system. To investigate drug toxicity, animal studies are performed. These studies also estimate mutagenicity, that is, whether the compound under investigation damages genetic material.

Rarely does a drug pass through a biological system unchanged. Most drugs undergo chemical transformations (in a process known as drug **metabolism**) before they are **excreted** from the body. The drug transformation products (**metabolites**) must be identified so that their toxicological profiles can be determined.

metabolism: the complete range of biochemical processes that take place within living organisms; comprises processes that produce complex substances from simpler components, with a consequent use of energy (anabolism), and those that break down complex food molecules, thus liberating energy (catabolism)

excrete: to eliminate or discharge from a living entity

metabolites: products of biological activity that are important in metabolism

Since the 1970s more attention has been given to drug formulation and methods of drug delivery. Historically, drugs have been administered orally, as a pill or a liquid, or in an injectable form. The goal of drug-delivery systems is to enable controlled and targeted drug release. Today, many medications are commonly introduced as inhalants or in a time-release formulation, either encapsulated in a biodegradable polymer or by means of a transdermal patch.

Once scientists and government regulatory agencies have determined the drug candidate to be relatively safe, it can enter into clinical trials. The clinical stage involves four phases of testing on human volunteers. Animal studies and in vitro testing continue during clinical investigations of a drug. Drug-therapy evaluation is very costly and time consuming. Phase I clinical trials evaluate drug tolerance and safety in a small group of healthy adult volunteers. Phase II trials continue to assess the drug's safety and effectiveness in a larger population. Volunteers participating in phase I trials understand that they are receiving experimental therapy. While those patients involved in phase II clinical trials are made aware of the medication and any known side effects, some of the volunteers may be administered a placebo (a compound with no pharmacological activity against the condition being treated) rather than the drug being studied. In a blind study, only the physician administering therapy knows whether the patient is receiving the drug or a placebo. Both groups of patients are monitored, and physicians or clinicians evaluate whether there is significant improvement in the condition of the group receiving the experimental drug, compared with those individuals who were administered a placebo. In a double-blind study, neither the physician nor the patient knows whether the drug, a placebo, or a related remedy has been administered. Therapy is monitored by an outside group.

Phase III and phase IV clinical trials involve larger populations. During phase III trials, which can last two to eight years, a drug is often brought to market. Phase IV studies continue after the drug is being marketed.

The field of pharmaceutical chemistry is diverse and involves many areas of expertise. Natural-product and analytical chemists isolate and identify active components from plant and other natural sources. Theoretical chemists construct molecular models of existing drugs to evaluate their properties.

These computational studies help medicinal chemists and bioengineers design and synthesize compounds with enhanced biological activity. Pharmaceutical chemists evaluate the bioactivity of drugs and drug metabolites. Toxicologists assess drug safety and potential adverse effects of drug therapy. When a drug has been approved for human studies, clinicians and physicians monitor patients' response to treatment with the new drug. The impact of pharmaceutical chemistry on the normal human life span and on the quality of life enjoyed by most people is hard to overestimate. SEE ALSO COMBINATORIAL CHEMISTRY; COMPUTATIONAL CHEMISTRY; MOLECULAR MODELING.

Nanette M. Wachter

Bibliography

Vogelson, Cullen T. (2001). "Advances in Drug Delivery Systems." *Modern Drug Discovery* 4(4):49–50, 52.

Williams, David A., and Lemke, Thomas L. (2002). *Foye's Principles of Medicinal Chemistry*, 5th edition. Philadelphia: Lippincott Williams & Wilkins.

Wolff, Manfred E., ed. (1996). *Burger's Medicinal Chemistry and Drug Discovery*, 5th edition. New York: Wiley.

Phospholipids

Phospholipids are an important class of biomolecules. Phospholipids are the fundamental building blocks of cellular membranes and are the major part of **surfactant**, the film that occupies the air/liquid interfaces in the lung. These molecules consist of a polar or charged head group and a pair of **nonpolar** fatty acid tails, connected via a glycerol linkage. This combination of polar and nonpolar segments is termed amphiphilic, and the word describes the tendency of these molecules to assemble at interfaces between polar and nonpolar phases.

surfactants: surface-active agents that lower the surface tension of water; e.g., detergents, wetting agents

nonpolar: molecule, or portion of a molecule, that does not have a permanent, electric dipole

The structure of the most common class of phospholipids, phosphoglycerides, is based on glycerol, a three-carbon alcohol with the formula $CH_2OH–CHOH–CH_2OH$. Two fatty acid chains, each typically having an even number of carbon atoms between 14 and 20, attach (via a dual **esterification**) to the first and second carbons of the glycerol molecule, denoted as the sn1 and sn2 positions, respectively. The third hydroxyl group of glycerol, at position sn3, reacts with phosphoric acid to form phosphatidate. Common phospholipids, widely distributed in nature, are produced by further reaction of the phosphate group in phosphatidate with an alcohol, such as serine, ethanolamine, choline, glyercol, or inositol. The resulting lipids may be charged, for example, phosphatidyl serine (PS), phosphatidyl inositol (PI), and phosphatidyl glyercol (PG); or dipolar (having separate positively and negatively charged regions), for example, phosphatidyl choline (PC), and phosphatidyl ethanolamine (PE). The term "lecithin" refers to PC-type lipids. A typical phospholipid arrangement is the presence of a saturated fatty acid, such as palmitic or stearic acid, at the sn1 position, and an unsaturated or polyunsaturated fatty acid, such as oleic or arachodonic acid, at sn2 (see Figure 1 for the structure of a phosphoglyceride).

esterification: chemical reaction in which esters (RCO_2R') are formed from alcohols (R_1OH) and carboxylic acids (RCO_2H)

Another class of phospholipids is the sphingolipids. A sphingolipid molecule has the phosphatidyl-based headgroup structure described above, but (in contrast to a common phospholipid molecule) contains a single fatty acid

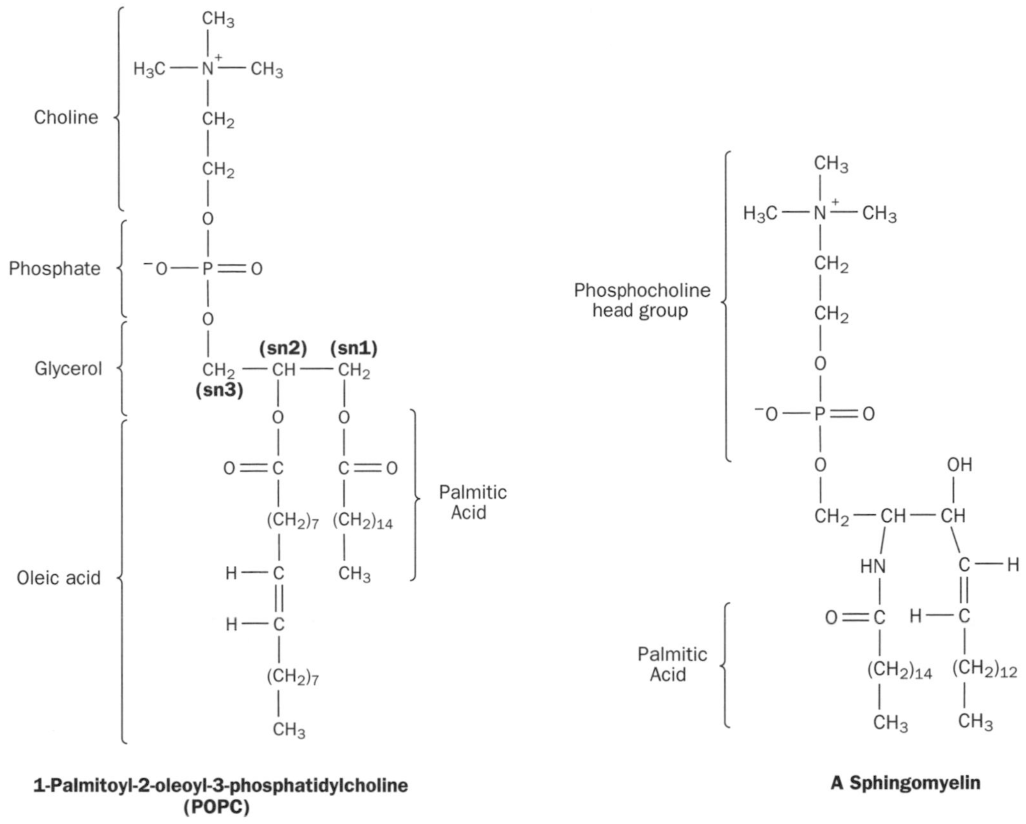

1-Palmitoyl-2-oleoyl-3-phosphatidylcholine (POPC)

A Sphingomyelin

and a long-chain alcohol as its **hydrophobic** components. Additionally, the backbone of the sphingolipid is sphingosine, an **amino alcohol** (rather than glyercol). The structure of a representative sphingolipid, sphingomyelin, is also shown in Figure 1. Sphingolipids, occurring primarily in nervous tissue, are thought to form cholesterol-rich domains within lipid bilayer membranes that may be important to the functions of some membrane proteins.

Phospholipids have many functions in biological systems: as fuels, as membrane structural elements, as signaling agents, and as surfactants. For example, **pulmonary surfactant** is a mixture of lipids (primarily dipalmitoyl phosphatidyl choline [DPPC]) and proteins that controls the surface tension of the fluid lining of the inner lung (the site of gas exchange), allowing rapid expansion and compression of this lining during the breathing cycle. Phospholipids are the major lipid constituent in cell membranes, thus maintaining structural integrity between the cell and its environment and providing boundaries between compartments within the cell. SEE ALSO LIPIDS; MEMBRANE; TRIGLYCERIDES.

Scott E. Feller
Ann T. S. Taylor

Figure 1. The structures of two phospholipids. Structure A represents a classic glycerophospholipid, POPC, and it is composed of choline, phosphate, glycerol, and two fatty acids. Structure B is an example of a sphingomyelin, and it is composed of choline, phosphate, sphingosine, and only one fatty acid.

hydrophobic: part of a molecule that repels water

amino alcohol: an organic molecule that contains both an amino ($-NH_3$) and an alcohol ($-OH$) functional group.

pulmonary surfactant: The protein-lipid mixture that prevents the collapse of alveoli during expir ation.

Bibliography

Berg, Jeremy M.; Tymoczko, John L.; and Stryer, Lubert (2002). *Biochemistry*, 5th edition. New York: W. H. Freeman.

Voet, Donald; Voet, Judith G.; and Pratt, Charlotte (1999). *Fundamentals of Biochemistry.* New York: Wiley.

Phosphorus

MELTING POINT: 44.1°C
BOILING POINT: 280°C
DENSITY: 1.82 g/cm^3
MOST COMMON IONS: PH_4^+, P^{3-}, $H_2PO_3^-$, PO_4^{3-}

DNA: deoxyribonucleic acid—the natural polymer that stores genetic information in the nucleus of a cell

The element phosphorus is essential to living organisms. It is part of the backbone of **DNA** (deoxyribonucleic acid), the carrier and transmitter of genetic information in cells. The element and its compounds have many commercial applications.

Phosphorus was first isolated by the alchemist Hennig Brand of Hamburg around 1670. He prepared white phosphorus, one of two common forms (allotropes) of the element, by evaporating human urine and strongly heating the residual solids. White phosphorus distilled and was collected under water.

The two common forms of phosphorus are white, which is made up of P_4 molecules, containing four atoms of phosphorus arranged in a regular tetrahedral formation, and red, which is a noncrystalline polymer. White phosphorus glows in the dark and bursts into flame in air. Red phosphorus does not react rapidly with air.

Phosphorus makes up about 0.12 percent of Earth's crust. It is extracted from minerals that contain phosphate (PO_4^{3-}) groups. Large deposits of such minerals, of which the most important is fluorapatite, $Ca_5F(PO_4)_3$, are found in the United States, Morocco, Russia, and Tunisia. At the present rate of extraction, the known deposits of phosphate rock would be sufficient to supply the world's demand for phosphorus for the next 1,000 years.

ester: organic species containing a carbon atom attached to three moieties: an O via a double bond, an O attached to another carbon atom or chain, and a H atom or C chain; the R(C=O)OR functional group

More than 90 percent of commercial phosphorus production is in the form of calcium salts of phosphoric acid, H_3PO_4, used as fertilizers. Other significant uses of phosphorus compounds are in the manufacture of matches (phosphorus sulfides), food products and beverages (purified phosphoric acid and its salts), detergents (sodium polyphosphates), plasticizers for polymers (**esters** of phosphoric acid), and pesticides (derivatives of phosphoric acid). Related to the phosphorus pesticides are nerve gases, poisonous compounds that rapidly attack the central nervous system, initially developed during World War II. SEE ALSO Deoxyribonucleic Acid (DNA); Fertilizer; Pesticides.

Harold Goldwhite

Bibliography

Greenwood, Norman N., and Earnshaw, A. (1984). *Chemistry of the Elements.* New York: Pergamon Press.

Weeks, Mary Elvira, and Leicester, Henry M. (1968). *Discovery of the Elements,* 7th edition. Easton, PA: Journal of Chemical Education.

Photography

It has long been known that certain substances, when illuminated, undergo permanent visible changes. In the early part of the nineteenth century, these materials were sometimes used to make "photogenic drawings," for exam-

The first known photograph, made in 1826. It shows the courtyard outside the room of Joseph-Nicéphore Niepce.

ple, by exposing them to sunlight through patterned masks. The most light-sensitive compounds are silver salts, and the photography that prospered in the second half of the nineteenth and throughout the twentieth century was based almost entirely on the use of silver halides.

Early Photography

Practical photographic processes were devised in the 1830s by Louis-Jacques-Mandé Daguerre in France and by William Henry Fox Talbot in England. In Daguerre's method, a silver iodide-coated silver plate was exposed to light in a camera, whereby the exposed silver iodide was decomposed to metallic silver and iodine. A clear image was obtained by treating the plate with mercury vapor (which amalgamated the silver) and by rinsing it in a strong salt solution to remove the remaining silver iodide. A positive image could be viewed by holding this "Daguerreotype" in oblique lighting with a dark background, so that the amalgamated silver zones appeared bright and the silver plate appeared dark.

Talbot's procedure consisted of washing paper successively in baths of saltwater and silver nitrate solution, thus depositing silver chloride in the fibers of the paper. The still wet paper was then exposed in a camera until a dark silver image appeared in the light-struck regions, and the remaining silver chloride was removed by washing with a concentrated salt solution or a sodium thiosulfate solution. By waxing or oiling the negative sheet, Talbot made the paper transparent, and then by making an exposure of diffuse light through the negative onto another sensitized sheet, he produced a positive image. An unlimited number of copies of a photograph could thus be made from any one negative.

Improvements in Talbot's Method

Both Daguerre's and Talbot's methods were inconvenient because they required long exposures in the camera—sometimes as long as 60 minutes. In 1840 Talbot greatly improved his process. He found that a very short camera exposure (about 1/60 of that required to give a visible image) left an invisible "latent" image on the sensitized paper. The latent image was then "developed" into a visible image by treatment with a solution of gallic acid and silver nitrate. This modification, together with the negative/positive feature, made Talbot's process so superior that it has survived, in its general form, to the present day. The main difference between Talbot's process and modern photographic practice is that now the silver halide, in the form of approximately micron-sized crystals or "grains," is suspended in gelatin. The gelatin emulsion is coated as a thin film on glass plates or flexible sheets of plastic or paper.

Mechanism of the Photographic Process

When a photon is absorbed by a silver halide grain, an electron is ejected from a halide ion and temporarily held at some site in the crystal. A silver ion can migrate to the site and combine with the electron to form a silver atom. The atom is not stable; it can decompose back into a silver ion and a free electron. However, during its lifetime, the atom can trap a second electron if one becomes available. If this second electron remains trapped until the arrival of a second silver ion, a two-atom cluster forms. This buildup of a silver cluster can continue as long as photoelectrons are available. The smallest cluster corresponding to a stable latent image speck is believed to consist of three or four silver atoms. Specks of this size or greater on the crystal surface can catalyze the subsequent action of a developer.

Classic Processing

A common, well-established procedure for making photographic prints is as follows:

1. Exposure of the sensitive material, usually a gelatin emulsion of silver halides on a cellulose acetate film, in the camera.

2. Development in the darkroom by treating the film with a solution of organic reducing agents such as hydroquinone and N-methyl paraaminophenol. The reagents reduce to metallic silver those silver halide crystals that acquired latent-image silver clusters. The brighter the subject of a photograph, the

darker is the image that forms in this development, so that one obtains a negative picture.

3."Fixing" the image so that the film will not darken on further exposure to light. This is accomplished by dissolving the undeveloped silver halide grains in a solution of sodium thiosulfate:

$$AgBr + 2S_2O_3^{2-} \longrightarrow Ag(S_2O_3)_2^{3-}\ Br^-$$

4. Washing away the dissolved silver salts and drying the negative.

5. Printing, that is, shining diffuse light through the negative onto a sheet of sensitive photographic paper (a gelatin emulsion on paper).

6. Darkroom development of the exposed paper using developer solution much like that used in the film development step. This step produces a positive image, in which the tones are like those in the original scene.

7. Fixing, washing, and drying the print as in the analogous film processing steps.

Reversal Processing

Transparencies, or photographic prints on a transparent base, can be produced essentially as paper prints are, but with replacement of the photographic paper by photographic film. This procedure can be used for making motion picture films. However, positive transparencies are more easily prepared by reversal processing, in which the final image is formed on the same film as that used in the original exposure. Typical reversal processing is as follows:

1. Exposure of the film in the camera.

2. Development of the negative image.

3. Dissolution of the developed silver image by treatment with an oxidizing agent.

4. Exposure of the remaining silver halide to light or to a chemical fogging agent.

5. Development of the silver halide, producing a positive image.

6. Washing and drying of the film.

Reversal processing can also be accomplished using the Sabatier effect, in which the emulsion is given a brief exposure to diffuse light in the midst of development. Some emulsions, when subjected to very intense camera exposure, will yield a positive image by ordinary development—a process referred to as overexposure solarization.

Spectral Sensitization

The silver halides are sensitive mainly to blue, violet, and ultraviolet light; hence, without sensitization, positive photographs reproduce all other colors as dark grays or blacks. However, by the addition of certain dyes to the emulsion, increased sensitivity to the other colors is obtained. Thus, "panchromatic" films, in which sensitivity is extended throughout the visible spectrum, are possible, and the resulting photographs are much more realistic than those obtained using old-fashioned red-insensitive films.

Color Photography

The sensitization of emulsions to the three primary colors (blue, green, and red) is essential to conventional color photography. A common method for producing color prints uses a film containing three superimposed layers, each sensitive to one of the three primary colors. In the initial development, the deposition of silver is accompanied by the formations of a dye color complementary to the color sensitivity of the film layer. After removal of the silver and silver halides from all three layers, the image seen through the three layers is complementary in color to that of the original scene, that is, a color negative. This negative is then used to print a positive copy onto paper with similar layered emulsions, and development proceeds analogously to that of the film.

Instant Photography

In 1947 Edwin H. Land devised a diffusion transfer process for obtaining positive paper prints rapidly in the Polaroid Land camera. The negative is developed in the presence of a solvent for silver halides, which not only develops the negative, but also dissolves the nondeveloped silver halides. The silver halides dissolved out of the negative sheet are developed into an adjacent sheet (containing nuclei for development) to give a positive image. This principle was applied to color photography in the 1960s.

Digital Photography

Silver halide-based photography is being rapidly displaced by so-called digital photography, involving special cameras that contain no film, but rather charge-coupled devices (CCDs), consisting of rectangular arrays of millions of minute light sensors. Under exposure to light, each sensor produces an electric charge, and the enormous amount of information thereby produced (charge as a function of sensor position) is stored electronically as digital data in the camera. The CCD array can be reused indefinitely, the only limitation on the number of possible exposures being the amount of information that can be stored in the camera. However, this information can be downloaded from the memory bank of the camera to a computer, and the image can later be manipulated and printed out, for example, with an ink jet printer, a laser printer, or a dye sublimation printer. The CCD arrays are monochrome devices, but when combined with color filter arrays, they provide blue, green, and red data and thus yield color pictures.

W. L. Jolly

Bibliography

Carroll, B. H.; Higgens, G. C.; and James, T. H. (1980). *Introduction to Photographic Theory.* New York: Wiley.

Coe, Brian, and Haworth-Booth, Mark (1983). *A Guide to Early Photographic Processes.* London: Victoria and Albert Museum.

Rosenblum, Naomi (1984). *A World History of Photography.* New York: Abbeville Press.

Photosynthesis

No chemical process is more important to life on Earth than **photosynthesis**—the series of chemical reactions that allow plants to harvest sunlight and create carbohydrate molecules. Without photosynthesis, not only would there be no plants, the planet could not sustain life of any kind. In plants, photosynthesis occurs in the **thykaloid membrane** system of chloroplasts. Many of the enzymes that allow photosynthesis to occur are transmembrane proteins embedded in the thykaloid membranes. What then is the chemistry involved?

photosynthesis: process by which plants convert carbon dioxide and water to glucose

thykaloid membrane: part of a plant that absorbs light and passes the energy on to where it is needed

The most basic summary of the photosynthesis process can be shown with a net chemical equation

$$6CO_2(g) + 6\ H_2O(l) + h\nu \longrightarrow C_6H_{12}O_6(s) + 6O_2(g)$$

The symbol $h\nu$ is used to depict the energy input from light (in the case of most plants, sunlight). This chemical equation, however, is a dramatic simplification of the very complicated series of chemical reactions that photosynthesis involves. It also implies that the only product is **glucose**, $C_6H_{12}O_6$ (s), which is also a simplification.

glucose: common hexose monosaccharide; monomer of starch and cellulose; also called grape sugar, blood sugar, or dextrose

Still, take a moment to look at this chemical equation. If one were to guess where the various atoms in the reactants end up when products are produced, it would be reasonable to suggest that the oxygen atoms in the O_2 (g) were those originally associated with carbon dioxide. Most scientists believed this to be true until the 1930s when experiments by American biologist Cornelius van Niel suggested that oxygen-**hydrogen bonds** in water must be broken in photosynthesis. Further research confirmed his hypothesis and ultimately revealed that many reactions are involved in photosynthesis.

hydrogen bond: interaction between H atoms on one molecule and lone pair electrons on another molecule that constitutes hydrogen bonding

There are two major components of photosynthesis: the light cycle and the dark cycle. As implied by these names, the reactions in the light cycle require energy input from sunlight (or some artificial light source) to take place. The reactions in the dark cycle do not have to take place in the dark, but they can progress when sunlight is not present.

The critical step of the light cycle is the absorption of electromagnet radiation by a pigment molecule. The most famous pigment is **chlorophyll**, but other molecules, such as β-carotene, also absorb light (see Figure 1). Together, these pigment molecules form a type of light harvesting antennae that is more efficient at interacting with sunlight than would be possible with

chlorophyll: active molecules in plants undergoing photosynthesis

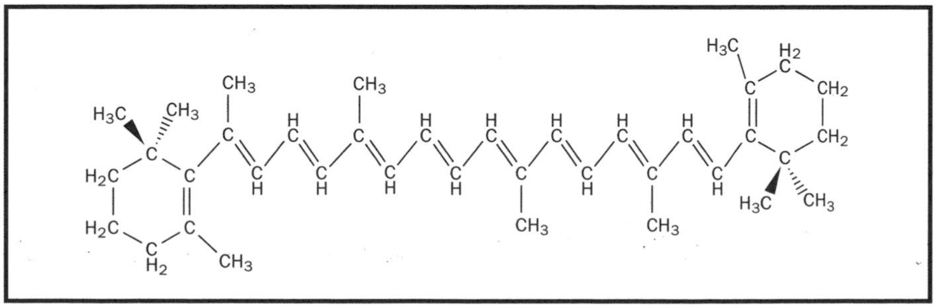

Figure 1a. Structure of β-carotene.

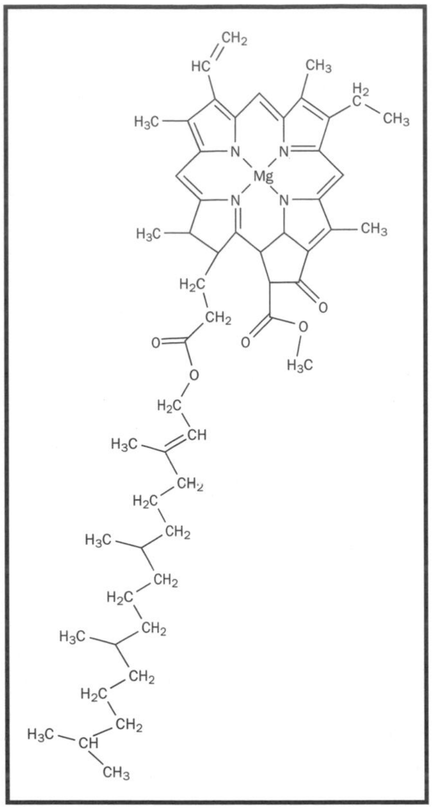

Figure 1b. Structure of GAP.

adenine: one of the purine bases found in nucleic acids, $C_5H_5N_5$

oxidation: process that involves the loss of electrons (or the addition of an oxygen atom)

photon: a quantum of electromagnetic energy

synthesis: combination of starting materials to form a desired product

the pigments acting alone. When the light is absorbed, electrons in the pigment molecule are excited to high energy states. A series of enzymes called electron transport systems help channel the energy present in these electrons into reactions that store it in chemical bonds.

For example, one major chemical reaction that results from the absorbed light energy (and excited electrons) involves water and nicotinamide **adenine** dinucleotide phosphate ($NADP^+$). The net reaction is shown by the chemical equation

$$2\ NADP^+ + 2\ H_2O \rightarrow NADPH + O_2 + 2H^+$$

This is an example of an **oxidation**–reduction reaction, and it shows that the light cycle is the stage of photosynthesis when water breaks up. The amount of energy required to make this reaction proceed is greater than what can be provided by a single **photon** of visible light. Therefore, there must be at least two ways that plants harvest light energy in photosynthesis. These two systems are referred to as photosystem I (PSI) and photosystem II (PSII), although the numbers associated with these names do not imply which one happens "first."

At the same time that NADPH is being produced, the combination of the photo systems also produces a concentration gradient of protons. Enzymes in the cell use this proton gradient to produce ATP from ADP. Thus, the light cycle produces two "high energy" molecules: NADPH and ATP.

With the high energy products provided by the light cycle, plants then use reactions that do not require light to actually produce carbohydrates. The initial steps in the dark cycle are collectively called the Calvin cycle, named after American chemist Melvin Calvin who along with his coworkers determined the nature of these reactions during the late 1940s and early 1950s.

The Calvin cycle essentially has two stages. In the first part of the cycle, several enzymes act in concert to produce a molecule called glyceraldehyde-3-phosphate (GAP). (See Figure 2). Note in the illustration that this molecule has three carbon atoms. Each of these carbon atoms comes originally from carbon dioxide molecules—so photosynthesis completes the amazing task of manufacturing carbohydrates out of air (the source of the carbon dioxide). This stage of the Calvin cycle is sometimes called carbon fixing. In order to carry out this **synthesis** of GAP, the Calvin cycle consumes some of the NADPH and ATP that was produced during the light cycle.

The carbon dioxide needed for this step enters through pores in the photosynthetic leaf (called stromata). Plants close these pores during hot, dry times of the day (to prevent water loss) so the details of carbon fixing vary for plants from different climates. In hot climates, where stomata are closed for a higher percentage of time, the trapping of carbon dioxide has to be more efficient than in cooler climates. This biochemical difference in photosynthesis helps explain why plants from one climate do not grow as well in warmer (or cooler) places.

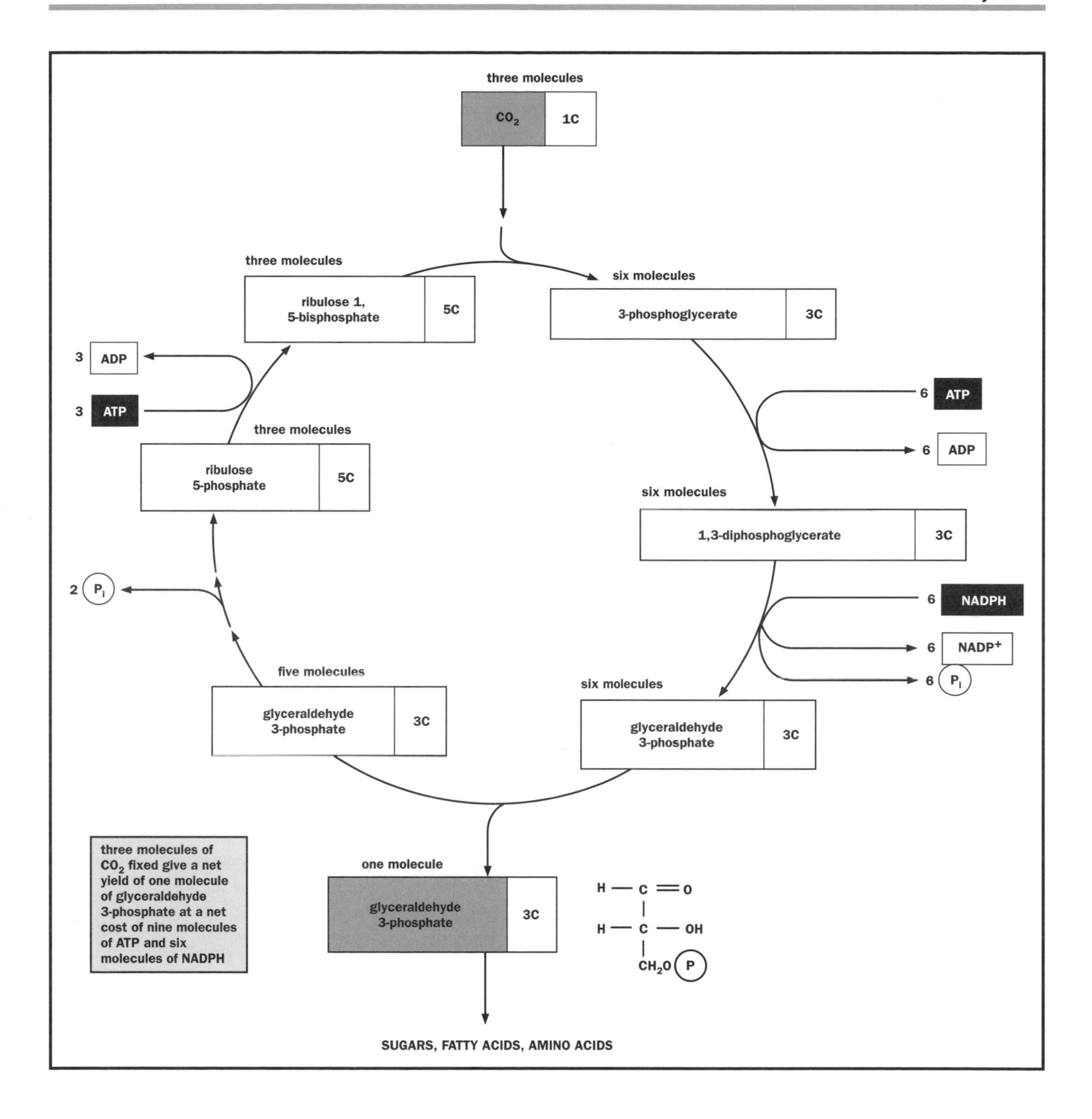

Figure 2. The Calvin cycle.

The second stage of the cycle builds even larger carbohydrate molecules. With more than half a dozen enzyme-catalyzed reactions in this portion of the dark cycle, five- and six-carbon carbohydrates are produced. The five-carbon molecules continue in the cycle to help produce additional GAP, thus perpetuating the cyclic process.

Photosynthesis is central to all life on the planet and has been for many thousands of years. As a result, there are numerous variations in the way it occurs in different cells. The efficient collection of carbon dioxide mentioned earlier is one example of variation in photosynthesis. Other differences occur when the process takes place in bacteria rather than plants. Nonetheless, the description provided here outlines the basic concepts that would be noted in all photosynthesis. These differences pose the research questions that continue to challenge scientists today. SEE ALSO CALVIN, MELVIN; CONCENTRATION GRADIENT.

Thomas A. Holme

Bibliography

Foyer, Christine H. (1984). *Photosynthesis.* New York: Wiley.

Govindjee, and Coleman, W. J. (1990). "How Plants Make Oxygen." *Scientific American* 262:50–59.

Internet Resources

Wong, Kate (2000). "Photosynthesis's Purple Roots." *Scientific American.* Available from <http://www.sciam.com>.

Physical Chemistry

Physical chemistry is the branch of chemistry concerned with the interpretation of the phenomena of chemistry in terms of the underlying principles of physics. It lies at the interface of chemistry and physics, inasmuch as it draws on the principles of physics (especially quantum mechanics) to account for the phenomena of chemistry. It is also an essential component of the interpretation of the techniques of investigation and their findings, particularly because these techniques are becoming ever more sophisticated and because their full potential can be realized only by strong theoretical backing. Physical chemistry also has an essential role to play in the understanding of the complex processes and molecules characteristic of biological systems and modern materials.

Hot water is vaporizing as it is thrown into air that is −37.2°C (−35°F).

Physical chemistry is traditionally divided into a number of disciplines, but the boundaries between them are imprecise. Thermodynamics is the study of transformations of energy. Although this study might seem remote from chemistry, in fact it is vital to the study of how chemical reactions yield work and heat. Thermodynamic techniques and analyses are also used to elucidate the tendency of physical processes (such as vaporization) and chemical reactions to reach **equilibrium**—the condition when there is no further net tendency to change. Thermodynamics is used to relate bulk properties of substances to each other, so that measurements of one may be used to deduce the value of another. **Spectroscopy** is concerned with the experimental investigation of the structures of atoms and molecules, and the identification of substances, by the observation of properties of the electromagnetic radiation absorbed, emitted, or scattered by samples. Microwave spectroscopy is used to monitor the rotations of molecules; infrared spectroscopy is used to study their vibrations; and visible and ultraviolet spectroscopy is used to study electronic transitions and to infer details of electronic structures. The enormously powerful technique of nuclear magnetic resonance is now ubiquitous in chemistry. The detailed, quantitative interpretation of molecular and solid-state structure is based in quantum theory and its use in the interpretation of the nature of the chemical bond. Diffraction studies, particularly x-ray diffraction and neutron diffraction studies, provide detailed information about the shapes of molecules, and x-ray diffraction studies are central to almost the whole of molecular biology. The scattering of neutrons, in inelastic neutron scattering, gives detailed information about the motion of molecules in liquids. The bridge between thermodynamics and structural studies is called statistical thermodynamics, in which bulk properties of substances are interpreted in terms of the properties of their constituent molecules. Another major component is chemical kinetics, the study of the rates of chemical reactions; it examines, for example, how rates of reactions respond to changes in conditions or the presence of a **catalyst**. Chemical kinetics is also concerned with the detailed mechanisms by which a reaction takes place, the sequences of elementary processes that convert reactants into products, including chemical reactions at solid surfaces (such as electrodes).

equilibrium: condition in which two opposite reactions are occurring at the same speed, so that concentrations of products and reactants do not change

spectroscopy: use of electromagnetic radiation to analyze the chemical composition of materials

catalyst: substance that aids in a reaction while retaining its own chemical identity

There are further subdivisions of these major fields. Thermochemistry is a branch of thermodynamics; its focus is the heat generated or required by chemical reactions. Electrochemistry is the study of how chemical reactions can produce electricity and how electricity can drive chemical reactions in "reverse" directions (electrolysis). Increasingly, attention is shifting from equilibrium electrochemistry (which is of crucial importance in interpreting the phenomena of inorganic chemistry) to dynamic electrochemistry, in which the rates of electron-transfer processes are the focus. Chemical kinetics has divisions that are based on the rates of reaction being studied. Special techniques for studying atomic and molecular processes on ever shorter time scales are being developed, and physical chemists are now able to explore reactions on a femtosecond (10^{-15} second) timescale. Chemical kinetics studies are theoretical as well as experimental. One goal is to understand the course of reactions in step-by-step (and atomic) detail. Techniques are available that allow investigators to study collisions between individual molecules.

X-ray diffraction gives detailed information about shapes of molecules and is the basis of molecular biology.

Physical chemistry is essential to understanding the other branches of chemistry. It provides a basis for understanding the thermodynamic influences (principally, the entropy changes accompanying reactions) that drive chemical reactions forward. It provides justifications for the schemes proposed in organic chemistry to predict and account for the reactions of organic compounds. It accounts for the structures and properties of **transition metal** complexes, **organometallic compounds**, the microporous materials known as zeolites that are so important for **catalysis**, and biological macromolecules, such as proteins and nucleic acids (including **DNA**). It is fair to say that there is no branch of chemistry (including biochemistry) that can be fully understood without interpretations provided by physical chemistry.

transition metals: elements with valence electrons in d-sublevels; frequently characterized as metals having the ability to form more than one cation

organometallic compound: compound containing both a metal (transition) and one or more organic moieties

catalysis: the action or effect of a substance in increasing the rate of a reaction without itself being converted

DNA: deoxyribonucleic acid—the natural polymer that stores genetic information in the nucleus of a cell

There is a distinction between physical chemistry and chemical physics, although the distinction is hard to define and it is not always made. In physical chemistry, the target of investigation is typically a bulk system. In chemical physics, the target is commonly an isolated, individual molecule.

Theoretical chemistry is a branch of physical chemistry in which quantum mechanics and statistical mechanics are used to calculate properties of molecules and bulk systems. The greater part of activity in quantum chemistry, as the former is commonly termed, is the computation of the electronic structures of molecules and, often, their graphical representation. This kind of study is particularly important to the screening of compounds for potential pharmacological activity, and for establishing the mode of action of enzymes. SEE ALSO CATALYSIS AND CATALYSTS; ELECTROCHEMISTRY; EQUILIBRIUM; KINETICS; QUANTUM CHEMISTRY; SPECTROSCOPY; THEORETICAL CHEMISTRY; THERMODYNAMICS.

Peter Atkins

Bibliography

Atkins, Peter, and de Paula, Julio (2002). *Atkins' Physical Chemistry*, 7th edition. New York: Oxford University Press.

Berry, R. Stephen; Rice, Stuart A.; and Ross, John (2000). *Physical Chemistry*, 2nd edition. New York: Oxford University Press.

Laidler, Keith James (1995). *The World of Physical Chemistry*. New York: Oxford University Press.

Pigments

Pigments and dyes are called colorants. The ways in which colorants are used determines whether they are pigments or dyes. Pigments are water- and oil-insoluble natural and synthetic products that impart color to materials such as paper and plastics. Dyes, by contrast, are water-soluble colorants, although some are converted into insoluble lake pigments by coprecipitating onto an inorganic base. Artists' colors are pigments that are spread on a surface suspended in a suitable medium, such as oil. The mass coloration of textile fibers, polymers, plastics, and rubber takes place when pigments exist in the form of dispersions. A convenient way of classifying pigments is into organic and inorganic pigments.

Pigments are used in the production of paint to add color.

Organic Pigments

Natural organic pigments were used in cave paintings and for decoration from the earliest times. The ancient Britons obtained indigo from the woad plant *Isatis tinctoria*, and used the extract to color their bodies. Here the insoluble blue is used as a pigment rather than as a dye. Other organic pigments found in nature include **chlorophyll**, the green coloring matter of leaves responsible for **photosynthesis**, and heme, which gives blood its red color and, when bound to proteins in hemoglobin, transports oxygen around the body. These biochemical pigments are members of the **porphyrin** family.

chlorophyll: active molecules in plants undergoing photosynthesis

photosynthesis: process by which plants convert carbon dioxide and water to glucose

porphyrin: common type of heterocyclic ligand that has four five-member rings with a nitrogen, all of which lie in a plane; nitrogen atom lone pairs are available for coordinate covalent bonds

During the early 1930s synthetic organic pigments, called phthalocyanines, were developed in Britain and manufactured by Imperial Chemical Industries (ICI). Academic researchers showed that phthalocyanines are coordination complexes that mimic the structures of porphyrins. The coordination concept, involving groups of atoms called chelates attached to a central **metal**, was developed beginning in 1893 by the Swiss chemist Alfred Werner and confirmed in 1911, when he collaborated in Zurich with Victor L. King, later a leading technical expert in the U.S. colorant industry. Phthalocyanines represent the only structurally novel class of synthetic colorants invented in the twentieth century. Copper phthalocyanine, known as Monastral blue, and its congeners are used in automobile finishes, printing inks, and plastics.

metal: element or other substance the solid phase of which is characterized by high thermal and electrical conductivities

Lake pigments made from the first synthetic dyes, such as mauve, were used during the 1860s for printing postage stamps and wallpaper. Red lake pigments, such as that derived from the madder dye, or after 1870 from synthetic dyes, were highly regarded by the Impressionist painters. Other synthetic organic pigments are made from azo dyes, those containing the $-N{=}N-$atomic grouping, introduced during the 1870s. The use of synthetic organic pigments in printing inks increased perhaps threefold in the last two decades of the twentieth century, as color became the norm in newspapers, magazines, advertising, and packaging. Careful standardization of the **microcrystalline** form, the crystal habit, is required for pigments employed in printing inks and paints. In order to add color to synthetic polymers, the plastics and resins, pigments are mixed in bulk with other chemicals during the manufacturing process.

microcrystalline: relating to the structure of crystals of very small size, typically a micron (μm) in a dimension

Inorganic Pigments

Ultramarine, or lazurite, is a natural blue pigment derived from lapis lazuli, a semiprecious mineral of the sodalite group, found in Afghanistan. From around 1000 C.E. it was used as a pigment for illumination and later in murals and paintings. In 1271, the explorer Marco Polo visited the site where it was found. Artificial ultramarine, a blue pigment of variable composition, became available in Europe during the late eighteenth and nineteenth centuries. The German poet and dramatist Johann Wolfgang von Goethe, who was interested in color, visited one of the factories in 1787. Although the natural and synthetic forms of ultramarine are chemically similar, they differ in their particle forms. Prussian blue, iron hexacyanoferrate (III) or **ferric** ferrocyanide, is a synthetic dark blue pigment discovered in Berlin in 1704. It was originally made from animal matter, including blood, and the salts of iron.

ferric: older name for iron in the +3 oxidation state

Chromium pigments followed the discovery of the element in a rare Siberian mineral by the French chemist Louis-Nicolas Vauquelin in 1797. Lead chromate gives yellows and oranges, including chrome yellow, invented in 1809, and extensively used by artists, though often modified by addition of other ingredients. Also in 1809 George Field in England produced the dye lemon yellow, based on barium chromate.

Lead has been known and used since ancient times, in part because of the ease with which it can be isolated in the free state. It was converted into the dense pigment white lead (lead carbonate), an essential component of artists' palettes, including the Italian masters of the Renaissance. In the past lead was often used in white paint for the external protection of homes. However, lead is a toxic metal that induces swelling of the brain (lead encephalophy), and causes madness and death. Young children are particularly susceptible to its scourge. In 1910 the pioneer of American industrial hygiene, Alice Hamilton, selected as her first assignment at the Bureau of Labor an investigation of conditions in the white lead industry. Lead-based paints fell out of favor during the 1950s and were banned by the federal government during the 1970s. In modern paints, the toxic lead has been replaced by titanium dioxide.

Theories of Color

Color arises from the way in which colorants interact with light. Colored organic compounds contain groups of atoms whose bonds are unsaturated, such as C=C, C=O, and N=N. These are part of an extended **delocalized** system of electrons called a **chromophore**. A sequence of alternating double bonds through which the electrons are spread is termed a conjugated system. The presence of salt-forming groups of atoms such as –OH and NH_2 modify the color. They are called auxochromes and contain lone pairs of electrons that become part of the delocalized electron system. The nomenclature is based on the first successful theory of color and constitution, that of German chemist Otto N. Witt.

delocalized: of a type of electron that can be shared by more than one orbital or atom

chromophore: part of the molecule that yields characteristic colors

Colored inorganic compounds often contain **transition metals** in which the *d* **subshell** of electrons is split by attached groups, the **ligands**. The extent of this splitting is responsible for the color. The **oxidation** state of metals also affects the *d* electrons and determines color. When absorbed light brings about the transfer of an electron from the ground state of an atom to the excited state of a nearby atom, the process is electron or charge transfer. This accounts for the colors of Prussian blue and chrome yellow.

transition metals: elements with valence electrons in d-sublevels; frequently characterized as metals having the ability to form more than one cation

subshell: electron energy sublevel, of which there are four: s, p, d, and f

ligand: molecule or ion capable of donating one or more electron pairs to a Lewis acid

oxidation: process that involves the loss of electrons (or the addition of an oxygen atom)

Although pigments have been associated with artists' paints, printing inks, and the coloration of synthetic polymers, they are also used in electronics and telecommunications, for the absorption of light, especially at semiconductor wavelengths, and in ink-jet printers, in addition to xerography (electrophotography) and thermography. SEE ALSO DYES; PERKIN, WILLIAM HENRY; WERNER, ALFRED.

Anthony S. Travis

Bibliography

Ashok, Roy, ed. (1993). *Artists' Pigments: A Handbook of Their History and Characteristics,* Vol. 2. Washington, DC: National Gallery of Art.

German physicist Max Karl Ernst Ludwig Planck, recipient of the 1918 Nobel Prize in physics, "in recognition of the services he rendered to the advancement of Physics by his discovery of energy quanta."

Bomford, David; Kirby, Jo; Leighton, John; and Roy, Ashok (1990). *Art in the Making: Impressionism*. New Haven, CT: Yale University Press.

Feller, Robert L., ed. (1986). *Artists' Pigments: A Handbook of Their History and Characteristics*, Vol. 1. Washington, DC: National Gallery of Art.

Fitzhugh, Elisabeth W., ed. (1997). *Artists' Pigments: A Handbook of Their History and Characteristics*, Vol. 3. Washington, DC: National Gallery of Art.

Internet Resources

"Paint & Pixel." Available from <http://www.total.net/~daxx/pap.shtml>.

"Pigments through the Ages." Available from <http://webexhibits.org/pigments>.

Planck, Max

GERMAN PHYSICIST
1858–1947

Max Karl Ernst Ludwig Planck was born into a family of lawyers and clergymen, and he became the fourth generation of university professors from his family. As a child, he exhibited considerable talent in mathematics, music, and philology (the scientific study of language). By 1874, when he entered the University of Munich, the sixteen-year-old Planck had decided to study mathematics. Very quickly, however, he became interested in physics and the application of mathematics to the physical world. The university's professor of physics, Philip von Jolly, discouraged the young student from studying physics because—as Jolly told him—it was very nearly a closed subject with little left to discover. Luckily, Planck disregarded his professor's advice.

Planck also studied at the University of Berlin with such notable physicists as Gustav Kirchhoff, whom he thought brilliant, but a dry and boring teacher. He also became familiar with the thermodynamics research of Rudolf Clausius, and in 1879 Planck received his Ph.D. from the University of Munich, only three months after his twenty-first birthday. His dissertation explored the second law of thermodynamics.

After holding posts at the universities of Munich and Kiel, Planck succeeded Kirchhoff at the University of Berlin in 1888 after the latter's death. Planck continued his research in thermodynamics, including attempts to connect heat with the Scottish physicist James Clerk Maxwell's theory of electromagnetic radiation. He also addressed a problem suggested by Kirchhoff, who had earlier established that the energy of radiation emitted by a blackbody depends on temperature and the frequency of the radiation.

A blackbody is any object that absorbs all the radiation falling on its surface. Thus, it appears black. A perfect absorber, a blackbody is also a perfect emitter of radiation, and Kirchhoff had challenged physicists to find the mathematical equation relating the energy to temperature and frequency.

The German physicist Wilhelm Wien had proposed such an equation, which worked well only for high frequencies, and Lord Rayleigh (born John William Strutt) proposed another equation, which worked well only at low frequencies. In 1900 Planck was able to develop a single expression that combined these two earlier equations and accurately predicted the energy over the entire range of frequencies.

Subsequently, Planck tried to provide a theoretical basis for his equation. He found that to do so, it was necessary to reject the idea from classical physics that electromagnetic radiation is wavelike and continuous and instead to make the bold assumption that it is particle-like and discrete. Planck assumed that radiation can occur in discrete packets of energy, which Albert Einstein called "quanta." This radical idea is expressed in the equation

$$E = hv$$

in which the energy E is directly proportional to the frequency v, and the proportionality constant h, now known as Planck's constant, has the value 6.62×10^{-34} joule per second.

Planck's revolutionary idea about energy provided the basis for Einstein's explanation of the photoelectric effect in 1906 and for the Danish physicist Niels Bohr's atomic model of the hydrogen atom in 1913. Their success, in turn, lent support to Planck's theories, for which he received the Nobel Prize in physics in 1918. In the mid-1920s the combination of Planck's ideas about the particle-like nature of electromagnetic radiation and French physicist Louis de Broglie's hypothesis of the wavelike nature of electrons led to the formulation of quantum mechanics, which is still the accepted theory for the behavior of matter at atomic and subatomic levels.

By the second decade of the twentieth century, Planck was less active in quantum theory research, taking on, in addition to his teaching responsibilities, various administrative duties, including the presidency of the Kaiser Wilhelm Gesellschaft during the years 1930 through 1937 and again after World War II from 1945 until 1946. Planck suffered many personal losses during this part of his life. His first wife died in 1909; his elder son was killed in World War I; and his two daughters both died in childbirth during the early part of the century. In World War II his younger son was executed after being accused of helping plot the assassination of Adolf Hitler, and his home and library were destroyed by Allied bombing. Planck spent the last few years of his life in Göttingen, living long enough to witness the establishment of the Max Planck Gessellschaft from the earlier Kaiser Wilhelm Gesellschaft, to which he had devoted so much of his professional life. SEE ALSO BOHR, NIELS; DE BROGLIE, LOUIS; EINSTEIN, ALBERT; MAXWELL, JAMES CLERK.

Richard E. Rice

Bibliography

Heathcote, Niels H. de V. (1971). *Nobel Prize Winners in Physics, 1901–1950.* Freeport, NY: Books for Libraries Press.

Heilbron, J. L. (1986). *The Dilemmas of an Upright Man: Max Planck as Spokesman for German Science.* Berkeley: University of California Press.

Planck, Max (1949). *Autobiography and Other Papers.* New York: Philosophical Library.

Weber, Robert L. (1988). *Pioneers of Science: Nobel Prize Winners in Physics,* 2nd edition Bristol, U.K.: Adam Hilger.

Plastics

The term "plastic" can be broadly defined as any inherently formless material that can be molded or modeled under heat and/or pressure. It is derived from the Greek word *plastikos,* meaning a shaped or molded substance.

The term "plastics" first included only natural polymers—usually animal proteins (horn and tortoise shell), tree resins, or insect secretions called shellac—that were subsequently mixed with fillers such as wood flour to yield substances having better molding properties. (A polymer, from the Greek word *poly*, meaning "many," and *mer* meaning "unit," is a molecule with an extremely high molecular weight.)

The use of natural polymers to make plastic products started as early as 1760, when Enoch Noyes opened a business making combs out of keratin and albuminoid organic proteins derived from animal horns and horse hoofs. However, the first commercially successful plastic material, celluloid, would not come about for another hundred years.

In the 1840s German chemist Christian Schönbein developed cellulose nitrate from a mixture of cotton, nitric acid, and sulfuric acid. Cellulose nitrate is a highly flammable doughlike substance primarily used in the manufacture of explosives. Schönbein's innovation represents the beginning of the modification of natural polymers by chemists so as to increase their processibility and functionality. Cellulose nitrate's properties as a molding substance interested other scientists of the time, and in 1855 an Englishman named Alexander Parkes developed a form of cellulose nitrate he named Parkesine. From this material, Parkes manufactured a number of buttons, pens, medallions, and combs. In 1862 he displayed this material officially at the Great International Exhibit in London. Parkes made small commercial gains with Parkesine and eventually sold the rights to Daniel Spill, who subsequently began production of the substance under the names Xylonite and Ivoride, around 1865. Spill received British patents for Xylonite and Ivoride in 1867 and 1869, respectively.

At around the same time in the United States, a billiard ball company advertised a $10,000 reward for the discovery of an alternate material to ivory. John Wesley Hyatt developed collodion, a mixture of cellulose nitrate and alcohol. Like cellulose nitrate, collodion was highly flammable and would produce a small explosion upon agitation. Hyatt reported: "[W]e had a letter from a billiard saloon proprietor in Colorado mentioning this fact . . . saying he did not care so much about it, but that instantly every man in the room pulled a gun." To avoid melee, camphor, a derivative of the laurel tree, was added, and in 1870 Hyatt received a U.S. patent for celluloid. In 1871 Hyatt and his brother Isaiah formed the American Celluloid Company, which is today the Plastics Division of the Celanese Corporation.

A more common perception of plastic is that it is a synthetic or manmade material, with highly engineered properties and product designs. Dr. Leo Baekeland engineered the first totally synthetic plastic in 1907. Patented in 1909 and named Bakelite after its inventor, the material was the first thermoset plastic. The term "thermoset" refers to a plastic that under initial heat and pressure can be molded into form. After cooling, the material sets and cannot be remelted or re-formed. This setting is due to the cross-linking of polymer chains, wherein strong covalent bonds form between separate oligomers, short chains of polymer units called monomers. The most common thermoset resin is vulcanized rubber, created by Charles Goodyear in the United States in 1839. Vulcanized rubber utilizes natural hevea rubber made from the gutta percha tree, and therefore is not totally synthetic (like Bakelite). Ironically, the first use of Bakelite was as a replacement for

HERMAN MARK (1895–1992)

The influence of Herman Mark, the so-called father of polymer science, on the plastics industry still echoes today in a legacy of education and research. His work in the 1920s on the structure of cellulose opened the door for the development of synthetic fibers such as acrylic, nylon, polyester, polystyrene, and PVC.

—***Valerie Borek***

natural rubber in electrical insulations. Bakelite is formed via the reaction of phenol and formaldehyde under high heat. Initially, formaldehyde is added to the reaction mixture in small amounts (forming a resin); the mixture is then poured into a mold, into which more formaldehyde is added; and pressure is applied to create the final product.

Over the next several decades, many varieties of synthetic thermoplastic materials would be developed in Germany, England, and the United States. Thermoplastic materials such as vinyls, nylons, and acrylics are polymers that can be molded or formed under heat and pressure, and if necessary can be reheated and re-formed (and will retain most of their original mechanical properties).

Eugen Baumann created today's most common vinyl, polyvinyl chloride (PVC), in 1872. However, Friedrich Heinrich August Klatte did not patent it until 1913. At that time PVC was not well received, as illustrated by Waldo Semon's comment, "People thought of PVC as worthless back then; they would throw it in the trash." Semon was responsible for creating plasticized PVC. He had been attempting to dehydrohalogenate PVC in a high boiling solvent when he realized that the molten material was exhibiting greater flexibility and elasticity. The exposure of PVC to a boiling solvent introduced a plasticizer, or low molecular weight molecule, to the PVC matrix. Today plasticizers are commonly added to polymers (especially PVC) to enhance flexibility, prevent stress cracking, and enhance processability. This has enabled the use of PVC in diverse commercial applications, including the manufacture of rigid tubing and flexible car seats.

In 1920 German scientist Hermann Staudinger published his theories on polyaddition polymerization, the formation of long-chain molecules. (Previously, the manner in which long-chain molecules were formed was unexplained.) Nine years later, in a publication that detailed the polymerization of styrene, this method of chain formation would be laid out. During this time period Staudinger developed polystyrene into a commercial product. A division of the German chemical company IG Farben, known as Badische Anilin- und Soda-Fabrik, or BASF, produced polystyrene in 1930. The Dow Chemical Company introduced the American public to polystyrene in 1937.

In 1928 directors at E. I. du Pont de Nemours & Company (Du Pont) placed Dr. Wallace H. Carothers in charge of fundamental research into what are now classic studies on the formation of polymer chains. During his years at Du Pont, Carothers published his theory on polycondensation, and discovered both neoprene and nylon.

Nylon, not publicly announced until 1938, was first used for bristles on combs, but made headlines in 1939 when nylon stockings debuted at the World's Fair in New York City. Nylon is known by its chemical name, poly(hexamethylene) adipamide, but more often simply as nylon. The first nylon manufacturing plant went into production at Seaford, Delaware, in 1940. Commercial production of nylon 6 by IG Farben in Germany began in 1941. These two plants would go on to produce millions of pounds of nylon annually. This mass production was essential to the World War II effort, as nylon was used for everything from belts, ropes, and straps to tents and parachutes.

Another polymer that came into use during World War II was polytetrafluoroethylene (PTFE), which received the trademark Teflon. Dr. Roy J. Plunkett and his assistant Jack Rebock at Du Pont discovered PTFE accidentally on April 6, 1938. They had been conducting research on alternate refrigeration methods when they discovered the polymerization of tetrafluoroethylene. Plunkett received a patent for PTFE in 1941. It was found that the material was resistant to corrosion by all the solvents, acids, and bases that were available for testing at that time. This led to the U.S. military's interest in PTFE, and its subsequent use as a cover for proximity fuses on the nose cones of artillery shells. It was not until the material was declassified in 1946 that the public learned of the material Du Pont had named Teflon two years earlier. Teflon has since become a household name; its best-known use being its contribution to nonstick surfaces on pots and pans.

Today's most widely produced and perhaps most versatile plastic, polyethylene, was discovered at the Imperial Chemical Industries (ICI) in England in 1933. E. W. Fawcett and R. O. Gibson set off a reaction between ethylene and benzaldehyde under 2,000 atmospheres of pressure, resulting in the polymerization of ethylene and the birth of polyethylene. By 1936, ICI had developed a larger volume compressor that made the production of useful quantities of polyethylene possible. Among polyethylene's first applications were its uses as underwater cable coatings and as insulation for radar during World War II.

In 1943, Karl Ziegler began work that would drastically alter the production of polyethylene. Ziegler used organometallic compounds, which have both metallic and organic components, as catalysts. At very modest pressures, these catalysts generated a linear, more rigid, high molecular weight polyethylene, and the innovation increased the number of the polymer's applications. Today polyethylene is used in the production of detergent bottles and children's toys, and is even replacing Kevlar as a bulletproof material.

In 1957, at the Montecatini Laboratories in Italy, Giulio Natta continued the work of Ziegler and used what is now termed Ziegler–Natta polymerization to create polypropylene. When Natta reported the polymerization of ethylene with a titanocene catalyst, it became clear that polymer chains with specific tacticities, or specific ordered structures, were possible. Polypropylene rose to become a substitute for polyethylene in products in which slightly higher temperature stability was necessary, for example, dishwasher-safe cups and plates.

Polycarbonate, a popular plastic used originally to make eyeglass lenses, was first discovered by A. Einhorn in 1898. But it would be more than fifty years before further research was performed on the material. In the 1950s Dr. Herman Schnell, working at Bayer, a division of IG Farben, along with Daniel Fox of General Electric's Corporate Laboratory in Schenectady, New York, conducted concurrent research on the synthesis of polycarbonate. Schnell and Fox each achieved a polymerization that produced polycarbonate via different methods, and received patents in 1954 and 1955, respectively. Upon his achievement of polymerization, Fox described his attempts to remove the newly formed polymer from the reaction vessel: "The remnants of the glass were broken away to yield a hemispherical, glass fragment embedded, glob of plastic on the end of a steel stirrer shaft. The glob was pounded on the cement floor and struck with a hammer in abortive attempts to remove the remaining glass, and/or, shatter the plastic. The pseudo plastic mallet was even used to drive nails into wood." That glob would eventually be developed into bulletproof glass and provide General Electric and Bayer with billions of dollars in revenue.

Means to improve the material properties of plastics have been sought for decades. Improvement has sometimes come in the form of compounds such as mineral fillers, antioxidants, and flame-retardants. One of the first searches for an improved material was centered on cellulose nitrate. Cellulose nitrate is colorless and transparent, which enabled it to be used as photographic film. However, it is extremely flammable, and its early use in motion picture film and concomitant exposure to hot lights led to numerous fires. In 1900, Henri Dreyfus substituted acetic acid for nitric acid in the synthesis of cellulose nitrate, and created instead a less flammable material, cellulose acetate. Today, polymers are often halogenated in order to achieve flame-retardation.

Plastics have been designed to be chemically resistant, stable compounds, and have been extremely successful in these regards. In fact, they have been so successful that an environmental problem has been created. Plastic products discarded in landfills decay slowly. They sometimes contain heavy metal additives. In addition, the millions of pounds of plastic discarded annually have engendered a crisis over landfill space. In the early 1980s plastic recycling programs began to spring up across the United States in response to the large number of polyethylene terephthalate (PET or PETE) bottles being discarded. In 1989, 235 million pounds of PET bottles were recycled. The number rose to 1.5 billion pounds in 1999.

Most plastics can be recycled. Even mixed plastic waste can be recycled into artificial lumber or particleboard. Plastic "wood" is easy to saw, and it has better resistance to adverse weather and insects than real wood. SEE ALSO BAEKELAND, LEO; CAROTHERS, WALLACE; GOODYEAR, CHARLES; STAUDINGER, HERMANN; NYLON; POLYMERS, SYNTHETIC.

Paul E. Koch

Bibliography

DiNoto, Andrea (1984). *Art Plastic Designed for Living.* New York: Abbeville Press.

Morawetz, Herbert (1985). *Polymers: The Origins and Growth of a Science.* New York: John Wiley.

Seymour, Raymond B. (1986). *High Performance Polymers: Their Origin and Development.* New York: Elsevier Science.

metal: element or other substance the solid phase of which is characterized by high thermal and electrical conductivities

Platinum

MELTING POINT: 1,739°C
BOILING POINT: 4,170°C
DENSITY: 21.45 g/cm^3
MOST COMMON IONS: Pt^{2+}, $Pt(Cl)_6{}^{2-}$, $Pt(CN)_4{}^{2-}$, $Pt(CN)_6{}^{2-}$

The first reports of the discovery of platinum were the papers of Antonio de Ulloa, who found an unworkable **metal**, *platina* (Spanish for "little silver"), in the gold mines of Colombia in 1736. Charles Wood provided the first

samples in 1741. Platinum has a concentration of approximately 10^{-6} percent in Earth's crust. Platinum crystallizes in the face-centered cubic structure. The pure metal is malleable and **ductile**, and lustrous and silvery in appearance. It is capable of absorbing gaseous hydrogen. Platinum is found in nature in alluvial deposits and in association with copper, iron, and nickel sulfide ores. The metal is soluble in aqua regia, isolated as $(NH_4)_2PtCl_6$ from aqua regia, and obtained as a sponge or powder by ignition of $(NH_4)_2PtCl_6$.

Platinum is used as a **catalyst** in a wide variety of chemical reactions. Some of the more common **catalytic** uses are the **oxidation** of organic vapors in automobile exhaust, the oxidation of ammonia in the production of nitric acid, and the rearrangement of atoms in petroleum reforming. Most of the halides are formed by direct combination of the **halogen** elements with platinum, resulting in PtF_6, $[PtF_5]_4$, PtX_4 (where X = F, Cl, Br, or I), and PtX_3 and PtX_2 (where X = Cl, Br, or I). The two oxides, PtO and PtO_2, are unstable and decompose upon heating. In the +2 and +3 oxidation states, platinum forms coordination complexes bonded to carbon, nitrogen, phosphorous, oxygen, and sulfur donor atoms. Perhaps the most well known coordination complex is *cis*-platin, $Pt(NH_3)_2Cl_2$, used in chemotherapy treatments of cancer.

ductile: property of a substance that permits it to be drawn into wires

catalyst: substance that aids in a reaction while retaining its own chemical identity

catalysis: the action or effect of a substance in increasing the rate of a reaction without itself being converted

oxidation: process that involves the loss of electrons (or the addition of an oxygen atom)

halogen: element in the periodic family numbered VIIA (or 17 in the modern nomenclature) that includes fluorine, chlorine, bromine, iodine, and astatine

D. Paul Rillema

Bibliography

Cotton, F. Albert, and Wilkinson, Geoffrey (1988). *Advanced Inorganic Chemistry*, 5th edition. New York: Wiley.

Greenwood, Norman N., and Earnshaw, A. (1984). *Chemistry of the Elements*. New York: Pergamon Press.

Plutonium

MELTING POINT: 640°C
BOILING POINT: 3,228°C
DENSITY: 19.84g/cm^3
MOST COMMON IONS: Pu^{3+}, Pu^{4+}, PuO_2^{+}, $PuO_2^{2,+}$ PuO_5^{3-}

Plutonium was discovered by Glenn Seaborg, Edwin McMillan, Joseph Kennedy, and Arthur Wahl in 1940. They prepared a new **isotope** of neptunium, ^{238}Np, which decayed by β-emission to ^{238}Pu.

$$^{238}U(d,2n)^{238}Np \xrightarrow{\beta-} {}^{238}Pu(t_{1/2} = 87.7y)$$

isotope: form of an atom that differs by the number of neutrons in the nucleus

Manhattan Project: government project dedicated to creation of an atomic weapon; directed by General Leslie Groves

Their work as part of the **Manhattan Project** was kept secret and was finally reported in 1946, after World War II, although the existence of plutonium had been revealed to the world earlier, when the atomic bomb was dropped over Nagasaki, Japan. There are sixteen isotopes of plutonium, having mass numbers ranging from 232 to 247. The principal isotopes of Pu are those having mass numbers 238, 239, 240, 241, 242, and 244. Ton quantities of ^{239}Pu (having a half-life of 2.4×0^4 y) are available. The isotope ^{239}Pu is the source material for nuclear weapons and is produced via neutron capture reactions on ^{238}U in nuclear reactors.

$$^{238}U(n,\gamma)^{239}U \xrightarrow{\beta-} {}^{239}Np \xrightarrow{\beta-} {}^{239}Pu$$

About 110 tons of ^{239}Pu are generated in nuclear power plants each year, with approximately 40 percent of the energy produced in the nuclear fuel cycle coming from ^{239}Pu. About three times as much electricity is generated from ^{239}Pu in the United States as from oil-fired electrical generating plants. The ground state (outer orbital) electronic configuration of Pu is $[Rn]5f^67s^2$. The most stable **oxidation** state for plutonium ions in solution is +4, although appreciable amounts of plutonium in its +3, +5, and +6 oxidation states can exist. The aqueous chemistry of plutonium is further complicated by the successive, stepwise hydrolysis of Pu(IV) compounds to form polymers of colloidal dimensions. Plutonium is the transuranium element that is most abundant in the environment, due to the atmospheric testing of nuclear weapons during the 1950s and 1960s that deposited approximately 4.2 tons of plutonium in the environment. Most of this plutonium is in the soil, in which it has no discernable effects. SEE ALSO Actinium; Berkelium; Einsteinium; Fermium; Lawrencium; Mendelevium; Neptunium; Nobelium; Protactinium; Rutherfordium; Seaborg, Glenn Theodore; Thorium; Uranium.

oxidation: process that involves the loss of electrons (or the addition of an oxygen atom)

Walter Loveland

Bibliography

Seaborg, Glenn T., and Loveland, Walter (1990). *The Elements beyond Uranium.* New York: Wiley.

Photo and Illustration Credits

Volume 1

p. 13: Acne on young woman's face, photograph by Biophoto Associates. National Audubon Society Collection/Photo Researchers, Inc. Reproduced by permission; p. 18: Macrophotograph of adhesive, photograph. © Astrid & Hanns-Frieder Michler/Science Photo Library, The National Audubon Society Collection/Photo Researchers, Inc. Reproduced with permission; p. 25: Protesters pulling up genetically modified crops from a field, sign: "genetix snowball," Banbury, Oxfordshire, England, c. 1990-1999, photograph. © Adrian Arbib/Corbis. Reproduced by permission; p. 26: Lab technician removing tissue from a sunflower plant for genetic engineering projects, photograph by Lowell Georgia. Corbis. Reproduced by Corbis Corporation; p. 31: Interior of alchemy laboratory used by John Dee and Edward Kelley, ca. 1585, Powder Tower, Prague, Czechoslovakia, photograph. © Guy Lynon Playfiar/Fortean Picture Library; p. 41: Rhazes, woodcut. The Granger Collection, Ltd. Reproduced by permission; p. 46: Urine drug testing lab, Kansas City, MO, 1995, photograph by Edward L. Lallo. The Picture Cube. Reproduced by permission; p. 50: Bottle of Amoxil (100ml) and a spoon, photograph. © 2003 Kelly A. Quin. Reproduced by permission; p. 58: Arrhenius, Svante, photograph. © Hulton-Deutsch Collection/Corbis. Reproduced by permission; p. 63: Diet Pepsi in a glass, sitting next to a Pepsi can, two packets of Equal sweetener in front, photograph by Kelly A. Quin. Reproduced by permission; p. 70: Gas pillars in Eagle Nebula, view from Hubble Space Telescope (HST), photograph. National Aeronautics and Space Administration (NASA); p. 73: Saints and Sinners, sculpture by Marshall M. Fredericks, photograph. © 2003 Kelly A. Quin. Reproduced by permission; p. 95: Tungsten crystal, magnified 2,7000,000 times, through a super-powered Muller Field Ion Microscope invented by Dr. Erwin W. Muller, photograph. © Bettmann/Corbis. Reproduced by permission; p. 97: Avery, Oswald (with petri dish), photograph. AP/Wide World Photos, Inc. Reproduced by permission; p. 101: Baekeland, Leo, standing in laboratory, photograph. Brown Brothers, Sterling, PA. Reproduced by permission; p. 110: Becquerel, Antoine Henri, photograph. © Bettmann/Corbis. Reproduced by permission; p. 113: Berthollet, Claude Louis, photograph. © Stefano Bianchetti/Corbis. Reproduced by permission; p. 115: Microtechnology, circuit board, satellite spacecraft transmitter, photograph by Brownie Harris. Stock Market. Reproduced by permission; p. 118: Jellyfish floating in dark water (glowing yellow/orange/red) off Cape Hatteras, North Carolina, 1991, photograph by M. Youngbluth. OAR/National Undersea Research Program (NURP)/National Oceanic and Atmospheric Administration; p. 127: Woman pouring chlorine bleach into washing machine, photograph by Robert J. Huffman. FieldMark Publications. Reproduced by permission; p. 129: Danish physicist Niels Bohr, leaning on his right forearm, wearing dark suit, white shirt with wing collar, some physics apparatus partially visible at left, c. 1940s, photograph. © Hulton-Deutsch Collection/Corbis. Reproduced by permission; p. 131: Boltzmann, photograph. Corbis. Reproduced by permission; p. 143: Robert Boyle, standing in laboratory with Denis Papin, engraving. UPI/Corbis-Bettmann. Reproduced by permission; p. 145: Bragg, Sir William, photograph. © Hulton-Deutsch Collection/Corbis. Reproduced by permission; p. 147: Bragg, Sir Lawrence, demonstrating an experiment. © Hulton-Deutsch Collection/Corbis. Reproduced by permission; p. 150: Bunsen, Robert Wilhelm, photograph. © Bettmann/Corbis. Reproduced by permission; p. 154: Espresso machine and cups of coffee, photograph. © Patrik Giardino/Corbis. Reproduced by Corbis Corporation; p. 155: Worker upon a flotation device sprays an acid lake with agricultural lime, done in attempt to counteract the continuing inflow of acidifying materials in waterways, photograph. © Ted Spiegel/Corbis. Reproduced by permission; p. 158: Inspectors viewing contents of parcel on X-ray monitors, photograph. AP/Wide World Photos. Reproduced by permission; p. 160: Cannizzaro, Stanislao, photograph. © Corbis. Reproduced by permission; p. 166: Allotropes, a pencil sitting next to a diamond, photograph by Paul Silverman. Photo Researchers, Inc. Reproduced by permission; p. 168: An atmospheric chemist with evapotranspiration measuring instrument, photograph. National Center for Atmospheric Research/University Corporation for Atmospheric Research/National Science Foundation 1016; p. 169: Chemist, standing at lab bench, photograph by Martha Tabor. Working Images Photographs. Reproduced by permission; p. 171: Researcher spraying decontamination foam in a laboratory to test neutralization of chemicals used as chemical weapons, photograph by © Randy Montoya/Sandia National Laboratories/Science Photo Library. Photo Researchers. Reproduced by permission; p. 173: Manufacturing of nylon carpet fibers, photograph by Ted Horowitz. The Stock Market. Reproduced with permission; p. 175: Carver, George Washington (blowing into tube that is inserted into a beaker), photograph. The Library of Congress; p. 180: Cavendish, Henry, photograph. Photo Researchers, Inc. Reproduced by permission; p. 183: Woman, painting ceramic figure, photograph. © AP/Wide World Photos. Reproduced by permission; p. 187: Top of cigarette lighter, photograph by Robert J. Huffman. FieldMark Publications. Reproduced by permission; p. 189: Chadwick, James, photograph by A. Barrington Brown. Photo Researchers, Inc. Reproduced by permission; p. 192: Chappelle, Emmet W., with bacteria detector, photograph. Corbis. Reproduced by permission; p. 193: Chardonnet, Louis Marie Hilaire Bernigaud, photograph. © Corbis. Reproduced by permission; p. 199: Molecular graphics, photograph by Ken Eward. BioGrafx/Science Source, National Audubon Society Collection/Photo Researchers, Inc. Reproduced by permission; p. 200: Combustion of sulfur, photograph. © Lawrence Migdale/Science Source/National Audubon Society Collection/Photo Researchers, Inc. Reproduced with permission; p. 207: Woman, wearing a uniform, holding three Chemiluminescence lightsticks, photograph. © Photo Researchers. Reproduced by permission; p. 211: High dose chemotherapy (boy in baseball cap and glasses), photograph. Custom Medical Stock Photo. Reproduced by permission; p. 217: Chevreul, Michel Eugene, portrait. The Library of Congress; p. 231: Two calves cloned from cells found in cow's milk, Japan, photograph. AP/Wide World Photos. Reproduced by permission; p. 235: Hands holding pieces of coal, coal-fired power plant in background., photograph. ©Tim Wright/CORBIS. Reproduced by permission; p. 242: Fibroblast and collagen fibers, photograph. © Cmeabg-ucbl. Photo Researchers. Reproduced by permission; p. 244: An Electronic sign on Interstate 287, in Parsippany, New Jersey, photograph. AP/Wide World Photos. Reproduced by permission; p. 249: Light passing through NaOh (left), and colloidal mixture (right), photograph by Charles D. Winters. Photo Researchers, Inc. Reproduced by permission;

p. 253: Robot adds solution and stirs DNA samples from New York Terrorism victims, photograph by Douglas C. Pizac. AP/Wide World Photos. Reproduced by permission; p. 255: Pople, John, Sugar Land, Texas, 1998, photograph Brett Coomer. AP/Wide World Photos. Reproduced by permission; p. 268: Cori, Carl F. and Gerty T. Cori, photograph. The Library of Congress; p. 270: Rust, oxidation of ferrous metal bolt, photograph by Matt Meadows. Peter Arnold, Inc. Reproduced by permission; p. 283: Multi-celled human embryo 2.5 days after it was removed from the womb and stored cryogenically at the Bourn Hall Fertility Clinic, Cambridgeshire, England, photograph. AP/Wide World Photos. Reproduced by permission; p. 285: CAT scan, photograph. © 1990 Terry Qing/FPG International Corp. Reproduced by permission; p. 287: Human brain with Parkinson's Disease showing atrophy, Computerized Axial Tomography, CAT, scan. © GJLP/CNRI/Phototake. Reproduced by permission; p. 289: Curie, Marie, photograph. AP/Wide World Photos. Reproduced by permission.

Volume 2

p. 2: Dalton, John, engraving. The Library of Congress, Prints and Photographs Division; p. 3: Davy, Sir Humphrey, portrait. Corbis-Bettmann. Reproduced with permission; p. 6: Broken egg, photograph by Arthur Beck. Corbis. Reproduced by permission; p. 7: Sunny side up egg, photograph by Japack Company/Corbis. Reproduced by permission; p. 12: Dewar, Sir James, portrait. The Library of Congress; p. 14: A female patient undergoing a cerebral angiography, photograph by Laurent. Photo Researchers. Reproduced by permission; p. 18: Baby wearing a disposable diaper, photograph by Elizabeth Thomason. Copyright © 2003 Elizabeth Thomason. Reproduced by permission; p. 23: DNA, computer graphics, photograph by Clive Freeman. Photo Researchers, Inc. Reproduced by permission; p. 25: Dupont, Eleuthere Irenee, engraving. Archive Photos, Inc. Reproduced by permission; p. 27: Textiles, fabric dyes, photograph by Randy Duchaine. Stock Market. Reproduced by permission; p. 31: Ehrlich, Paul, in laboratory, 1908, photograph. Liaison Agency/Hulton Getty. Reproduced by permission; p. 32: Einstein, Albert, photograph. Archive Photos. Reproduced by permission; p. 40: Hitchings, George, Gertrude Elion, photograph. AP/Wide World. Reproduced by permission; p. 43: Wheelchair racers participating in a marathon in Washington, D.C., photograph by Joe Sohm. Chromosohm/Stock Connection/PictureQuest. Reproduced by permission; p. 45: Electronic display (concentric octagons) of energies from subatomic particles, photograph. Newman Laboratory of Nuclear Science, Cornell University, National Audubon Society Collection/Photo Researchers, Inc. Reproduced by permission; p. 51: Leachate from the Dowelanco plant in Lafayette, Indiana, photograph by Sandra Marquardt. Greenpeace. Reproduced by permission; p. 52: Chemicals being sprayed on agricultural land on the Everglades Park border, Florida, photograph. Earth Scenes/© Arthur Gloor. Reproduced by permission; p. 55: Enzyme, lock-and-key model of interaction, drawing by Hans & Cassidy. Gale Research; p. 58: Chemical reaction, nitric acid, photograph. © LevyYoav/Phototake NYC. Reproduced by permission; p. 69: Reed, Ken and Peter Mckay, taking part in a demonstration against embryonic stem cell research outside of George Washington University Hospital, photograph by Kamenko Pajic. AP/Wide World Photos. Reproduced by permission; p. 70: Harman, Jane, holding up a support sign of stem cell research, at a press conference, outside the Capitol building, in Washington D.C., photograph by Shawn Thew. Getty Images. Reproduced by permission; p. 75: Nuclear explosion at sea, photograph by PHOTRI. Stock Market. Reproduced by permission; p. 78: Engraving of Professor Faraday delivering a lecture at the Royal Institution, London, photograph. © Corbis. Reproduced by permission; p. 86: Fermi, Enrico, photograph. © Bettmann/Corbis. Reproduced by Corbis Corporation; p. 88: Truck applying fertilizer in rice field, photograph by Thomas Del Brase. Stock

Market. Reproduced by permission; p. 92: Carpet trader in Sarajevo. Photograph. David Reed/Corbis. Reproduced by permission; p. 94: Dacron polyester fibers, photograph. © Science Photo Library, The National Audubon Society Collection/Photo Researchers, Inc. Reproduced with permission; p. 95: Colored scanning electron micrograph of eyelashes growing from the surface of human skin, photograph. Photo Researchers, Inc. Reproduced by permission; p. 100: Torchlight parade, photograph by Byron Hertzler. AP/Wide World Photos. Reproduced by permission; p. 103: Fleming, Sir Alexander, photograph. Corbis-Bettmann. Reproduced with permission; p. 107: Supermarket deli worker waiting on a customer, photograph by Jeff Zaruba. Corbis. Reproduced by permission; p. 109: Forensic scientist, removing a fragment of material from a bloodstained garment recovered from a crime scene for DNA testing, photograph by Dr. Jurgen Scriba. Photo Researchers, Inc. Reproduced by permission; p. 111: Lab Agent Kathy Dressel works on DNA evidence in the murder of JonBenet Ramsey at the Colorado Bureau of Investigation, Boulder, Colorado, photograph. AP/Wide World Photos. Reproduced by permission; p. 118: Motorist filling up his automobile gas tank at a Union 76 gas station in San Francisco, California, photograph. © Reuters Newsmedia Inc./Corbis. Reproduced by permission; p. 124: Franklin, Rosalind, British X-ray Crystallographer, photographer. Photo Researchers, Inc. Reproduced by permission; p. 127: Goddard Space Flight Center, National Aeronautics and Space Administration; p. 129: Buckministerfullerene, photograph by Laguna Design. Photo Researchers, Inc. Reproduced by permission; p. 132: Open microwave oven on kitchen counter, photograph by Robert J. Huffman. FieldMark Publications. Reproduced by permission; p. 137: Photograph by Kelly A. Quin. Copyright © 2003 Kelly A. Quin. Reproduced by permission; p. 138: Soil tilth test using compost (open palm with dirt), photograph by Robert J. Huffman/Field Mark Publications. Reproduced by permission; p. 149: Gay-Lussac, Joseph Louis, photograph. Archive Photos, Inc. Reproduced by permission; p. 153: Diamonds, a mixture of natural and synthetic, shapes and sizes, photograph by Sinclair Stammers. Photo Researchers, Inc. Reproduced by permission; p. 154: Handful of gems that were found in a bucket of dirt at a mine near Franklin, North Carolina, photograph by Raymond Gehman. Corbis. Reproduced by permission; p. 158: Lab technician removes corn embryos to be grown in controlled conditions for desired genetic traits at the Sungene Technologies Lab, Palo Alto, CA, photograph. © Lowell Georgia/Corbis. Reproduced by permission; p. 161: Gibbs, Josiah Willard (three-quarter view from his right, wide bow tie), engraving. The Library of Congress; p. 167: The Artsgarden (glass vault), Indianapolis, Indiana, photograph by Shawn Spence. The Indianapolis Project. Reproduced by permission; p. 177: Goodyear, Charles, sketch. The Library of Congress; p. 179: Zion National Park, Utah, photograph by Robert J. Huffman. Field Mark Publications. Reproduced by permission; p. 183: Haber, Fritz, photograph. © Austrian Archives/Corbis. Reproduced by permission; p. 188: Karen getting hair highlighted by stylist Patti (unseen), sheet of foil under lock of hair, at Maximillian's Salon, photograph by Kelly A. Quin. Reproduced by permission; p. 192: The first blobs of isolated aluminum, photograph. © James L. Amos/Corbis. Reproduced by permission; p. 193: Chlorine gas, in a clear jar, photograph by Martyn F. Chillmaid. Photo Researchers, Inc. Reproduced by permission; p. 197: Heisenberg, W., (second from left) discussing with the staff of the Russian Stand, the adiabatic trap "Ogra", photograph. © Bettmann/Corbis. Reproduced by permission; p. 202: Worker with the California Transportation Department riding on the front of a truck and spraying herbicide, photograph by Ed Young. Corbis. Reproduced by permission; p. 207: Heyrovsky, Jaroslav, portrait. The Library of Congress; p. 208: Hodgkin, Dorothy, photograph. Archive Photos, Inc./Express Newspapers. Reproduced by permission; p. 216: Chlorine plant in Louisiana, photograph.

© Royalty Free/Corbis. Reproduced by permission; p. 223: Petrochemical plant at dusk, photograph by Jeremy Walker. Photo Researchers, Inc. Reproduced by permission; p. 233: Protease inhibitor, photograph by Leonard Lessin. Photo Researchers, Inc. Reproduced by permission; p. 239: Tent caterpillars, (making cocoon), photograph by Edward S. Ross. Copyright E. S. Ross. Reproduced by permission; p. 241: Ladybug on a green stalk, photograph by Robert J. Huffman. Field Mark Publications. Reproduced by permission; p. 253: Sandwich meats in a supermarket, photograph by Todd Gipstein. Corbis. Reproduced by permission.

Volume 3

p. 1: Kekule, Friedrich, photograph. Photo Researchers, Inc. Reproduced by permission; p. 3: Kelsey, Frances Oldham, photograph. The Library of Congress; p. 5: Khorana, Har Gobind, photograph. AP/Wide World Photos. Reproduced by permission; p. 8: Acid dehydration of sugar, photograph by David Taylor. Science Photo Library, National Audubon Society Collection/Photo Researchers, Inc. Reproduced by permission; p. 10: Krebs, Hans Adolf, photograph. The Library of Congress; p. 16: Lavoisier, Antoine-Laurent, photograph. The Library of Congress; p. 17: Instruments from Lavoisier's lab, portrait. The Library of Congress; p. 18: Lawrence, Ernest (with atom smasher), photograph. Corbis-Bettmann. Reproduced by permission; p. 23: Le Bel, Joseph Achille, photograph. Photo Researchers, Inc. Reproduced by permission; p. 24: Leblanc, Nicholas, photograph. Bettmann/Corbis. Reproduced by permission; p. 28: Leloir, Luis F., photograph. UPI/Corbis-Bettmann. Reproduced by permission; p. 29: Lewis, Gilbert N., photograph. Corbis-Bettmann. Reproduced by permission; p. 34: Liebig, Justes von, portrait. The Library of Congress; p. 40: Transmission electron micrograph of lipid droplets in a rat fat cell, magnified x 90,000, photograph by David M. Phillips, The Population Council/Photo Researchers, Inc. Reproduced by permission; p. 41: Woman holding a digital fishfinder, photograph. © AFP/Corbis. Reproduced by permission; p. 45: Waterstrider bug, photograph. © Hermann Eisenbeiss, National Audubon Society Collection/Photo Researchers, Inc. Reproduced by permission; p. 46: Liquids, viscosity, photograph. © Yoav Levy/Phototake NYC. Reproduced by permission; p. 48: Lister, Joseph, photograph. © Bettmann/Corbis. Reproduced by permission; p. 50: Lonsdale, Kathleen, photograph. AP/Wide World Photos. Reproduced by permission; p. 52: False-color angiogram showing stenosis in the carotid artery, photograph by Alfred Pasteka. © CNRI/ Science Photo Library, National Audubon Society Collection/Photo Researchers, Inc. Reproduced by permission. p. 56: Bar magnet with iron filings showing field of force (bright blue background), photograph. Grantpix/Photo Researchers, Inc. Reproduced by permission; p. 60: Oppenheimer, J. Robert, 1946, photograph. © Bettmann/Corbis. Reproduced by permission; p. 62: Martin, Archer, photograph by Nick Sinclair. Photo Researchers, Inc. Reproduced by permission; p. 63: Red sports car, photograph. © Royalty Free/Corbis. Reproduced by permission; p. 67: Maxwell, James Clerk, engraving. The Library of Congress; p. 69: Post-mounted sundial in a Bolivian town, photograph. © John Slater/Corbis. Reproduced by permission; p. 72: Teenage girl being weighed on large scale, photograph by Robert J. Huffman. Field Mark Publications. Reproduced by permission; p. 74: Hahn, Otto, standing with Lise Meitner, photograph. Photo Researchers, Inc. Reproduced by permission; p. 77: Mendeleev, Dmitri Ivanovich, in the chemical factory of the University of St. Petersburg, photograph. © Bettmann/Corbis. Reproduced by permission; p. 82: Landfill with aluminum drum for mercury waste, photograph. Recio/Greenpeace. Reproduced by permission; p. 83: Ritalin tablets, photograph. © Photo Researchers, Inc. Reproduced by permission; p. 85: Lothar von Meyer, Julius, photograph. Photo Researchers, Inc. Reproduced by permission; p. 86: Millikan, Robert A., photograph. The Library of Congress; p. 88: Stratas of banded liassic limestone and shale on a seacliff by

Chinch Gryniewicz. Ecoscene/Corbis. Reproduced by permission; p. 89: Matrix of diamond, photograph. U. S. National Aeronautics and Space Administration (NASA); p. 104: Chaney, Michael wearing special 3-D glasses to view a computer model of Fluoxetine, or Prozac. Lilly Research Labs, Indianapolis, February 22, 1994, photograph. Corbis. Reproduced by permission; p. 115: Three-dimensional computer model of a protein molecule of matrix porin found in the E. Coli bacteria, photograph. © Corbis. Reproduced by permission; p. 118: Two models which represent the arrangement of atoms in a chemical molecule, illustration. Tek Image/Science Photo Library. Photo Researcher, Inc. Reproduced by permission; p. 122: Moseley, Henry, photograph. Photo Researchers, Inc. Reproduced by permission; p. 124: Six-legged frog, photograph. JLM Visuals. Reproduced by permission; p. 130: Coffee shop sign at the Pike Place Public Market, Seattle, Washington, ca. 1990, photograph by Stuart Westmorland. CORBIS/Stuart Westmorland. Reproduced by permission; p. 132: Professor Walther Nernst of Berlin University, Nobel Prize winner for physics, in his laboratory, photograph. © Bettmann/Corbis. Reproduced by permission; p. 137: Diagram showing transmitting and receiving neurons and the role of serotonin in communication between one neuron and the next, illustration by Frank Forney; p. 143: Photovoltaic cell powered desalination system, Saudi Arabia, photograph. Mobil Solar Energy Corporation/Phototake NYC. Reproduced by permission; p. 144: Newton, Sir Isaac (oval frame, allegorical elements surrounding), engraving. The Library of Congress; p. 153: Hand pouring liquid nitrogen into beaker, photograph. © David Taylor/Photo Researchers. Reproduced by permission; p. 155: Nobel, Alfred (wearing suit, checked bow tie), drawing. The Library of Congress; p. 158: Scientist from University of California—Berkley studying helium gas being released from a hot spring in Yellowstone National Park, photograph. © James L. Amos/Corbis. Reproduced by permission; p. 168: Lasers focus on a small pellet of fuel in attempt to create nuclear fusion reaction, photograph. © Roger Ressmeyer/Corbis. Reproduced by permission; p. 169: Seaborg, Glenn, photograph by Lawrence Berkeley. Photo Researchers, Inc. Reproduced by permission; p. 172: Nuclear reactors, interior of the containment building, photograph by Mark Marten. Photo Researchers, Inc. Reproduced by permission; p. 174: Nuclear fusion, computer simulation, photograph by Jens Konopka and Henning Weber. Photo Researchers, Inc. Reproduced by permission; p. 178: Magnetic Resonance Imaging System (room with equipment and viewing machine), photograph by Mason Morfit. FPG International Corp. Reproduced by permission; p. 181: Colored positron emission tomography (PET) scans comparing the brain of a depressed person (top) with the brain of a healthy person, illustration. © Photo Researchers, Inc. Reproduced by permission; p. 192: Oppenheimer, Dr. J. Robert, atomic physicist and head of the Manhattan Project, photograph. National Archives and Records Administration; p. 209: Ostwald, Friedrich Wilhelm, portrait. The Library of Congress; p. 211: Doctor communicating with a patient lying in a pressure chamber and undergoing hyperbaric oxygen therapy, photograph by James King-Holmes/Science Photo Library. Reproduced by permission; p. 214: Woodcut from "Works", of Paracelsus ca. 1580 edition, of Mercury in its alchemical form as Azoth, photograph. Fortean Picture Library. Reproduced by permission; p. 218: © Kevin Fleming/Corbis. Reproduced by permission; p. 220: Pasteur, Louis, photograph. Hulton-Deutsch Collection/Corbis. Reproduced by permission; p. 222: Pauli, Wolfgang, photograph by Zurich Bettina. Courtesy of CERN-European Organization for Nuclear Research. Reproduced by permission; p. 223: Pauling, Linus, proponent of heavy vitamin C use, tossing an orange, photograph. © Roger Ressmeyer/Corbis. Reproduced by permission; p. 225: Penicillin, photograph by Andrew McClenaghan, photograph. Photo Researchers, Inc. Reproduced by permission; p. 229: Early Periodic Table, portrait. The Library of Congress; p. 234: Worker in protective photograph by Ed Young. Corbis. Reproduced by permission; p. 238: Smith, Billy, Drilling Foreman, seated at Drake Well, photograph. AP/Wide World. Reproduced by permission; p. 241: Man in red hardhat standing on bluish oil refinery equipment, yellow flames from burning natural gas at top right, photograph. © Lowell Georgia/Corbis. Reproduced by permission; p. 249: The first known photograph of a courtyard outside Niepce's room, photograph. The Bettmann Archive/Corbis. Reproduced by permission; p. 256: Hot water vaporizing as it is thrown into air during minus 35-degree Fahrenheit temperature, Boundary Waters Canoe Area Wilderness, Minnesota, USA, photograph © Layne Kennedy/Corbis. Reproduced by permission; p. 258: X-ray diffraction crystallography, photograph by James King-Holmes. OCMS/Science Photo Library/ Photo Researchers, Inc. Reproduced by permission; p. 259: Several open cans of paint in an assortment of colors, photograph by Matthias Kulka. Corbis. Reproduced by permission; p. 262: Planck, Max, photograph. © Bettmann/Corbis. Reproduced by permission.

Volume 4

p. 5: Travolta, John, with Karen Lynn Gorney in a scene from "Saturday Night Fever," photograph. The Kobal Collection. Reproduced by permission; p. 12: Ghost crab walking among a sandy shoreline, photograph. Sapelo Island National Estuarine Research Reserve. Reproduced by permission; p. 30: Priestley, Joseph, engraving. International Portrait Gallery; p. 38: Scientist adjusts equipment used in fast protein liquid chromatography, photograph. Photo Researchers, Inc. Reproduced by permission; p. 50: Spectrum of electromagnetic radiation, (chart) illustrations by Robert L. Wolke. Reproduced by permission; p. 52: Radioactive emission of alpha particles from radium, yellow sunburst, dark blue background, photograph. Photo Researchers, Inc. Reproduced by permission; p. 58: International mobile radiological radiologists measuring soil radioactivity levels near Ukraine's Chernobyl Nuclear Plant, wearing protective clothing, from International Mobile Radiological Laboratories, near Chernobyl, Ukraine, photograph. © Reuters NewsMedia Inc./Corbis. Reproduced by permission; p. 60: Radon test kit and carbon monoxide detector, photograph by Robert J. Huffman. Field Mark Publications. Reproduced by permission; p. 63: Ramsay, William, portrait. The Library of Congress; p. 65: Reaction rate, illustration. GALE Group. Studio: EIG; p. 74: Bragg, Kevin, photograph. © James A. Sugar/Corbis. Reproduced by permission; p. 80: Open electric oven showing the glowing red heating element, photograph by Robert J. Huffman. FieldMark Publications. Reproduced by permission; p. 86: Robinson, Robert, portrait. The Library of Congress; p. 89: Captured German V-2 rocket cuts through the clouds during experiment, 1947, photograph. AP/Wide World Photos. Reproduced by permission; p. 90: Full-scale operational model of the Mars Viking Lander, photograph by Lowell Georgia/Corbis. Reproduced by permission; p. 97: Rubidium, photograph. © Russ Lappa/Science Source, National Audubon Society Collection/Photo Researchers, Inc. Reproduced by permission; p. 99: Rutherford, Ernest, Walton (left to right), Ernest Rutherford, and J.D. Cockcroft, standing together outside, photograph. © Hulton-Deutsch Collection/Corbis; p. 103: Large pile of salt at a salt mine, October 29, 1996, Upington, South Africa, photograph. © Charles O'Rear/Corbis; p. 106: Sanger, Frederick looking up at model of DNA molecule, photograph. © Bettmann/Corbis. Reproduced by Corbis Corporation; p. 108: Scheele, Carl Wilhelm, portrait. The Library of Congress; p. 110: Schrodinger, Professor Erwin of Magdalen College, Oxford, winner of the 1933 Nobel Prize in Physics along with Professor P.H.M. Dirac of Cambridge University, England, November 18, 1933, photograph. © Bettmann/Corbis. Reproduced by permission; p. 111: Seaborg, Glenn T., photograph. © Roger Ressmeyer/Corbis. Reproduced by permission; p. 115: Seibert, Florence, photograph. AP/Wide World Photos. Reproduced by permission; p. 116: Dish of selenium, photograph. © Rich

Treptow, National Audubon Society Collection/Photo Researchers, Inc. Reproduced by permission; p. 117: Person holding a conventional tub amplifier, photograph. © Bettmann/Corbis. Reproduced by permission; p. 120: Micro wires, silicon chip, photograph by Andrew Syred. The National Audubon Society Collection/ Photo Researchers, Inc. Reproduced by permission; p. 122: Worker testing silicon wafers at the Matsushita Semiconductor plant in Puyallup, Washington, c. 1994, photograph. © Kevin R. Morris/Corbis. Reproduced by permission; p. 131: Soddy, Frederick, drawing. The Library of Congress; p. 132: Sodium acetate hot pack reacting, photograph by Charles D. Winters. Photo Researchers, Inc. Reproduced by permission; p. 149: Solution of water and oil, droplets are exposed to polarized light, magnification, photograph by David Spears. © Science Pictures Limited/Corbis; p. 154: Electron spectroscopy, chemical analysis, photograph by Joseph Nettis. Photo Researchers, Inc. Reproduced by permission; p. 157: Stanley, Wendell Meredith, photograph. The Library of Congress; p. 159: Staudinger, Hermann, portrait. The Library of Congress; p. 161: Steel worker sampling molten metal, photograph by John Olson. Stock Market. Reproduced by permission; p. 164: Steel Bridge, Portland, Oregon, photograph by James Blank. © James Blank. Reproduced by permission; p. 167: Hand holding 50 mg bottle of the injectable anabolic steroid Durabolin, photograph. Custom Medical Stock Photo. Reproduced by permission; p. 169: Razor blade cutting powder form of the drug Speed into lines for snorting, photograph. Science Photo Library. Reproduced by permission; p. 174: Germinating garden pea Pisum Sativum, showing embryonic root growing downward, macrophotograph, photograph. © J. Burgess/Photo Researchers. Reproduced by permission; p. 176: Rayleigh, 3rd Baron (John W. Strut, III), drawing. The Library of Congress; p. 179: Vat of sulfur, photograph by Farrell Grehan. Corbis. Reproduced by permission; p. 180: Sumner, James B., portrait. The Library of Congress; p. 182: Meissner effect, cube floating above a superconductor, photograph by Yoav Levy. Phototake NYC. Reproduced by permission; p. 187: Scanning tunneling microscope image of sulfur atoms on rhodium surface, photograph by Dr. James Batteas. National Institute of Standards and Technology; p. 187: Topographic image of a reconstructed silicon surface showing outermost atomic positions, photograph of Dr. James Batteas. National Institute of Standards and Technology. Reproduced by permission; p. 189: Infrared near-field scanning optical microscope (NSOM) image of blended polystyrene/polyethylacrylate film on silicon, photograph by f Dr. Chris Michaels and Dr. Stephan Stranick, National Institute of Standards and Technology. Reproduced by permission; p. 189: Nanowriting of a mercaptbenzoic acid layer into a background layer of dodecanethiol using nanografting, photograph by Dr. Jayne Gerano, National Institute of Standards and Technology. Reproduced by permission; p. 191: Tractor riding past wind generators, photograph. © AP/Wide World Photos. Reproduced by permission; p. 192: Cutaway of the ITER tokamak device, photograph. Photo Researchers, Inc. Reproduced by permission; p. 193: Svedberg, Theodor, portrait. The Library of Congress; p. 195: Synge, Richard L.M., photograph. The Library of Congress; p. 198: Organic chemist synthesizing chemotherapy agent, photograph by Geoff Tompkinson. Photo Researchers, Inc. Reproduced by permission; p. 200: Szent-Gyorgyi, Dr. Albert, Nobel Prize winner, discoverer of Vitamin C and bioflavinoids, photograph. © Bettmann/CORBIS. Reproduced by permission; p. 205: Photocopier in office, photograph by Robert J. Huffman. FieldMark Publications. Reproduced by permission; p. 207: © Royalty-Free/Corbis. Reproduced by permission; p. 209: Yendell, Thomas, a Thalidomide baby, picks up a toy with his feet, photograph. AP/Wide World Photos. Reproduced by permission; p. 215: Thallium heart scan, photograph by Peter Berndt, M.D. Custom Medical Stock Photo. Reproduced by permission; p. 226: Thomson, Sir Joseph John, photograph. UPI/Corbis-Bettmann.

Reproduced by permission; p. 227: Man, writing, being flanked by two other men watching. AP/Wide World Photos. Reproduced by permission; p. 228: Army medic, taking an x-ray of a fellow soldier simulating an injury, photograph. © Bettmann/Corbis. Reproduced by permission; p. 236: Stanford Linear Accelerator Center, photograph by David Parler. National Audubon Society Collection/Photo Researchers Inc. Reproduced by permission; p. 241: Alchemist (seated left wearing eyeglasses) and his assistants, engraving. The Granger Collection, New York. Reproduced by permission; p. 244: Computer generated graphic of a molecule of hemoglobin-carrier of oxygen in the blood, photograph. © Science Photo Library. Photo Researchers. Reproduced by permission; p. 249: Urey, Harold, photograph. The Library of Congress; p. 253: Van Der Waals, Johannes Diderik, photograph. Science Photo Library. Photo Researchers, Inc. Reproduced by permission; p. 255: Helmont, Jan (Baptise) Van (standing), print; p. 256: Hoff, Jacobus Henricus vant, photograph. Photo Researchers, Inc. Reproduced by permission; p. 258: Banded gila monster, photograph by Renee Lynn. The National Audubon Society Collection/Photo Researchers, Inc. Reproduced by permission; p. 261: Allesandro Volta, standing, body turned slightly to his left, painting. UPI/Corbis-Bettmann. Reproduced by permission; p. 263: Waksman, Selman A.(in laboratory), photograph. The Library of Congress.; p. 265: Molecular graphic of water molecules evaporating from a solution, photograph. © K. Eward/Photo Researchers. Reproduced by permission; p. 270: Dead fish, killed from water pollution, photograph. © U.S. Fish & Wildlife Service; p. 273: Newly treated water in the Orange County Water Treatment Works, Florida, photograph by Alan Towse. © Ecoscene/CORBIS. Reproduced by permission; p. 275: Watson, Dr. James Dewey, with DNA model, 1962, photograph. UPI/Corbis-Bettmann. Reproduced by permission; p. 277: Werner, Alfred, portrait. The Library of Congress; p. 279: Willstatter, Richard, photograph. Photo Researchers, Inc. Reproduced by permission; p. 280: Wohler, Friedrich, portrait. The Library of Congress; p. 282: Woodward, Robert Burns (standing with unknown man, viewing book), photograph. The Library of Congress; p. 284: Cylindrical map of the eastern half of the surface of Venus, as viewed from the spacecraft Magellan, photograph. NASA/JPL/CALTECH. Reproduced by permission; p. 285: Yalow, Rosalyn, 1978, photograph. Corbis-Bettmann. Reproduced by permission; p. 288: Yukawa, Hideki (holding chalk, pointing to writing on board), photograph. The Library of Congress; p. 293: Kangaroo on the beach in Australia, photograph. © Australian Picture Library/Corbis. Reproduced by permission; p. 294: Zsigmondy, Richard, portrait. The Library of Congress.

Glossary

acetylcholine: neurotransmitter with the chemical formula $C_7H_{17}NO_3$; it assists in communication between nerve cells in the brain and central nervous system

acid rain: precipitation that has a pH lower than 5.6; term coined by R. A. Smith during the 1870s

activation analysis: technique that identifies elements present in a sample by inducing radioactivity through absorbtion of neutrons

adenine: one of the purine bases found in nucleic acids, $C_5H_5N_5$

adenosine triphosphate (ATP): molecule formed by the condensation of adenine, ribose, and triphosphoric acid, HOP(O)OH–O–(O)OH–OP(O)OH–OH; it is a key compound in the mediation of energy in both plants and animals

adrenalin: chemical secreted in the body in response to stress

alchemy: medieval chemical philosophy having among its asserted aims the transmutation of base metals into gold

aldehyde: one of the characteristic groups of atoms in organic compounds that undergoes characteristic reactions, generally irrespective of where it occurs in the molecule; the RC(O)H functional group

aliphatic: having carbon atoms in an open chain structure (as an alkane)

aliquot: specific volume of a liquid used in analysis

alkaloid: alkaline nitrogen-based compound extracted from plants

alloy: mixture of two or more elements, at least one of which is a metal

α subunit: subunit that exists in proteins that are composed of several chains of amino acids, the first unit in the "counting" of the units

α-particle: subatomic particle with 2+ charge and mass of 4; a He nucleus

amalgam: metallic alloy of mercury and one or more metals

amine functional group: group in which nitrogen is bound to carbon in an organic molecule in which two other groups or hydrogen atoms are bound to nitrogen; major component of amino acids

amino acid residue: in a protein, which is a polymer composed of many amino acids, that portion of the amino acid that remains to become part of the protein

amino acid sequence: twenty of the more than five hundred amino acids that are known to occur in nature are incorporated into a variety of proteins that are required for life processes; the sequence or order of the amino acids present determines the nature of the protein

amphetamine: class of compounds used to stimulate the central nervous system

anabolism: metabolic process involving building of complex substances from simpler ones, using energy

analgesic: compound that relieves pain, e.g., aspirin

androgen: group of steroids that act as male sex hormones

angiotensin: chemical that causes a narrowing of blood vessels

anhydrous compound: compound with no water associated with it

anion: negatively charged chemical unit, like Cl−, CO_{32}−, or NO_3−

anthcyanin: antioxidant flavanoid that makes raspberries red and blueberries blue

antibody: protein molecule that recognizes external agents in the body and binds to them as part of the immune response of the body

anticoagulant: molecule that helps prevent the clotting of blood cells

antiscorbutic: substance that has an effect on scurvy

apoenzyme: the protein part of an enzyme that requires a covalently bound coenzyme (a low molecular weight organic compound) or a cofactor (such as a metal ion) for activity

aqueous solution: homogenous mixture in which water is the solvent (primary component)

aromatic: having a double-bonded carbon ring (typified by benzene)

asparagine residue: amino acid asparagine unit as it occurs in a polypeptide chain

atomic mass units: unit used to measure atomic mass; 1/12 of the mass of a carbon-12 atom

atomic number: the number of protons in an atomic nucleus, expressed in terms of electric charge; it is usually denoted by the symbol Z

atomic orbital: mathematical description of the probability of finding an electron around an atom

atomic spectrum: electromagnetic array resulting from excitement of a gaseous atom

atomic theory: concept that asserts that matter is composed of tiny particles called atoms

atomic weight: weight of a single atom of an element in atomic mass units (amu)

attraction: force that brings two bodies together, such as two oppositely charged bodies

axial bond: covalent bond pointing along a molecular axis

azo dye: synthetic organic dye containing a –N=N– group

bacteriophage multiplication: process by which immune system cells responsible for battling bacterial infections reproduce

basal metabolism: the process by which the energy to carry out involuntary, life-sustaining processes is generated.

***β* subunit:** subunit that exists when two or more polypeptide chains associate to form a complex functional protein, the chains are referred to as "subunits"; these subunits are often identified as α, β, etc.

biological stain: dye used to provide contrast among and between cellular moieties

biomass: collection of living matter

biosynthesis: formation of a chemical substance by a living organism

boat conformation: the arrangement of carbon atoms in cyclohexane, C_6H_{12}. In which the spatial placement of the carbon atoms resembles a boat with a bow and a stern

brine: water containing a large amount of salts, especially sodium chloride (NaCl)

Brownian motion: random motion of small particles, such as dust or smoke particles, suspended in a gas or liquid; it is caused by collisions of the particle with gas or solvent molecules which transfer momentum to the particle and cause it to move

calc: calcium carbonate

calcine: to heat or roast to produce an oxide (e.g., CaO from calcite)

capacitor plate: one of several conducting plates, or foils, in a capacitor, separated by thin layers of dielectric constant, an insulating material

carboxylate: structure incorporating the –COO– group

carboxyl group: an organic functional group, –C(O), found in aldehydes, ketones, and carboxyl acids.

carboxylic acid: one of the characteristic groups of atoms in organic compounds that undergoes characteristic reactions, generally irrespective of where it occurs in the molecule; the $-CO_2H$ functional group

catabolism: metabolic process involving breakdown of a molecule into smaller ones resulting in a release of energy

catalysis: the action or effect of a substance in increasing the rate of a reaction without itself being converted

catalyst: substance that aids in a reaction while retaining its own chemical identity

catalytic conversion: catalytic oxidation of carbon monoxide and hydrocarbons in automotive exhaust gas to carbon dioxide and water

cell culture: artificially maintained population of cells, grown in a nutrient medium and reproducing by asexual division

cephalosporin C: family of antibiotics obtained from a fungus acting in a manner similar to penicillin

chain of custody: sequence of possession through which evidentiary materials are processed

chair conformation: arrangement of atoms in a cycle structure (usually a six-membered ring) that appears to be arranged like a chair (as opposed to the other conformation which is described as a "boat")

chemical-gated: of a membrane protein whose action to open a pore in the membrane occurs only after a substrate has been binded to the protein or a cofactor

chlorofluorocarbon (CFC): compound containing carbon, chlorine, and fluorine atoms that remove ozone in the upper atmosphere

chlorophyll: active molecules in plants undergoing photosynthesis

chromatography: the separation of the components of a mixture in one phase (the mobile phase) by passing in through another phase (the stationary phase) making use of the extent to which the components are absorbed by the stationary phase

chromophore: part of the molecule that yields characteristic colors

cladding: protective material surrounding a second material, which is frequently tubes filled with uranium dioxide pellets in a nuclear reactor

cleave: split

cobrotoxin: polypeptide toxin containing sixty-two residues that is found in the venom of cobras

code: mechanism to convey information on genes and genetic sequence

cofactor: inorganic component that combines with an apoenzyme to form a complete functioning enzyme

coherent mass: mass of particles that stick together

color fastness: condition characterized by retention of colored moieties from a base material

combustion: burning, the reaction with oxygen

competitive inhibitor: species or substance that slows or stops a chemical reaction

complementarity: basis for copying the genetic information, where each nucleotide base has a complementary partner with which it forms a base-pair

congener: an element or compound belonging to the same class

constitutional isomer: form of a substance that differs by the arrangement of atoms along a molecular backbone

contact activity: process involving the touching of different surfaces

contraction: the shortening of a normal trend of a quantity

coordinate covalent bond: covalent bond in which both of the shared electrons originate on only one of the bonding atoms

coordination chemistry: chemistry involving complexes of metal ions surrounded by covalently bonded ligands

corrosive gas: gas that causes chemical deterioration

covalent bond: bond formed between two atoms that mutually share a pair of electrons

crystal lattice: three-dimensional structure of a crystaline solid

crystallization: process of producing crystals of a substance when a saturated solution in an appropriate solvent is either cooled or some solvent removed by evaporation

culture: living material developed in prepared nutrient media

cyanobacterium: eubacterium sometimes called "the blue-green alga"; it contains chlorophyll (the pigment most abundant in plants), has very strong cell walls, and is capable of photosynthesis

cyclopentadienyl ring: five-membered carbon ring containing two C–C double bonds; formula C_5H_6

cysteine residue: sulfhydryl-containing cysteine unit in a protein molecule

cytosine: heterocyclic, pyrimidine, amine base found in DNA

dedifferentiation: the opposite of the biological process of differentiation by which a relatively unspecialized cell undergoes a progressive change to a more specialized form or function

degradative: relating to or tending to cause decomposition

degrade: to decompose or reduce the complexity of a chemical

delocalized: of a type of electron that can be shared by more than one orbital or atom

denitrification: process of removing nitrogen

density-functional theory: quantum mechanical method to determine ground states

depolarization: process of decreasing the separation of charge in nerve cells; the opposite of hyperpolarization

deterministic: related to the assumption that all events are based on natural laws

deuteron: nucleus containing one proton and one neutron, as is found in the isotope deuterium

dialcohol: organic molecule containing two covalently-bonded –OH groups

diamagnetic: property of a substance that causes it to be repelled by a magnetic field

diamine: compound, the molecules of which incorporate two amino groups (–NH_2) in their structure, such as 1,2 diamino ethane (sometimes called ethylenediamine) and the three diamine benzene compounds

dibasic acid: acidic substance that incorporates two or more acidic hydrogen atoms in one molecule, such as sulfuric (H_2SO_4) and phosphoric (H_3PO_4) acids

dihydroxy compound: compound with molecules that incorporate two hydroxyl groups (–OH) in their structure, such as 1,2 dihydroxy ethane (sometimes called glycol) and the three dihydroxy benzene compounds

directing effect: ability of a functional group to influence the site of chemical reaction, such as substitution, for a molecule

discharge display tube: glass tube containing gas at low pressure through which a beam of electrons is passed

disperse system: two-phase system in which one phase, the disperse phase, is distributed in the second phase, the dispersion medium

disulfide bond: bond in a complex substance that involves two bonding sulfur atoms, –S–S–

disulfide bridge: covalent –S–S– linkage that provides cross-links in protein molecules

DNA: deoxyribonucleic acid—the natural polymer that stores genetic information in the nucleus of a cell

dope: to add a controlled amount of an impurity to a very pure sample of a substance, which can radically change the properties of a substance

drug resistance: ability to prevent the action of a particular chemical substance

ductile: property of a substance that permits it to be drawn into wires

Eighteen Electron Rule: rule noting that coordination complexes with eighteen electrons are stable; electrons from both metal species and ligand donor species are counted

electrolyte solution: a liquid mixture containing dissolved ions

electron correlation error: quantum mechanical method for studying atoms, ions, or molecules

electronegative: capable of attracting electrons

electrophoresis: migration of charged particles under the influence of an electric field, usually in solution; cations, the positively charged species, will move toward the negative pole and anions, the negatively charged species, will move toward the positive pole

electrostatic interaction: force that arises between electrically charged particles or regions of particles

elemental analysis: determination of the percent of each atom in a specific molecule

emulsifier: substance that stabilizes the formation of an emulsion—normally it contains molecules with both polar and non-polar functional groups

emulsion: immiscible two-phase mixture in which one phase is dispersed (as small droplets) in the other phase

enantiomorphic shape: mixture of molecules with the same molecular formulas but different optical characteristics

endohedral: descriptive term for a point within a three-dimensional figure

endoplasmic reticulum: internal membrane system that forms a net-like array of channels and interconnections of organelles within the cytoplasm of eukaryotic cells

Equation of State for Ideal Gases: mathematical statement relating conditions of pressure, volume, absolute temperature, and amount of substance; $PV = nRT$

equatorial bond: covalent bond perpendicular to a molecular axis

equilibrium: condition in which two opposite reactions are occurring at the same speed, so that concentrations of products and reactants do not change

erythromycin: antibiotic used to treat infections

ester: organic species containing a carbon atom attached to three moieties: an O via a double bond, an O attached to another carbon atom or chain, and an H atom or C chain; the R(C=O)OR functional group

esterification: chemical reaction in which esters (RCO_2R_1) are formed from alcohols (R_1OH) and carboxylic acids (RCO_2R')

estrogen: female sex hormone

eukaryotic cell: cell characterized by membrane-bound organelles, most notably the nucleus, and that possesses chromosomes whose DNA is associated with proteins

excitatory: phenomenon causing cells to become active

excitatory neurotransmitter: molecule that stimulates postsynaptic neurons to transmit impulses

exclusion principle: principle that states that no two electrons can have the same four quantum numbers

excrete: to eliminate or discharge from a living entity

expressed: made to appear; in biochemistry—copied

extracellular matrix: entity surrounding mammalian tissue cells, also called connective tissue; composed of structural proteins, specialized proteins, and proteoglycans

face centered cubic structure: close-packed crystal structure having a cubic unit cell with atoms at the center of each of its six faces

feedstock: mixture of raw materials necessary to carry out chemical reactions

Fermi conduction level: vacant or partially occupied electronic energy level resulting from an array of a large number of atoms in which electrons can freely move

ferric: older name for iron in the +3 oxidation state

ferrous: older name for iron in the +2 oxidation state

fibril: slender fiber or filament

fission: process of splitting of an atom into smaller pieces

fissionable: of or pertaining to unstable nuclei that decay to produce smaller nuclei

5′ end: situation in nucleic acids in which the phosphate group is attached at the fifth carbon atom from where the base is attached

folic acid: pteroylglutamic acid; one of the B complex vitamins

formaldehyde: name given to the simplest aldehyde HC(O)H, incorporating the –C(O)H functional group

fractional distillation: separation of liquid mixtures by collecting separately the distillates at certain temperatures

fulcrum: prop or support to an item as in a lever

functional group: portion of a compound with characteristic atoms acting as a group

galactose: six-carbon sugar

galvanic: relating to direct current electricity, especially when produced chemically

galvanometer: instrument used to detect and measure the strength of an electric current

gas density: weight in grams of a liter of gas

glucocorticoid: class of hormones that promotes the breakdown of proteins to make amino acids available for gluconeogenesis; this elevates the blood sugar level and leads to glycogen synthesis in the liver

glucose: common hexose monosaccharide; monomer of starch and cellulose; also called grape sugar, blood sugar, or dextrose

golgi apparatus: collection of flattened stacks of membranes in the cytoplasm of eukaryotic cells that function in the collection, packaging, and distribution of molecules synthesized in the cell

gram negative: bacteria that do not retain their color when exposed to basic dyes such as crystal violet and then exposed to organic solvents; named after Danish bacteriologist Hans Christian Joachim Gram

gram positive: bacteria that retain their color when exposed to basic dyes such as crystal violet and then exposed to organic solvents; named after Danish bacteriologist Hans Christian Joachim Gram

Gray: unit of radiation dose per second; 1 Gray = 1 J/kg

greenhouse effect: presence of carbon dioxide in the atmosphere prevents heat from escaping, thereby raising Earth's temperature

Griess reagent: solution of sulfanilic acid and a-naphthylamine in acetic acid; reagent for nitrites

guanine: heterocyclic, purine, amine base found in DNA

halogen: element in the periodic family numbered VIIA (or 17 in the modern nomenclature) that includes fluorine, chlorine, bromine, iodine, and astatine

heavy metal: by convention, a metal with a density greater than 5 g/cm^3; 70 elements are thus classified as heavy metals

helix: in the shape of a spiral or coil, such as a corkscrew

heme group: functional group formed by an iron atom interacting with a heterocyclic ligand called a porphyrin

hemiacetal: relating to organic compounds formed from an alcohol and a carbonyl-containing molecule

hemlock: poisonous herb of the genus Conium

Hippocrates: Greek physician of fifth century B.C.E. known as the "father of medicine"

homogeneous: relating to a mixture of the same materials

homogeneous solution: mixture of molecules that forms a single phase (solid, liquid, or gas)

hormonal signaling: collective processes by which hormones circulate in the blood stream to their target organs and trigger the desired responses

hydrogen bonding: intermolecular force between the H of an N–H, O–H or F–H bond and a lone pair on O, N or F of an adjacent molecule

hydrolyze: to react with water

hydrophilic: having an affinity with water

hydrophobic: water repelling

hyperbolic relationship: a geometric system in which two or more lines can be drawn through any point in a plane and not intersect a given line in that plane

hyperpolarization: process of causing an increase in charge separation in nerve cells; opposite of depolarization

hypertension: condition in which blood pressure is abnormally high

Ibn Sina: given name of an Islamic scientist known in the West as Avicenna (979–1037); reputed to be the author of more than 100 books that were Europe's most important medical texts from the 12th century until the 16th century

inert: incapable of reacting with another substance

inhibitory: preventing an action that would normally occur

integro-differential: complex mathematical model used to calculate a phase transition

interface tension: contractile force at the junction of two liquids

intermediate: molecule, often short-lived, that occurs while a chemical reaction progresses but is not present when the reaction is complete

intermolecular force: force that arises between molecules, generally it is at least one order of magnitude weaker than the chemical bonding force

internuclear: distance between two nuclei

intestinal epithelium: layer of cells in the intestines that allows the passage of water and solutes

intramolecular force: force that arises within molecules—essentially the force associated with chemical bonds

invertebrate: category of animal that has no internal skeleton

in vitro: Latin, meaning "in glass" and applied to experiments done under artificial conditions, that is, in a test tube

in vivo: Latin, meaning "in life" and applied to experiments conducted in a living cell or organism

ion exchange chromatography: form of liquid-solid chromatography based on the reversible formation of bonds between the fixed ions bound to an insoluble matrix of an ion exchanger and mobile counter ions present in the liquid phase passing over the insoluble matrix

ionization: dissociation of a molecule into ions carrying + or − charges

isolate: part of a reaction mixture that is separated and contains the material of interest

isomer: molecules with identical compositions but different structural formulas

isoprene: common name for 2-methyl-1,3butadiene, the monomer of the natural rubber polymer

isostructural: relating to an arrangement of atomic constituents that is geometrically the same although different atoms are present

isotope: form of an atom that differs by the number of neutrons in the nucleus

ketone: one of the characteristic groups of atoms in organic compounds that undergoes characteristic reactions, generally irrespective of where it occurs in the molecule; the RC(O)R functional group

kinetic theory: theory of molecular motion

Kohlrausch drum: rotating cylinder used to mount a variable resistance slide wire for a polarograph

lanthanides: a family of elements (atomic number 57 through 70) from lanthanum to lutetium having from 1 to 14 4f electrons

lattice: systematic geometrical arrangement of atomic-sized units that describe the structure of a solid

ligand: molecule or ion capable of donating one or more electron pairs to a Lewis acid

lipid: a nonpolar organic molecule; fatlike; one of a large variety of nonpolar hydrophobic (water-hating) molecules that are insoluble in water

lipophilic: a molecule that tends to congregate in solution with lipids—it will be a nonpolar molecule or the nonpolar portion of a molecule

liposome: sac formed from one or more lipid layers that can be used for drug transport to cells in the body

liquefaction: process of changing to a liquid form

locomotor: able to move from place to place

Lucretius: Roman poet of first century B.C.E., also known as Titus Carus; author of *De Rerum Natura*

lysergic acid: one of the Ergot series of alkaloids, which constrict blood vessels so that the victim develops burning sensations in the limbs, gangrene, and, ultimately, convulsions; the diethylamide of this substance (LSD) induces visual perception disorders, delusion

lysis: breakdown of cells; also the favorable termination of a disease

macrolide: substance with a large ring lactone structure

macronutrient: one of a number of substances, needed in relatively large amounts, that provide nourishment for living organisms

macroscopic phenomena: events observed with human vision unassisted by instrumentation

mammalian toxicity: poisonous effect on humans and other mammals

Manhattan Project: government project dedicated to creation of an atomic weapon; directed by General Leslie Groves

manifold of ensemble states: a set of quantum states that meet the specific requirements (such as total energy) being considered in a calculation

mechanical energy: energy of an object due to its position or motion

mediate: to act as an intermediary agent

melting point: temperature at which a substance in the solid state undergoes a phase change to the liquid state

mentorship: the process by which a wise and trusted teacher guides a novice in the development of his/her abilities

metabolism: the complete range of biochemical processes that take place within living organisms; comprises processes that produce complex substances from simpler components, with a consequent use of energy (anabolism), and those that break down complex food molecules, thus liberating energy (catabolism)

metabolites: products of biological activity that are important in metabolism

metal: element or other substance the solid phase of which is characterized by high thermal and electrical conductivities

metal cation: positively charged ion resulting from the loss of one or more valence electrons

metalloenzyme: a protein enzyme that contains at least one metal atom, often in an active site of the protein

metalloid: elements that exhibit properties that are between those of metals and nonmetals; generally considered to include boron, silicon, germanium, arsensic, antimony, tellurium, and polonium

metallothionein: class of low molecular weight proteins and polypeptides with very high metal and sulfur content; thought to play a role in concentration and flow of essential elements, e.g., Cu and Zn, and in ameliorating the influence of toxic elements, e.g., Hg and Cd, in the body

metallurgy: the science and technology of metals

microchemistry: chemical investigation carried out on a microscopic level

microcrystalline: relating to the structure of crystals of very small size, typically a micron (μm) in dimension

micromolar: relating to a solution of a substance that is in the concentration range of micromoles per liter, or 10^{-6} moles per liter

mitochondrial matrix: soluble phase inside the inner mitochondrial membrane containing most of its enzymes

mitosis: process by which cells divide, particularly the division of the cell nucleus

molecular identity: "fingerprint" of a molecule describing the structure

monoclinic: one of several arrangements of atoms found in crystalline solids; characterized by a unit cell of three axes each of a differing length; two axes are mutually perpendicular while the third is at an oblique angle

monodentate: capable of donating one electron pair; literally, one-toothed

monosaccharide: one class of the natural products called carbohydrates with the general formula $C_x(H_2O)_y$; monosaccharides have a weak sweet taste, are readily soluble in water, and are called sugars

mordant dye: dye substance containing an acidic moiety, e.g., a carboxyl group, which binds metallic compounds, forming insoluble colored compounds

natural philosophy: study of nature and the physical universe

nesosilicate: any silicate in which the SiO_4 tetrahedra are not interlinked

net charge: total overall charge

neurologic: of or pertaining to the nervous system

neuropathy: degenerative state of the nerves or nervous system

neuropeptide: neurotransmitter released into the blood stream via nerve cells

neutron activation analysis: method for detecting traces of elements by bombardment with high-flux neutrons and then measuring the decay rate of the radioactive products

Newtonian: based on the physics of Isaac Newton

nicotine adenine dinucleotide (NAD): one compound of a group of coenzymes found in hydrogen-transferring enzymes

nitric oxide: compound, NO, which is involved in many biological processes; the drug Viagra enhances NO-stimulation of pathways to counteract impotence; may be involved in killing tumors

nitrotoluic acid: benzoic acid molecule with methyl and nitro groups attached

noble gas: element characterized by inert nature; located in the rightmost column in the Periodic Table

noncovalent: having a structure in which atoms are not held together by sharing pairs of electrons

noncovalent aggregation: non-specific interaction leading to the association of molecules

nonpolar: molecule, or portion of a molecule, that does not have a permanent, electric dipole

nuclear: (a) having to do with the nucleus of an atom; (b) having to do with the nucleus of a cell

nucleosynthesis: creation of heavier elements from lighter elements via fusion reactions in stars

octahedral: relating to a geometric arrangement of six ligands equally distributed around a Lewis acid; literally, eight faces

odd chain fatty acid: long chain carboxylic acid with an odd number of carbon atoms

oligomeric chain: chain that contains a few repeating units of a growing polymeric species

opioid: naturally produced opium-like substance found in the brain

optically active: capable of rotating the plane of plane-polarized light

organoleptic: effect of a substance on the five senses

organometallic compound: compound containing both a metal (transition) and one or more organic moieties

oxidation: process that involves the loss of electrons (or the addition of an oxygen atom)

oxidation state zero: condition characterized by an atom having neither lost nor gained electrons

oxidation–reduction reaction: reaction, sometimes called redox, that involves the movement of electrons between reactants to form products

oxide ion conductor: series of oxides of various compounds in perovskite structure—especially of interest in high-temperature fuel cells

parabolic curve: planar curve each point of which is equidistant from a straight line (axis)

paraffin: saturated aliphatic hydrocarbon (alkane), or hydrocarbon wax

partial pressure: portion of a total pressure of a gas mixture contributed by a single type of gas in the sample

passive diffusion: mechanism of transporting solutes across membranes

pasteurization: process of heating foods such as milk to destroy bacteria

peerage: a body of peers; dignitaries of equal standing

perpendicular: condition in which two lines (or linear entities like chemical bonds) intersect at a 90-degree angle

pH effect: effect caused by a change in the concentration of hydrogen ions

phase: homogenous state of matter

phenol: common name for hydroxybenzene (C_6H_5OH)

phosphorylation: the addition of phosphates into biological molecules

photodiode assembly: grouping of electronic devices which includes a photodiode—a photodetector; useful in medical diagnostics, bar code readers, and guidance systems

photon: a quantum of electromagnetic energy

photosynthesis: process by which plants convert carbon dioxide and water to glucose

physostigmine: alkaloid derived from the leaves of the Calabar bean, formula $C_{15}H_{12}N_3O_2$; salts used for anticholinesterase activity

pilot plant: intermediate stage of testing for chemical process, between bench-top and production scale

planar complex: arrangement of atoms in which all atoms lie within a common two-dimensional plane

plane polorized light: electromagnetic radiation (light) in which the electric (or magnetic) vectors are all vibrating in the same plane

platelet: smallest noncellular component of human blood

pneumatic chemist: early chemist who studied primarily the properties of gases

polynucleotide synthesis: formation of DNA or RNA

polypeptide: compound containing two or more amide units—C(O)NH—produced by the condensation of two or more amino acids

porphyrin: common type of heterocyclic ligand that has four five-member rings with a nitrogen, all of which lie in a plane; nitrogen atom lone pairs are available for coordinate covelent bonds

postsynaptic neuron: receptor nerve cell

potash: the compound potassium oxide, K_2O

precipitation: process of separating a solid substance out of a solution

precursor molecule: molecule that is the chosen starting point for the preparation of another molecule; a starting molecule in a chemical synthesis; a reactant molecule

primary electrochemical cell: voltaic cell based on an irreversible chemical reaction

principal oxidation state: oxidation state that is most important

prism: triangular-shaped material made from quartz or glass used to diffract light

prodrug: precursor of a drug that is converted into an active form by a metabolic process

progesterone: steroid found in the female reproductive system; formula $C_{21}H_{30}O_2$

prokaryotic: relating to very simple cells of the type found in bacteria

propagating: reproducing; disseminating; increasing; extending

protecting group: substance added to a functional group of a molecule preventing further reaction until the substance is removed by subsequent reactions

proximate percent: nearest percent of a population (e.g. people, substances)

purine base: one of two types of nitrogen bases found in nucleic acids

putative: commonly believed or hypothesized

pyramidal: relating to a geometric arrangement of four electron-donating groups at the four vertices of a pyramid

pyrimidine base: one of two types of nitrogen bases found in nucleic acids

pyruvate: anion of pyruvic acid produced by the reaction of oxygen with lactic acid after strenuous exercise

quantum: smallest amount of a physical quantity that can exist independently, especially a discrete amount of electromagnetic energy

quantum mechanical: theoretical model to describe atoms and molecules by wave functions

quantum physics: physics based on the fact that the energy of an electron is equal to its frequency times Planck's constant

radioactive decay: process involving emission of subatomic particles from a nucleus, typically accompanied by emission of very short wavelength electromagnetic radiation

radioelement: a radioactive element; one in which the nucleus spontaneously decomposes (decays) producing α (alpha) and β (beta) particles and γ (gamma) rays

rare earth elements: older name for the lanthanide series of elements, from lanthanum to lutetium

rate-limiting step: slowest step in a complex reaction; it determines the rate of the overall reaction; sometimes called the rate-determining step

reagent: chemical used to cause a specific chemical reaction

receptor: area on or near a cell wall that accepts another molecule to allow a change in the cell

reducing agent: substance that causes reduction, a process during which electrons are lost (or hydrogen atoms gained)

reducing potential: stored energy capable of making a chemical reduction occur

relativistic calculation: quantum mechanical model that includes the effects of relativity, particularly for core electrons

repulsive force: force that repels two bodies; charges of the same sign repel each other

reserpine: one of a group of alkaloids found naturally in the shrub *Rarewolfia serpentina*; has been used for centuries to treat hypertension, insomnia, and mental disorders; more recently it has been used to reduce blood pressure

retardation: to slow down a chemical reaction

retrosynthetic analysis: method of analyzing chemical reactions that starts with the product and works backward to determine the initial reactants

reverberator furnace: furnace or kiln used in smelting that heats material indirectly by deflecting a nearby flame downward from the roof

ribosome: large complex of proteins used to convert amino acids into proteins

RNA: ribonucleic acid, a natural polymer used to translate genetic information in the nucleus into a template for the construction of proteins

RNA polymerase: enzyme used to make RNA using DNA as a template

rough endoplasmic reticulum: regions of endoplasmic reticulum the outer surfaces of which are heavily studded with ribosomes, which make proteins for activities within membrane-bounded organelles

Royal Society: The U.K. National Academy of Science, founded in 1660

rutile: common name of TiO_2; also a common structural type for compounds with the general composition AB_2

saltpeter: potassium nitrate; chile saltpeter is sodium nitrate

screen: process of comparing multiple reagents simultaneously to provide information on reaction of one reagent with another

seed germination: beginning of the process by which a seed produces a new plant

selenium toxicity: condition created by intake of excess selenium (Se) from plants or seleniferous water; acute and chronic toxicity are known

semisynthetic: produced by synthesis from natural starting materials

serology: the study of serum and reactions taking place within it

sigma plus pi bonding: formation of a double bond within a molecule or ion

single Slater determinant: wave function used to describe atoms and molecules

size of the basis set: number of relatively simple mathematical functions (called the basis set) used to represent a more complicated mathematical function such as an atomic orbital

smelting: process by which ores are reduced in the production of metals

Socrates: Greek philosopher, c.470–399 B.C.E.

somatic cell: cells of the body with the exception of germ cells

spectral line: line in a spectrum representing radiation of a single wavelength

spectroscopy: use of electromagnetic radiation to analyze the chemical composition of materials

spinel: name given to a group of minerals that are double oxides of divalent and trivalent metals, for example, $MgO \cdot Al_2O_3$ or $MgAl_2O_4$; this mineral is called spinel; also a structural type

stacking interactions: one type of interaction that affects conformation of double-stranded DNA; caused by van der Waals forces

stereospecific: yielding one product when reacted with a given compound but the opposite product when reacted with its stereoisomer

steric repulsion: repulsive force that exists when two atoms or groups get too close together

sterol: steroid containing an alcohol group; derived from plants or animals; e.g., cholesterol

stigmasterol: sterol found in soybeans, $C_{29}H_{48}O$

stratosphere: layer of the atmosphere where ozone is found; starts about 6.2 mi (10 km) above ground

streptomycin: antibiotic produced by soil bacteria of genus *Streptomyces*

subcritical: mass of nuclear materials below the amount necessary to cause a chain reaction

subshell: electron energy sublevel, of which there are four: *s*, *p*, *d*, and *f*

sulfonamides: first of a series of drugs used for the prevention and cure of bacterial infections in humans; sulfanomides are amides of sulfuric acids and contain the $-SO_2NRR_1$ group

super-heavy elements: elements of atomic number greater than 103

superhelix: helical-shaped molecule synthesized by another helical-shaped molecule

surfactants: surface-active agents that lower the surface tension of water; e.g., detergents, wetting agents

synaptic cleft: tiny space between the terminal button of one neuron and the dendrite or soma of another

synthesis: combination of starting materials to form a desired product

synthon: in retrosynthesis, molecules are broken into characteristic sections called synthons

tetrachloride: term that implies a molecule has four chlorine atoms present

tetravalent oxidation state: bonding state of an atom that can form four bonds

theoretical physics: branch of physics dealing with the theories and concepts of matter, especially at the atomic and subatomic levels

3′ end: situation in nucleic acids in which the phosphate group is attached at the third carbon atom from where the base is attached

thykaloid membrane: part of a plant that absorbs light and passes the energy on to where it is needed

thymine: one of the four bases that make up a DNA molecule

toluic acids: methylbenzoic acids

torsion balance: instrument used to measure small forces (weights), based upon the resistance of a wire to be twisted

toxicology: division of pharmacology dealing with poisons, their identification, detection, effects on the body, and remedies

toxin: poisonous substance produced during bacterial growth

trace element: element occurring only in a minute amount

transcription: enzyme-catalyzed assembly of an RNA molecule complementary to a strand of DNA

transition metal complex: species formed when a transition metal reacts with ions or molecules, including water

transition metals: elements with valence electrons in d-sublevels; frequently characterized as metals having the ability to form more than one cation

translational process: transfer of information from codon on mRNA to anticodon on tRNA; used in protein synthesis

trigonal bipyramidal: geometric arrangement of five ligands around a central Lewis acid, with ligands occupying the vertices of two trigonal pyramids that share a common face; three ligands share an equatorial plane with the central atom, two ligands occupy an axial position

tropocollagen: fibers, or fibrils, consisting of three polypeptide sequences arranged in a helix

tyrosine: one of the common amino acids

ultraviolet radiation: portion of the electromagnetic spectrum with wavelengths shorter than visible but longer than x rays

uracil: heterocyclic, pyrimidine, amine base found in RNA

valence: combining capacity

vertabrates: animals that have a skeleton

vesicle: small compartment in a cell that contains a minimum of one lipid bilayer

vitamins: organic molecules needed in small amounts for the normal function of the body; often used as part of an enzyme catalyzed reaction

vitriol: sulfate of a metal; there are blue (Cu), white (Zn), green (Fe), and rose (Co) vitriols

volatile: low boiling, readily vaporized

voltage: potential difference expressed in volts

vulcanized rubber: chemical process of mixing rubber with other materials like sulfur; it strengthens rubber so it can be used under hot or cold conditions; discovered by Charles Goodyear

wetting agent: molecule that, when added to a liquid, facillitates the spread of the liquid across a surface

zoology: branch of biology concerned with the animal kingdom

zwitterion: molecule that simultaneously contains a positive and a negative charge

Index

*The number preceding the colon indicates the volume number; the number after the colon indicates the page number. Page numbers in **boldface type** indicate article titles; those in italic type indicate illustrations. The letter t indicates a table.*

B

D

E

F

I

J

M

N

P

Q

R

S

T

U

V

W

X

Y

Z